Sand Dune Conservation, Management and Restoration

Coastal Research Library

VOLUME 4

Series Editor:

Charles W. Finkl
Department of Geosciences
Florida Atlantic University
Boca Raton, FL 33431
USA

The aim of this book series is to disseminate information to the coastal research community. The Series covers all aspects of coastal research including but not limited to relevant aspects of geological sciences, biology (incl. ecology and coastal marine ecosystems), geomorphology (physical geography), climate, littoral oceanography, coastal hydraulics, environmental (resource) management, engineering, and remote sensing. Policy, coastal law, and relevant issues such as conflict resolution and risk management would also be covered by the Series. The scope of the Series is broad and with a unique crossdisciplinary nature. The Series would tend to focus on topics that are of current interest and which carry some import as opposed to traditional titles that are esoteric and non-controversial. Monographs as well as contributed volumes are welcomed.

For further volumes:
http://www.springer.com/series/8795

J. Patrick Doody

Sand Dune Conservation, Management and Restoration

Springer

J. Patrick Doody
National Coastal Consultants
Brampton, Huntingdon, UK

ISSN 2211-0577 ISSN 2211-0585 (electronic)
ISBN 978-94-007-4730-2 ISBN 978-94-007-4731-9 (eBook)
DOI 10.1007/978-94-007-4731-9
Springer Dordrecht Heidelberg New York London

Library of Congress Control Number: 2012948595

Printed on acid-free paper

Springer is part of Springer Science+Business Media (www.springer.com)

*To Norma who has fought the trials
and tribulations of life with fortitude
and humour and to Jean, for being there.*

Preface

Sandy beaches and inland dunes occupy an important place in the coastal ecosystem. They occur in moderately energetic environments where waves and then wind move sand grains towards the land. They are essentially terrestrial in character, although in the early stages of development the plant and animal communities colonising the sandy shore are tolerant of saline conditions. They provide coastal protection, buffering tides and waves, which may be particularly important in areas where relative sea level is rising, and during storms. They support a rich and varied fauna and flora with many species especially adapted to the habitat. Managing these assets in the face of continuing pressure from human populations on a sustainable basis is a major task.

The book is a guide introducing the sand dune and its main features, together with a summary of the changes brought about by human activities. Thereafter it provides a description of the various states in which the habitat exists, and information on their values. There are signposts to issues and activities, which alter the ecosystem services the sand dune system provides. Options for management are considered and the likely consequences of taking a particular course of action highlighted. These options include the traditional approaches to management (for the conservation of wildlife and landscapes) as well as habitat restoration.

This is an ecological textbook. However, coastal systems are highly dynamic. It is therefore important to consider the geomorphological context for the development of the sand dune system's biological attributes. Due to this, discussion includes the active sand-sharing system at the beach/foredune interface (Chaps. 4 and 6) and the inland[1] sand dune (Chaps. 5 and 7).

This book concentrates on sand dunes in temperate regions of the world using examples mainly from the British Isles, mainland Europe and North America. It includes information based on personal knowledge, published scientific papers, reports and the internet. It is for those with a special interest in the practical aspects of sand dune conservation, management and restoration and undergraduates.

[1] Note "inland" refers to the sand dune immediately behind the beach/foredune.

Plant names are those given in the International Plant Names Index (IPNI http://www.ipni.org/index.html). At first mention, English and Latin names are given with Latin names used thereafter. Similarly, animals have both Latin and English names but with English names where they appear in subsequent text.

Acknowledgments

This book represents a synthesis of research and information derived from the work of a large number of scientists, managers and policy advisors over the last 70 years or so. The studies of people such as Ranwell (1972) and work that is more recent (Packham and Willis 1997; Maun 2009) provide a foundation for understanding the ecology of coastal sand dunes. Carter (1989), Carter and Woodroffe (1994) and Psuty (2004) provide a geomorphological context.

Thanks to all friends and former colleagues from the United Kingdom Nature Conservancy Council and Joint Nature Conservation Committee for their help during my time as coastal specialist within those organisations. Dr Paul Rooney, Liverpool Hope University, played an important part in highlighting errors and omissions from an early draft of the book. Thanks to him for all his efforts. Dr Albert Salman, The Coastal and Marine Union (EUCC), commented on several chapters. Thanks also to Dr. Maike Isermann, Bremen University, for identifying omissions in Chap. 8 and Dr. Stewart Angus, Scottish Natural Heritage (Chap. 11). Special thanks to Prof. Norbert Psuty of the Institute of Marine and Coastal Sciences, Rutgers University, New Jersey, who provided important and critical comment on all aspects of the book, especially its geomorphological content. His help was invaluable.

Carter RWG (1989) Coastal environments. An introduction to the physical, ecological and cultural systems of coastlines. Academic, London, p 617

Carter RWG, Woodroffe CD (1994) Coastal evolution – late quaternary shoreline morphodynamics. Cambridge University Press, Cambridge, p 517

Maun MA (2009) The biology of coastal sand dunes. Oxford University Press, Oxford, p 265

Packham JR, Willis AJ (1997) Ecology of dunes, salt marsh and shingle. Chapman and Hall, London, p 335

Psuty NP (2004) The coastal foredune: a morphological basis for regional coastal dune development. In: Martínez M, Psuty NP (eds) Coastal dunes: ecology and conservation. Springer, Berlin, pp 11–27

Ranwell DS (1972) Ecology of salt marshes and sand dunes. Chapman and Hall, London, p 258

Contents

Chapter 1
Introduction

Abstract Sand dunes exist in a wide range of locations around the world. This book is largely concerned with coastal sand dunes that have had contact with the sea in the Holocene. The term Holocene literally means "completely recent". It refers to the present geological era. It marks the end of the Pleistocene (period of the Ice Ages) and begins around 12,000 years ago. It is marked by a climatic warming phase, with rapidly rising sea levels and is the latest interstadial (warm period between glaciations) which last approximately 1.5 million years. It also concentrates on those sand dunes developed in temperate regions, mostly from the northern hemisphere. This chapter introduces the habitat, its origins, geomorphologic development and vegetation. Using the physical condition as a backdrop, it discusses sand dunes from an ecological point of view. In particular, it describes primary succession and subsequent development above upper beach levels, into what is essentially a terrestrial environment.

1.1 Origins – Late Pleistocene – Holocene

Sand dunes, especially those associated with inland deserts can be very old. Areas of the dry, cold Taklamakan desert dunes of China, for example, probably date back at least 5.3 million years (Sun and Liu 2006). Coastal dunes are much younger than this, although they may develop through reworking of older sands. The 'upland' coastal dunes in Oregon, United States of America originated from aeolian sand transport by onshore winds when sea level was lower than today in the Late Pleistocene. During the middle to late Holocene, following the decline in the rate of sea level rise, onshore waves transported sand to create the beach. Wind subsequently moved the beach sediments landward forming the Holocene dune sheets present today (Peterson et al. 2007).

J.P. Doody, *Sand Dune Conservation, Management and Restoration*,
Coastal Research Library 4, DOI 10.1007/978-94-007-4731-9_1,

1.1.1 Northern Europe

Some of the oldest Holocene coastal sand dunes occur in Finland, where their active development took place along the coast from about 8,000 years ago. These are now above sea level and out of reach of modern coastal processes due to isostatic rebound following the end of the last Ice Age, a process that continues today (Hellemaa 1998). There are similar dates for aeolian activity along the southwest coast of Norway (Prøsch-Danielsen and Selsing 2011). Dune deposits dating up to 8,700 years ago occur on the Southern Isles of the Outer Hebrides of Scotland (Gilbertson et al. 1999). Also in Scotland, sand dune development on Orkney (near the site of the Neolithic settlement of Skara Brae) took place between 5,000 and 6,100 years ago (de la Vega Leinert et al. 2000). In England and Wales, most sand dunes originate from between 5,000 and 6,000 years ago (Pye et al. 2007). In the Netherlands, 'older' dunes began to develop from about 4,800 years ago becoming stabilised and forested (Issar 2003). The origin and formation of sand dunes in Northern Ireland dates from between 2,800 and 3,300 years ago (Wilson and Braley 1997).

1.1.2 Southern Europe

Away from the edge of the main glacial ice-sheets, there are examples of Late Pleistocene sand dune deposits. In Portugal, on the northwest coast, these include coastal sand dunes dating from 25,000 to 14,000 years ago when sea level was much lower than today. These sediments derive from much greater fluvial activity due to high rainfall and spring ice melting. The dunes became 'stranded' above present sea level due to rapid tectonic uplift. Reworking of these 'fossilised' dune cliffs, together with sediment from the transgression of the sea between 9,900 and 3,400 years ago, created a further series of sand dunes. These stabilised around 3,800–1,400 years ago when sea level finally stopped rising in this region (Dias et al. 2000; Thomas et al. 2008).

1.1.3 North America

On the east coast of United States of America there are older barrier islands that have coastal dunes on them. The dates for dune formation are slightly different to those in Europe. The maximum depression of sea level was about 20,000 years ago. The rapid rise in sea level occurred from about 15,000 to 8–10,000 years ago, slowing in stages to about 3,000 years ago. Virtually all the dune features, including those on barrier islands and coastal spits that are present today have formed since then.

1.1.4 Phases of Development

Three phases of sand dune development in response to sea level change are discernable. In Australia, for example, dunes developed along the seashore as sea levels fell (60,000 years ago). As sea levels rose (10,000 years ago), dunes migrated landwards but with greater stability. New, Holocene dunes (approximately 6,000 years ago to the present day) accumulated above the shore (Bird 2008). Pye and Tsoar (1990, page 149) have developed four alternative models for dune development, namely:

1. High sea level model – sand dunes develop and remain at or near the maximum elevation, even when sea level falls. New dunes only form when sea levels return to the higher levels;
2. Falling sea level model – as sea level falls landward sand movement is followed by seaward dune development;
3. Low sea level model – reworked, exposed marine deposits create sand dunes above the low water, which move progressively landward;
4. Rising sea level model – sand moves as sea level rises creating a dune, which migrates landwards.

A general picture emerges in the northern hemisphere of the release of copious amounts of sediment as continental ice retreated and mountain glaciers melted towards the end of the last glaciation and into the early Holocene. After the major period of change from about 15,000 to 10,000 years ago, at 7,000 years ago as the rapid rate of rise in sea level slowed, sand dunes began to form near their position on the present shoreline. The chronology of aeolian activity on the coastal area of Vejers, western Jutland, Denmark over a period of 7,000 years supports the notion that climatic change "strongly influenced dune field dynamics" (Clemmensen et al. 2006). Present day coastal sand dunes represent reworking of these sediments, particularly in the last 2,000 years, together with additional material from more recent erosion and sand movement.

1.1.5 Late Holocene Development

Changing sea levels and climatic variation including wind speed and direction, rainfall and temperature effect change. Over the last 2,000 years, two periods are especially significant, at least in northwest Europe:

1. The Medieval Warm Period (MWP) generally dated to approximately 1,200–800 years ago (800–1200 AD) when temperatures were close to, or a little above those of today;
2. The Little Ice Age (LIA), approximately 700–150 years ago (1300–1850 AD) when temperatures were on average 1–2° colder than at present.

Table 1.1 Periods of sand migration for different parts of northwest Europe related to the Medieval Warm Period and the Little Ice Age

Location	Dates AD	Reference
Outer Hebrides, Scotland	300–700 1400–1800	Gilbertson et al. (1999)
Western Ireland	– 1580–1880	Delaney and Devoy (1995)
Eastern England	500–1000 1500–1800	Orford et al. (2000)
Aquitaine, south western France	c700–1100 1450–1750	Clarke et al. (2002)
Portugal	200BC–500 1770–1905	Clarke and Rendell (2006)

These dates represent a maximum range within which the changes in climate occurred. The Medieval Warm Period spanned about 400 years, when the colder weather of the Little Ice Age began. The LIA lasted some 550 years, until about 1850. As early as 1203, northern sea ice reached as far south as Iceland for the first time since the last (Devensian) glaciation. Here Polar Bear *Ursus maritimus* skins carpeted some church floors in the late Middle Ages suggesting the ice persisted for some time (Lamb 1995). Work in the Outer Hebrides shows that there was a general increase in sand mobility associated with periods before and after the MWP and during the LIA. The combination of increased sea ice and with it a greater thermal gradient throughout the western European region, was related to increased storminess (Dawson et al. 2004). Other work suggests a similar pattern elsewhere (Table 1.1) although the evidence for greater sand movement prior to the MWP is less clear-cut than for the LIA.

In many places in the United Kingdom the present dune topography is only a few hundred years old (Pye et al. 2007). Information from more recent documented events clearly shows the extent of sand movement from the 1300s onwards, especially on the north and west coasts of Great Britain (Table 1.2). The 1880s were particularly significant as during this period, towards the end of the LIA, some of the most intense Atlantic storms developed (Lamb and Frydendahl 1991). This together with high rabbit numbers and human activities may explain why many sites still had extensive areas of bare sand as recently as the 1950s (Chap. 2).

Similarly, on the Sefton Coast, northwest England a major storm in 1739 caused sand to drift 1.5 km inland creating a landscape described by a traveller some years later as being like the "Sahara Desert" (Smith 1999).

Coastal sand dunes in Finland fit this general trend with 'modern' sand dunes having developed from about 1,000 years ago, with most originating in the last 500 years (Hellemaa 1998). In the Netherlands a sequence of 'younger dunes' dating from about 1,200 years ago (Klijn 1990) overlay the 4,800 year 'older' dunes and may be connected to the LIA (Issar 2003, page 49).

A review of documentary records, instrumental data and proxy records over the last 1,000 years for Western Europe, confirm the importance of the Little Ice Age to

Table 1.2 Dates of large scale historical sand movement in Great Britain

Site	Dates AD	Notes on sand movement
Margam (S Wales)	1300	Abbey reported overwhelmed by sand (Steers 1969)
Newborough Warren (Anglesey, Wales)	1331	186 acres noted as being rendered useless for agriculture by sand blow on 6th December (Ranwell 1959)
Aberffraw, (Anglesey, Wales)	1331	Historical documents record sand being mobilised in a strong storm (Steers 1969)
Kenfig (S Wales)	1316	Date of closure of medieval port due to sand dune development (Lamb 1995)
Morfa Harlech (NW Wales)	1385	Closure of the port of Harlech due to sand invasion (Lamb 1995)
Penard Burrows (South Wales)	1478–1528	Sand reported as advancing dangerously – church overwhelmed (Steers 1969)
Sands of Forvie (NE Scotland)	1413	Forvie village abandoned, following major storm. The date given is very near to an extreme astronomical tide (Lamb and Frydendahl 1991)
Culbin Sands (NE Scotland)	1695	Final inundation of agricultural estate following many years of sand blow (Ross 1992)

These examples are of loss of land due to catastrophic storms, by slow inundation from blowing sand, or a combination of both

sand dune development. This appears to have resulted from strong winds associated with Atlantic storms. There may have been up to 250,000 ha of drifting sand during this period (Clarke and Rendell 2010).

In North America, there are three identified periods of sand dune development for the east coast (North Carolina to Virginia) 740, 1260 and 1810 AD. These are associated with colder, dry and stormy conditions as well as sea level rise. Again, widespread development of more modern sand dunes appears related to the conditions that occurred within the more recent LIA (Havholm et al. 2004). Most of the east coast barrier islands have an interior belt of dunes derived from coastal processes operating during a period with a relatively stable sea level 2,500 years ago. Modern dunes developed over the last 400 years or so.

On a global basis, the MWP may not be as significant as suggested above, with the warmth confined to Europe and regions neighbouring the North Atlantic. Relatively colder worldwide conditions did appear around 1400 AD and continued into the nineteenth century. However, the coldest periods occurred at very different times in different regions (Houghton et al. 2001). Despite this, it is possible to identify a general picture of greater foredune development during periods of deteriorating weather when colder, drier and more stormy conditions predominate. Note, these features are more easily recognised from inland 'perched' sand dunes, which are unaffected by sea level change today (Haslett et al. 2000). The picture is complicated and even adjacent sand dunes may have had very different developmental histories (Wilson and Braley 1997). As we will see in the next chapter, human intervention could be as important as climate change in influencing active dune development in many areas. Chapter 12 considers the wider implications of climate change and the relationship with sea levels in determining coastal sand dune conservation, management and restoration.

1.2 Physical Development

Coastal sand dunes are features derived from sedimentary particles that depend on the action of waves to transport sand onshore and wind, which is sufficiently strong to drive sand grains inland. They develop where there is a supply of sediment in the range of 0.06–4 mm with an exposed beach wide enough for the wind to move the particles over the surface. The size range is not absolute and depends to some extent on exposure to wind and the wetness of the sand. When sediments are light enough to be moved by wind, but too heavy to be in suspension in the air, sand dunes may form. The wind speed threshold ranges from about 4 m per second (1 m above ground level) for 0.2 mm diameter dry sand, to in excess of 10 m per second for damp sand (Sherman and Nordstrom 1994). Higher wind speeds will move larger sand grains and 2–4 mm represents a range of overlap between the upper limit for sand deposition and the lower size limit for the definition of pebble/boulder (shingle) beaches. Below 0.06 mm (silt or mud), sedimentary particles are held in suspension until they fall from the seawater column to form mudflats and saltmarshes. Table 1.3 provides a summary of particle size across the range of coastal habitats.

1.2.1 Sediment Movement

Much of the sand available for dune formation in the northern hemisphere derives from material produced through glacial activity as described above, and eroded from the sea floor. To the south as these glacial sediments diminish, other sources

Table 1.3 Particle sizes, first proposed in 1898 by J.A. Udden, modified and extended by Wentworth (1922)

Size range (mm)	Size class
64–246	Cobbles
4–64	Pebbles
2–4	Granules
1–2	Very coarse sand
0.5–1	Coarse sand
0.25–0.5	Medium sand
0.125–0.25	Fine sand
0.0625–0.125	Very fine sand
0.0039–0.0625	Silt
0.00006–0.0039	Clays

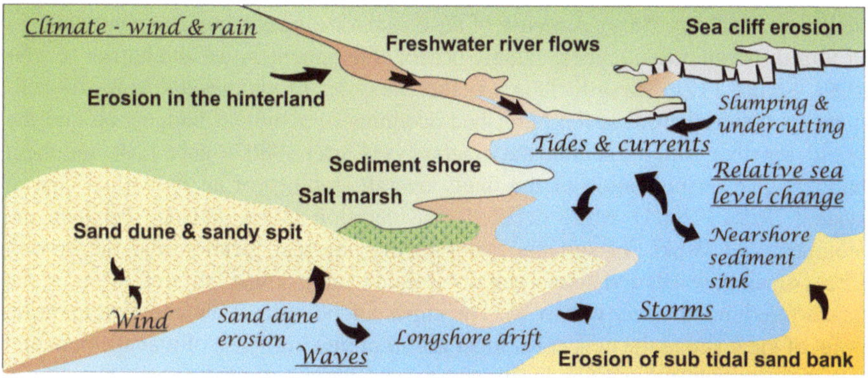

Fig. 1.1 Movement of sediment in the coastal zone in relation to the development of sand dunes. The *arrows* are indicative of the direction of sediment movement and do not imply quantity. Wind strength and direction, tidal range and exposure all affect the extent to which erosion and transportation make sediments available to the system. *Underlined words* represent the pressures forcing change

become progressively more important. These include fluvial material from erosion in the hinterland and coastal cliffs (such as those composed of sandstone, chalk and limestone). Sediment also comes from coral reefs or deposition and movement of shell fragments from marine animals, although these sources are more restricted. Volcanic activity can provide 'black sand' suitable for limited sand dune formation. Today the relative balance between inputs from the land and sea, and movement of sand grains inside and outside the littoral zone, are critical to the dynamic status of the sand dune (Fig. 1.1). Bird (1996) provides more detail on the provenance of beach sediments and Bagnold (1954) describes the way sand grains move.

Sediment availability helps to determine the size and location of an individual sand dune. Storms cause rapid rates of sediment movement. Winds move sand grains onshore where vegetation helps to trap them, creating large sand dunes in areas with abundant sediment. Over long time scales, sea level change moves the beach/foredune landwards or seawards depending on whether it is rising or falling. In the absence of human interference (Chap. 2), the balance between these forces determines the rates of change (erosion or accretion) and the direction of movement.

1.2.2 Sediment Budget

The concept of 'sediment budget' is useful in assessing the overall balance between the volume of sediment entering and leaving a section of coast. If there is an abundance of sand grains then the sand dune is likely to continue to expand laterally as well as vertically. If not the sand dune will erode, although this does not mean the feature will disappear. So long as there is space, the sand dune will migrate landward (Psuty and Silveira 2010). A mildly negative sediment budget may even favour foredune development (Psuty 2004). A review of the concept includes schematic representation of four different coastal types, namely beach, embayment, shoal and barrier (Carter 1989, Figure 117). These include specific examples of sources and sinks of sedimentary material from seven published accounts of sediment budgets around the world, together with the sedimentary pathways (Carter 1989, Figure 120). Sediment budgets give an indication of the likely overall stability of an individual coastal system (Rosati 2005). More detailed consideration of the balance between the 'sources' and 'sinks' of suitable material, the physical location and the driving forces of wind, waves and human activity is used to help define 'physical states' at the beach/foredune interface (Chap. 4). The important point from a nature conservation point of view is that the concept is most useful in making sense of management and restoration needs at a specific location.

1.2.3 Sedimentary Processes

The process by which sand grains move from one place to another provides the material for subsequent deposition to form sand dunes. Erosion followed by transport by water (river flows and tidal currents) represents the first stage. The critical factors determining whether a coastal sand dune develops or not, is the presence of beach wide enough for the sand to dry out and for there to be sufficient exposure to wind to move the grains across the beach. Textbooks describe the movement of sand grains as saltation.[1] Sand may also 'creep' along the surface or under high wind speeds be transported in 'suspension' through the air. Thereafter other factors come into play, notably the extent to which wind speed slows allowing the sand grains to fall to the ground.

This may be because of vegetation, objects on the upper beach or the simple fact that as the air mass moves inland, including over extant dunes, it slows down. In temperate regions of the world, vegetation plays an important role in determining whether sand dunes develop or not. There is a simple formula the '*Dune Mobility Index*', derived from studies of desert dunes in the Kalahari, South Africa. This describes the movement of dunes as being "directly proportional to the presence of strong winds and inversely proportional to the presence of vegetation." (Lancaster 1988).

In summary, if there is a lot of wind and not much vegetation, dunes erode; if there is good vegetation cover and not much wind, they do not. The formula is: $\mathbf{M} = \mathbf{W}/\mathbf{P}/\mathbf{PE}$ where **M** is dune mobility, **W** is the percentage of time that the wind blows above the threshold velocity for sand transport, **P** is annual rainfall, and **PE** is potential evapotranspiration (loss of water from the soil by evaporation and from the plant by transpiration). Wind, as expressed by **W**, may be most critical through

[1] Saltation derived from the Latin 'saltus' meaning leap.

its effect on vegetation. As the **P/PE** value decreases, the vegetation becomes more susceptible to desiccation by the wind and the dunes become more active. This is only part of the story in coastal situations, climate variability, atmospheric and oceanographic processes, alongshore transport, changes in sediment morphology, engineering activities, policy and politics combine to create a "cascade of uncertainties" when attempting to model sand movement on coastal beaches (Pilkey and Pilkey-Jarvis 2007). On vegetated inland sand dunes, the influence of land-use including grazing management is, from a nature conservation perspective especially important (Chap. 5).

Carter (1989, pages 305–320) provides an excellent summary of the processes involved and the nature of sand dune development, see also Bagnold (1954), Pye and Tsoar (1990) and Maun (2009).

1.3 'Natural' Vegetation Succession

Typically, vegetation plays a crucial role in the initiation and stabilisation of coastal sand dunes, especially in temperate regions. This section describes the sequence of development from the beach to stable forms of sand dune vegetation including woodland, using examples from the United Kingdom and the Netherlands.

Lying above sandy beaches, coastal sand dunes are essentially terrestrial in nature. However, they depend on the availability of sand moved onshore by wave action and subsequently driven inland by the action of wind, as described above. Burial by sand provides a stimulus for growth in early colonising vegetation, and the foundation for zonation by providing small amounts of nutrients and soil volume (Maun 1998, 2004). The most successful species survive rapid burial by sand in addition to sand blasting, exposure to salt spray and very low nutrient levels.

Different early colonising species of sand dunes around the world have similar characteristics producing rhizomes, stolons and suckers. These break up during storms and are transported back to the beach to form new colonies (Maun 2004). Arbuscular mycorrhizal fungi[2] also play an important role by improving the uptake of essential plant nutrients such as phosphorus (Bever et al. 2001). Seed germination is less important in the early stages of colonisation as seedling emergence, survival, and growth of seedlings and adult plants is impaired as the level of burial increases (Maun 1998).

Sand dunes accumulate organic matter although this is at a much lower percentage than other habitats. For example, even in a transition zone of an 800 year old dune, organic matter was 4.5, 2.5 and 1.0% at soil depths of 5, 10 and 25 cm respectively showing how relatively rapidly the amount of organic matter decreases with soil depth (Baldwin and Maun 1983). These values are much lower than garden soil, which typically has a 10–30% of organic material.

[2] Arbuscular mycorrhizal fungi have a symbiotic relationship with a host plant without which they cannot survive. They enhance a number of factors important to growth and survival (Smith and Read 2008).

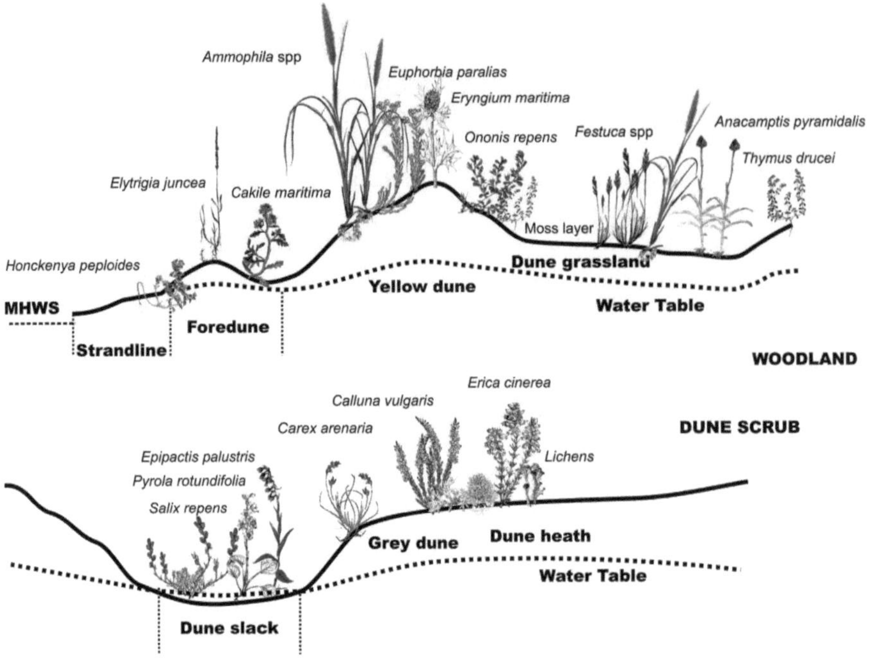

Fig. 1.2 A highly simplified depiction of succession along a landward gradient from the foreshore to dune heath. Slacks occur when the water table is at or near the dune surface, often the result of 'blowouts' within the dune. Dune heath develops when there is high silica or low calcium content in the soil. The final stage is native scrub and woodland

Sand dunes, perhaps because of their perceived 'naturalness' and apparent freedom from human interference were included in early ecological studies. The growth habit of individual plants, including the root system, soil water and mineral nutrient relationships, were the subject of classic studies (e.g. Salisbury 1952). These related the autoecology and ecophysiology of the vegetation succession to the physical development of the sand dune. The early descriptions emphasised the pattern of plant communities and the process through which these patterns developed. Thus, the description of coastal vegetation was of a series of types progressing from early pioneer stages to dune heath (Fig. 1.2).

Tansley (1949) in describing sand dune vegetation in Great Britain also does so in a sequence beginning with foreshore communities, through mobile dunes and fixed dune, including grassland and heath. This is the approach generally adopted throughout the literature. The further studies of Ranwell (1972), for example, continued the description of the ecological principles of sand dune development. These studies recognised the importance of wind in the movement of sand grains and the key role of vegetation in helping to stabilise mobile sand dunes in temperate regions of the world. The British National Vegetation Classification (Rodwell 2000) and the manual interpreting the habitats for Natura 2000 (European Commission 2007) follow a similar linear approach.

Fig. 1.3 Strandline with *Cakile maritima* and *Elytrigia juncea*, west Wales, United Kingdom in July

The summary descriptions that follow also use this general approach. However, it is possible to make a distinction between the beach/foredune, which is essentially a 'sand-sharing system' and the dunes that lie inland. The former depend on inputs of sediment and active interplay between objects on the beach and vegetation. Inland of this active dune front many other factors affect development and distribution of vegetation. They include the height of the water table, precipitation or lack of it, sand movement and rate of deposition or erosion within the body of the dune. As a result, most sand dunes are much more complex than the above linear description suggests.

1.3.1 Strandline (Drift Line, Strandwall)

In the early stages of dune growth, there are a few species able to tolerate inundation with seawater or spray, rapid changes in rates of sand deposition, desiccation and exposure. Pioneer plants of the strandline are the first colonisers. The communities that develop are by their nature patchy and ephemeral, forming strips near the high water mark, often rooted in buried organic drift material. The development of a drift line thus depends on the presence of debris and colonising plants on the beach, which help trap sand. The precise location varies with the incidence of storms, tides and time of year. Two of the main species in western Europe, Sea Rocket *Cakile maritima* and Sand Couch *Elytrigia juncea*, are able to colonise sand close to the high water mark (Fig. 1.3).

Fig. 1.4 'Yellow' mobile dunes with *Ammophila arenaria* Coto Doñana, Spain in October 2006

Other species fulfil a similar function in other parts of the world. For example, Spinifex *Spinifex sericeus* in Australia and New Zealand, or creeping species such as Beach Morning Glory *Ipomoea pes-caprae* or Beach Bean *Canavalia rosea* in the tropics (Hesp 2004).

1.3.2 Mobile Foredune (Yellow Dune)

Foredune development in northwest Europe depends on the ability of plants such as *Elytrigia juncea* and Marram Grass *Ammophila arenaria* to withstand burial by sand of up to 1 m in a single year. Together with the other stresses such as lack of water, this is an inhospitable environment for plants (Ranwell 1972). Foredune plants are specially adapted, occurring in the zone above high water, in areas with abundant sediment and relatively high accretion rates. The term 'yellow dune' reflects the 20% or more bare sand that gives this stage of habitat development its 'yellow' appearance. In locations with abundant sediment, they can help create large areas of dune (Fig. 1.4). This community is often a simple mixture of the grasses with a few other species, such as Sea Bindweed *Calystegia soldanella* or Sea Holly *Eryngium maritimum*. It will continue to exist so long as there is active movement and deposition of sand. The sediment budget (Sect. 1.2.2) determines whether there is a seaward or landward migration of this mobile dune front.

Another species tolerant of burial by sand is Sea Lyme-grass *Leymus arenarius*, which has a northerly distribution in Europe. It is the primary coloniser of the sand dunes of Iceland, both inland and where glacial melt-water delivers large volumes of sediment to the coast. Similar species occupy the same niche along the eastern shoreline of much of the eastern United States of America, where American Beachgrass *Ammophila breviligulata* replaces the ubiquitous *Ammophila arenaria* of Europe.

1.3.3 Dune Grassland and Dune Heath

Inland the mobile dune stabilises as organic matter accumulates and soil formation takes place. A further sequence of communities develops in situ, with three key factors influencing this development. These are the calcium carbonate content of the sand, soil moisture and grazing (dealt with in later Chapters). The original calcium content of the sand, and the age of the dune soil (including the degree of leaching) helps to determine whether succession is to calcareous dune grassland or dune heath. Sand dunes with a low calcium carbonate content of between 1 and 2% tend towards the early development of dune heath. Above 3%, calcareous sand dune grassland develops and it can take several hundred years for the carbonate content to drop to levels that favour heathland vegetation.

In Western Europe, with greater stability *Ammophila arenaria* gradually disappears from the sward. In calcareous dune grassland, under the influence of grazing, a richer variety of plants including species such as Common Restharrow *Ononis repens* or *Calystegia soldanella* occur in a short turf with abundant mosses. Sometimes these include patches on the sides of hollows with conspicuous species such as Pyramidal Orchid *Anacamptis pyramidalis* (Fig. 1.5) and Common Centaury *Centaurium erythraea* as well as a variety of grasses including *Vulpia membranacea* and Sea Barley *Hordeum marinum.*

Acid dune grassland dominated by Wavy Hair-grass *Deschampsia flexuosa*, Red Fescue *Festuca rubra* and Sheep's Fescue *F. ovina,* or dune heath with Ling *Calluna vulgaris* and Bell Heather *Erica cinerea* develop on dunes where the calcium carbonate content of the soil is low. In Europe, these also include Grey Hair-grass *Corynephorus canescens* and Sand Sedge *Carex arenaria,* with abundant mosses such as *Tortula ruralis* ssp. *ruraliformis* and lichens, which are often characteristic (Rhind et al. 2006). On the Wadden Sea island Terschelling, the Netherlands lichen diversity in several plant communities is very high, with a total of 45 species. These include usually epiphytic species, such as *Bryoria fuscensens*, *Evernia prunastri*, *Hypogymnia physodes*, *H. tubulosa*, *Pseudevernia furfuracea* and *Usnea* spp., growing on a moss carpet, as well as on open sites with *Corynephorus canescens* (Ketner-Oostra and Sýkora 2004). Lichen-rich dune heath (Fig. 1.6) is rare. Lee slopes may also include semi-fixed dune vegetation, sometimes interspersed with large patches of Burnet Rose *Rosa pimpinellifolia.* Northern European examples include Crowberry *Empetrum nigrum* and *Calluna vulgaris* within dry dune heath. In wetter areas Cross-leaved Heath *Erica tetralix* is also present. Dunes developed on sand with high silica or low calcium content and hence acid soils tend to be more frequently encountered in northern Europe.

Fig. 1.5 Calcareous dune grassland with *Anacamptis pyramidalis* and Ladies Bedstraw *Galium verum*, an example from the island of Islay, western Scotland, in July

Fig. 1.6 Lichen-rich dune heath with *Calluna vulgaris* and scattered *Ammophila arenaria*, Winterton Dunes, National Nature Reserve, Norfolk, England in July

1.3.4 Dune Slacks (Swales[3])

Although sand dunes are inherently dry habitats, they often include areas that become periodically wet (slack) habitat. Slacks develop in several different ways that relate to the physical conditions of sand movement and water relations. 'Primary' slacks occur when accreting sand spits or parallel ridges enclose a sandy beach with the water table near the surface of the sand. These may or may not become completely cut off from the sea. In the southern North Sea these include dune valleys, which have open access to the sea and are influenced by the tide and known as 'slufters' (Verwaest et al. 2005). These will have gradients from halophytic vegetation to freshwater species depending on the extent and frequency of tidal inundation, or over longer time scales, soil compaction and sea level rise (or fall).

'Secondary' slacks occur in hollows in the sand dune resulting from wind-eroded 'blowouts[4]' or as a result of the landward movement of the dune over wet or seasonally wet sand. They occur on many dunes and are often rich in species, particularly when associated with calcareous sand. In Western Europe Fen Orchid *Liparis loeselii* var. *ovata*, present in sand dunes in south Wales, only occurs in the early stages of colonisation (Jones 2008). Later stages also include other uncommon plants such as Grass of Parnassus *Parnassia palustris*, Large Wintergreen *Pyrola rotundifolia* (Fig. 1.7) and Marsh Helleborine *Epipactis palustris*. They can represent a significant proportion of the plant diversity in a sand dune. On the Sefton Coast, northwest England, for example of 222 vascular plants recorded in a survey of 1983, 150 (68%) were plants of dune slacks, with one slack having 78 species (Smith 1999).

Vegetation succession in dune slacks are closely related to the underlying soil processes (acidification) and build up of plant material (Sýkora et al. 2004; Sect. 7.4.5). As the vegetation develops, the surface becomes elevated above the water table and other more vigorous species replace the plants present in the earlier stages of colonisation. In Europe, these include species of Willow, especially Creeping Willow *Salix repens* together with Common Reed *Phragmites communis* and Wood Small-reed *Calamagrostis epigejos*, which become dominant in wetter areas of older slacks (Davy et al. 2006).

1.3.5 Dune Scrub

In the absence of grazing, or under a light grazing regime, the natural progression is to scrub and woodland. Often equated with invasive species, many of these vegetation

[3] Low-lying land, especially when moist or marshy. Often used as an alternative name for dune slacks outside Europe.

[4] Blowouts occur when the destabilising forces create mobile sand within the body of the sand dune. Sand grains move under the influence of wind until the ground water surface is reached, creating a stable surface on which vegetation can develop.

Fig. 1.7 Dune slack vegetation with *Parnassia palustris* and inset *Liparis loeselii*, *Pyrola rotundifolia* and *Ophioglossum vulgatum*

types in Europe occur because of a reduction in grazing pressure where their presence is a cause for concern (Chap. 8). In the United Kingdom *Salix repens*, Hazel *Corylus avellana*, Silver Birch *Betula pendula*, and on the east coast and in the Netherlands Sea Buckthorn *Hippophaë rhamnoides* are probably the main natural colonisers. Other native shrubs in Europe include Elder *Sambucus nigra*, Privet *Ligusticum vulgare* and English Elm *Ulmus procera* all of which occur on sand dunes. In the north, species include Juniper *Juniperus communis* even on some heavily grazed dunes.

In many areas, dune scrub is a ubiquitous component of the succession. Mediterranean vegetation includes sclerophyllous (drought and fire tolerant) and evergreen shrubs. Also in the Mediterranean a number of *Juniper* spp. (*Juniperus turbinata* ssp. *turbinata, J. macrocarpa, J. navicularis, J. communis, J. oxycedrus*) occur and form associations on the Iberian peninsula with other dune scrubs such as Portuguese Crowberry *Corema album* and Rockrose *Halimium halimifolium*. Many of these species are 'fire-adapted' and even 'fire-dependent', having survived the extensive burning carried out by humans, probably from prehistoric times.

Species with the same characteristics occur in similar climatic regions especially in California, which has several related native plants (Grove and Rackham 2001). Coastal dune scrub was once extensive here with up to 14 square miles said to have

Fig. 1.8 Dune scrub, Salisbury Beach dunes, Massachusetts, United States of America in September

occurred on the Pacific coast near San Francisco (USA National Park Service 2008 http://www.nps.gov/prsf/naturescience/coastal-dune-scrub-community.htm). Common plants include:

- Yellow Bush Lupine, *Lupinus arboreus*;
- Sticky Monkeyflower, *Mimulus aurantiacus*;
- California Coffeeberry, *Rhamnus californica*;
- Chamisso's Lupine, *Lupinus chamissonis*;
- Coyote Bush, *Baccharis pilularis*;
- Lizard Tail, *Eriophyllum staechadifolium*;
- Mock Heather, *Ericameria erucoides*;
- Poison Oak, *Toxicodendron diversilobum.*

The area also has a number of rare endemic species including the San Francisco Campion *Silene verecunda* and San Francisco Wallflower *Erysimum franciscanum.* On the east coast, extensive areas occupy the dunes immediately behind the foredune. The dune scrub communities can be dense (Fig. 1.8) and have transitions to low-growing pine woodland.

Other "Mediterraneoid" regions of the world with similar types of scrub are Mid Chile, Cape of Good Hope, South Africa and Southwest Australia (Grove and Rackham 2001).

Fig. 1.9 Open, presumed native woodland with Aleppo Pine *Pinus halepensis* on the coast of central Albania in August

1.3.6 Woodland

Climax woodland is the final stage in succession. This is scarce because of the extensive deforestation that took place in historical times (Chap. 2). In areas relatively free from human interference, natural woodland or probably more frequently naturalised 'secondary' woodland on sand dunes, will reflect the native species associated with the surrounding area. Scots Pine *Pinus sylvestris* probably dominated dune forests in northwest Europe, accompanied by Oak *Quercus* spp., *Betula pendula*, and *Corylus avellana* (Provoost et al. 2009). Today *B. pendula* is the most frequently encountered woodland tree on sand dunes in Great Britain and the Netherlands.

Further south around the Mediterranean, the evergreen oak forests that were the natural climax vegetation are now low-growing scrub (<5 m high) because of human activity (notably fire). Today Stone Pine *Pinus pinea* (mostly occurring as a planted species) and/or Maritime Pine *P. pinaster* predominate. These woodlands are scattered around the margins of some dune systems providing an open and often diverse habitat (Fig. 1.9).

1.4 Complex Systems

Sand dunes are much more complex than the straightforward succession depicted as a linear zonation in Sect. 1.2 above implies. It is possible to illustrate several different forms based on the natural interplay of geomorphological driving forces and ecological processes.

1.4.1 Settings for Coastal Dunes

The size and shape of the dune depends on the balance between the amount of sediment available, exposure to wind and physical location. In areas with a positive sediment budget, the beach builds seaward. A negative sediment budget will result in landward movement as the foredune erodes. In both situations, in temperate regions of the world vegetated sand dune (Sects. 1.3.1 and 1.3.2 above) accompanies the beach migration.

Behind the beach or beach/foredune complex, sand dunes vary in size and shape depending on the landform in which they develop. In some areas with persistent onshore winds to drive sediment landward, the dune can become a dominant feature stretching several km into the hinterland. These inland dunes become 'stranded' above the foreshore where sea level is falling relative to the land. They may have a typical undulating dune topography reflecting their original formation and subsequent reprofiling of dune ridges and blowouts. In a few of the most exposed locations, beaches supply sediment blown onto and over sea cliffs or other rising ground. These can have a veneer of blown sand or occasionally more typical undulating dunes. On less exposed shores, wind speeds are lower and bays fill with sediment but do not form such extensive inland dunes. Occasionally offshore winds can result in a 'ness' like formation. Here the active dune derives from sediment eroded from the side of the dune, accumulating near the ness. In most of these examples, the active dune front, where it is present, is relatively small by comparison to the inland dune area (Fig. 1.10a).

Sand dunes also develop on barrier islands, spits and in deltas where conditions are favourable (onshore winds, and/or alongshore drift and a suitable supply of beach sediment). Figure 1.10b shows some examples of the complex coastal systems within which they can occur.

1.4.2 Trophic Levels

Within these complex geomorphological systems, there is also an equally complex biological exchange. This includes the beach/foredune (Chap. 4) and the stabilised inland sand dune (Chap. 5). The primary sources of organic material are from the growth and decay of dune vegetation, marine debris deposited on the beach by tides and river-borne material moved along the shore. These support animals at various levels within the sand dune. There are also exchanges beyond the sand dune into the hinterland (Fig. 1.11).

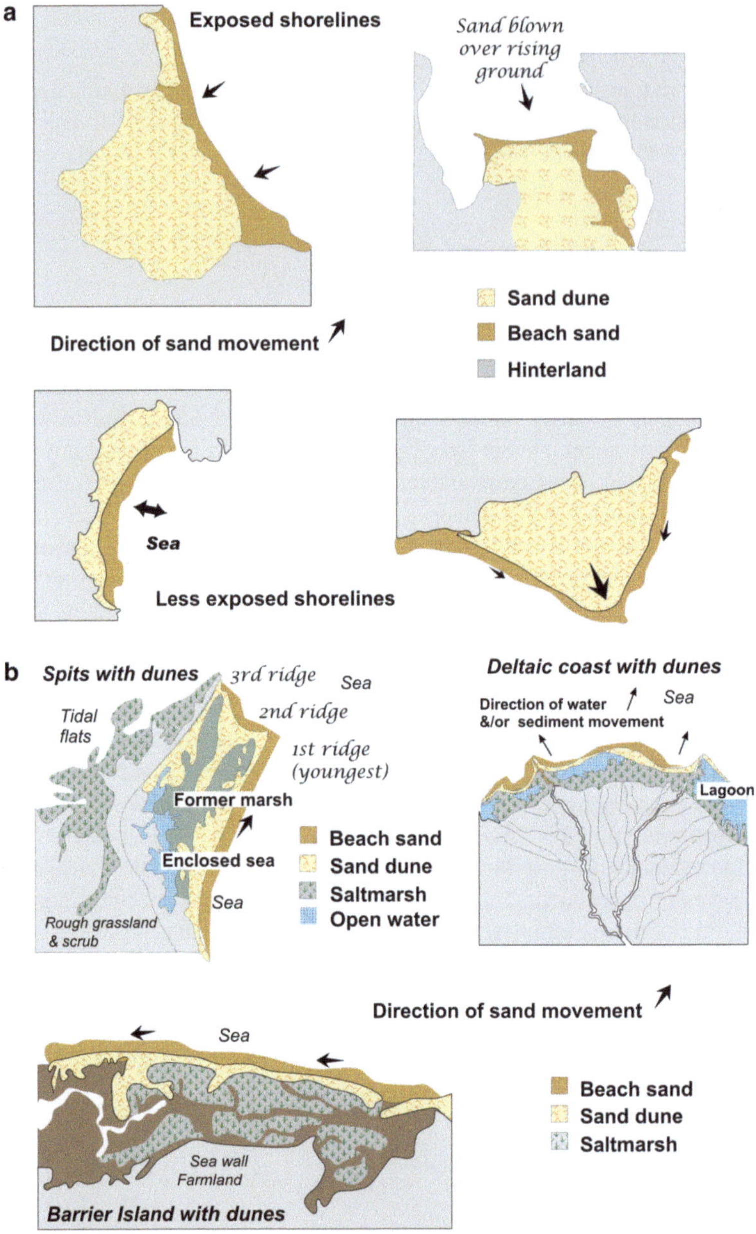

Fig. 1.10 (**a**) Places where inland sand dunes are the dominant feature. (**b**) Some situations where sand dunes develop in association with other habitats, based on examples from temperate regions of the northern hemisphere (After Ranwell and Boar 1986), not to scale

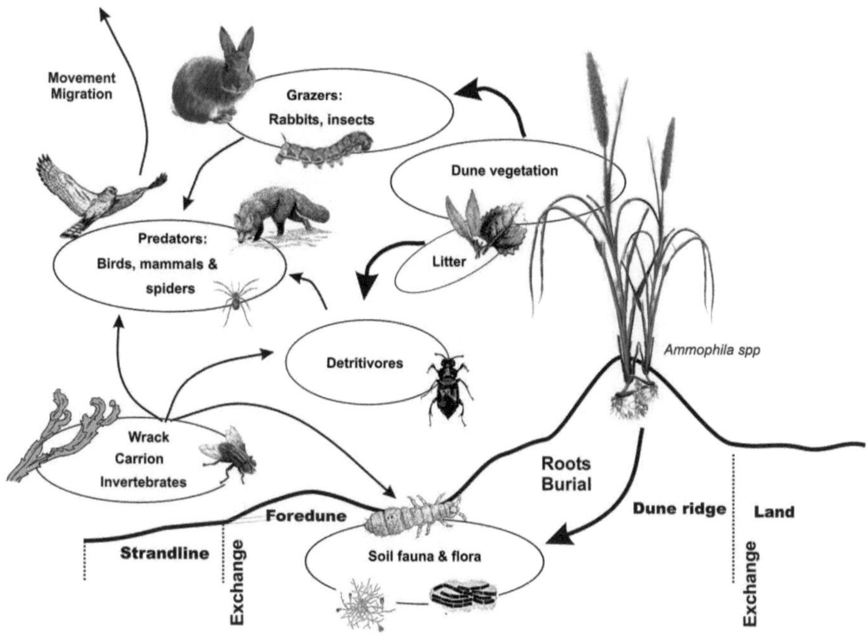

Fig. 1.11 Three basic trophic levels in a vegetated sand dune where exchange takes place. Interactions occur between the beach, sand dune and hinterland, within the soil and vegetation, and between predators and prey (Adapted from McLachlan 1991)

1.5 Geographical Location and Scale

Sand dunes occur extensively around the coastlines of the world in both temperate (including cold) and tropical (including subtropical) regions. A map showing the world distribution derived from information in the two volumes of Dry Coastal Ecosystems (van der Maarel 1993a, b) appears in Martinez and Psuty (2004). Two of the largest dune fields with a coastal element are the Atacama Desert on the west coast of South America, Peru, which has a great variety and quantity of windblown sand (Gay 2005) and the Namib Desert on the southwest coast of Southern Africa (Seely 1992).

By contrast, along the Low and High Arctic sand dunes tend to be small, narrow and isolated. There are active dunes in Alaska, again mainly small and isolated. In Greenland, sand dunes are mostly restricted to the southern part of the island in small sheltered sites. There do not appear to be any coastal dunes in the Antarctic or on sub-Antarctic islands (van der Maarel 1993a).

Extensive beaches and dunes develop onshore where there is abundant sediment, especially on very exposed coasts where the sand migrates landward, sometimes for tens of kilometres. In exposed locations, they may reach 100 m in height and exceptionally 200–300 m (Short 2005b). They also occur as narrow linear features, broken by cliffed coast forming dunes in embayments or in association with spits, bars and barriers. This section provides summary information on the geographical range and variation of the habitat mainly in temperate regions of the northern hemisphere.

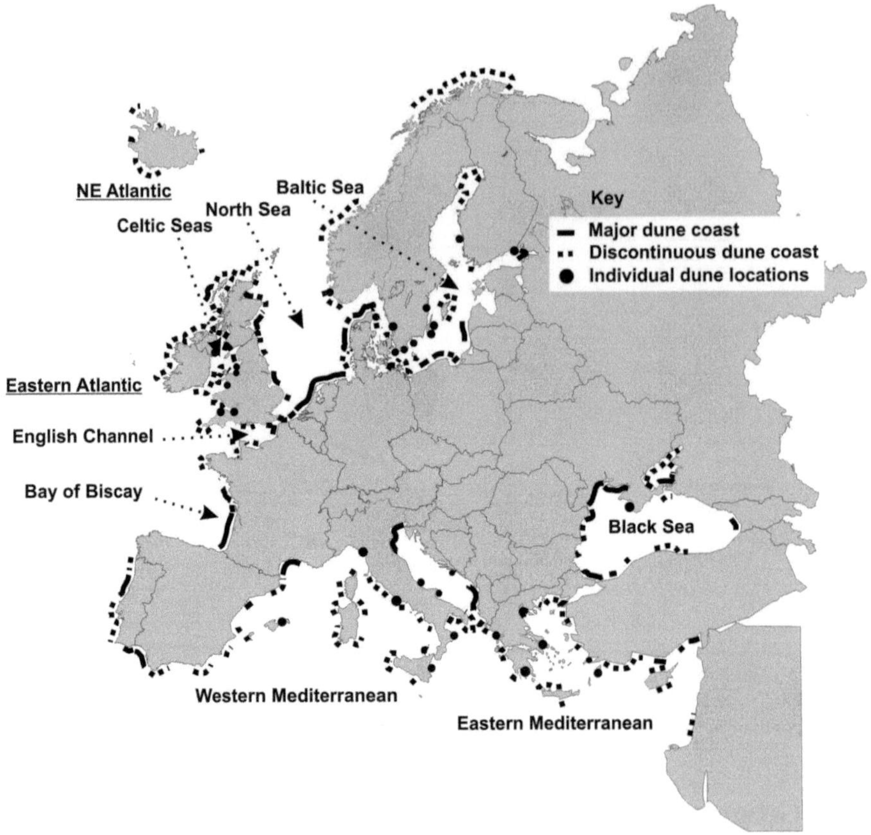

Fig. 1.12 Sand dune distribution in Europe, (Updated from Doody (2008))

1.5.1 Habitat Distribution in Europe

The overall distribution and size of the dunes in Europe (Fig. 1.12) is a reflection of the key influences, namely the physical nature of the coast, exposure, sediment availability, vegetation succession and climatic conditions described above.

1.5.2 North East Atlantic, Celtic and North Seas and the Baltic Sea

Much of the coastline of the exposed North East Atlantic has rocks resistant to erosion. In Iceland, retreating glaciers and volcanic eruptions constantly provide material for the growth of new sand dunes. Glacial rivers bring these sediments to the coast resulting in large dune systems totalling 120,000 ha (Doody 2008).

The paucity of sedimentary material for the development of sand dunes elsewhere means that there are a relatively large number of small and scattered sites. Many of these are associated with embayments between rocky outcrops.

In Great Britain on west facing coasts, dominant winds reinforce the prevailing westerly winds and where there is an abundance of sediment derived from offshore sources of shell sand; large inland dunes have developed. In the Outer Hebrides in Scotland, these include some of the best and largest examples of the extensive cultivated sandy plain or 'machair', also present in the west of Ireland (Chap. 11). In northern areas where coastlines that are rising relative to sea level, due to isostatic uplift, prograding sand dune ridges form, sometimes interspersed with damp hollows in which dune slack vegetation develops. Examples include Magilligan dunes in Northern Ireland, showing 700 years of development (Wilson and Farrington 1989).

Scotland, northeast England and Wales have the greater part of the sand dunes in Great Britain (Fig. 1.13). On the east coast of northern England and Scotland, the coastline tends to be sheltered from the westerly prevailing winds, which are in opposition to the dominant winds. This results in a tendency for the dunes to be narrower often nestling in bays or forming a 'ness'.

In the southern North Sea and southern Baltic sand dunes are very extensive, for example they make up 80% of the coastline of Poland (Doody 2008). There are also major dune systems in Denmark and the Netherlands, derived from accumulations of glacial sand moved onshore under the action of coastal currents and the prevailing wind. In the Netherlands most of the coastline has a sand dune frontage varying in width from a narrow strip to 11 km wide (Doing 1995). The Wadden Sea (Denmark, Germany and the Netherlands), includes a series of barrier islands lying parallel to the coast with extensive sand dunes. Sand dunes continue to evolve as material moves from west to east creating new complexes with saltmarsh see for example, the Ostplate on the island of Spiekeroog, Germany.

1.5.3 Eastern Atlantic, Channel Coast and the Bay of Biscay

Along the east Atlantic, dunes or dune remnants occur along many stretches of the coastline from southwest Ireland, southwest Britain, northern France and Spain. The coast has extensive cliffs and the availability of sediment to form beaches where dunes may subsequently form, is restricted. Hence, as with the coastal cliff landscapes farther north, dunes tend to develop in sheltered embayments. In a few exposed areas, blown sand creates a veneer over the cliffs, covering the underlying rock. In the north of France, some of the botanically richer areas occur where dunes composed of calcareous sand lie adjacent to chalk cliffs, as at le Nord de la Baie de Canche.

To the south, along the west facing coasts of France and Portugal, there is greater availability of material from fluvial sources. Combined with strong erosive action and alongshore drift large sediment transport systems develop. Both countries have

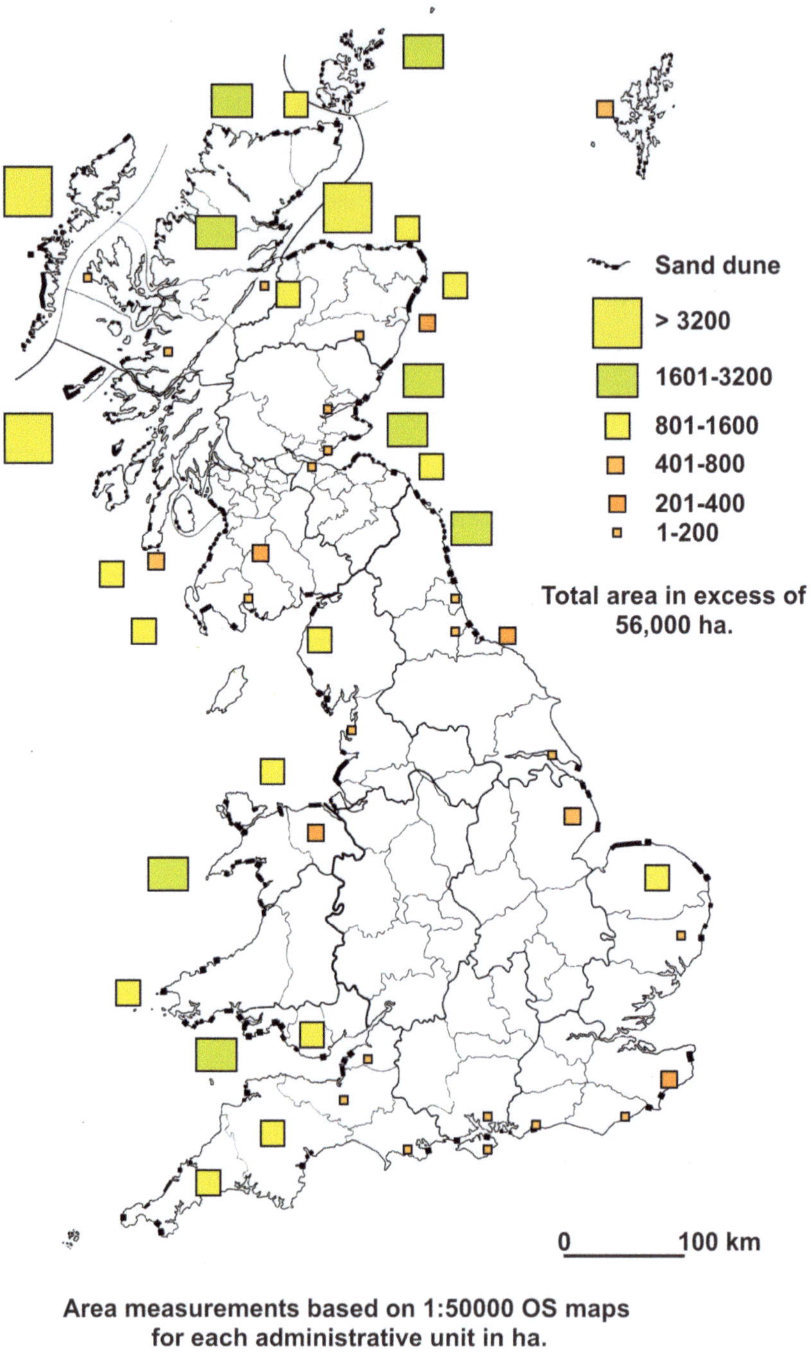

Fig. 1.13 Sand dune distribution in Great Britain and approximate extent based on data circa 1990. Note: figures (ha) are from 1:50,000 maps and meant to facilitate comparisons between local authority administrations. Since 1990, field survey has increased the knowledge of the resource, which is more extensive than indicated by these figures

strong, prevailing westerly winds and the dunes may stretch many kilometres inland. In France, for example, the area known as 'Les Landes' has a special significance and is one of the most extensive dune systems in Europe. Some 60% of the coastline of Portugal has sand dunes, mostly parabolic in form, extending from a few metres to up to 6 km inland.

1.5.4 Western Mediterranean

On the Mediterranean coast, dunes are narrower and often found in association with deltas. Erosion in the hinterland is the main source of sand, which is then transported by rivers to the sea. The Coto Doñana National Park includes sandy beaches, foredunes, high mobile dunes (up to 30 m) and stabilised dunes enclosing a major wetland. Other deltas include the Rhône delta, France within which lies the Carmargue Regional Nature Park and the Ebro Delta in Spain, both with low sand dunes. Elsewhere, dunes occur around much of the coastline, but only in a few protected areas, like the National Park of Circeo, Italy are they relatively free from human influence (Chap. 2).

1.5.5 Eastern Mediterranean and Black Sea

There are many places in the region where sand dunes cannot develop because the land shelves steeply into the sea, leaving nowhere for sediment to accumulate. The coast of Croatia, southern Albania and many of the Greek islands are such. Throughout this area, sand dunes either occupy a narrow fringe bordering flat areas of land or exceptionally, form extensive dunes up to 10 m in height as in Western Peloponnesus, in Greece.

In Turkey, and along the northern shores of Albania, almost all dune systems occur in the immediate vicinity of rivers sometimes within large delta systems. They may also form part of a barrier system enclosing a lake or series of lakes (lagoons). The maximum height is about 50 m and they have a width of up to 4.3 km in southwest Turkey. Many different dune forms are present, including beach plains with embryo dunes, parabolic dunes, blowouts, dune slacks, secondary barchans and dune fields. In the Mediterranean, the calcium carbonate content is very high, whereas siliceous (silica) sands prevail along the Black Sea, including the Danube Delta and Marmara coasts. There are significant areas of sand dune along the coast of Israel. The dune zone stretches for approximately 190 km and is up to 7 km wide in the south narrowing to 1 km wide in the north (Kutiel 2001).

1.5.6 North America

In North America, sand dunes occur along both the west and east coasts. On the Pacific coast they are distributed at irregular intervals, partly due to its steep rocky nature. Large dunes occur in coastal California. For example, of the 27 dune fields, the Monterey Bay dunes cover about 10,000 ha and the Nipomo Dune complex, north and south of the Santa Maria River a further 4,600 ha. There are extensive spits and barrier islands further north in Humboldt Bay. Large systems are also present along the coasts of Oregon and Washington (Cooper 1958). These extend into British Columbia, Canada. There are also dune features in the Gulf of Alaska (Bird 2010).

Coastal dunes occur along much of the Atlantic coast. In Canada these include Sable Island, 160 km due east of Nova Scotia, 42 km long and roughly 1.5 km wide, largely dominated by sand dunes (de Villiers and Hirtle 2006). In the United States of America, sand dunes extend from Maine to the southern tip of Florida. They form on barrier islands, as narrow dynamic sandy beaches and on the ocean side of many bays and estuaries. They also occur along the Gulf of Mexico.

The largest sand dune system in the United States of America is not on the coast at all. It is the area known as the Great Dunes, located in central Colorado. They originate from sand deposited in a glacial lake, which dried up nearly 500,000 years ago (Madole et al. 2008) leaving vast quantities of sediment to blow and form dunes rising up to 230 m from the valley floor. There are also extensive sand dunes around the Great Lakes.

1.5.7 South Africa

Dune fields comprise over 70% (roughly 3,000 km) of the South African coastline (Tinley 1985). Some of the dunefields date back 10 million years and have had many cycles of development as climate and sea levels have changed. The Kwaaihoek region of the Eastern Cape coast, South Africa, consists of an 8 km long stretch of largely unspoiled dunefield (La Cock et al. 1992).

For a detailed description of the location and succession of dunes on the eastern coast of South Africa, see Lubke (2004). Illenberger and Burkinshaw (2008) describe the coastal dunes and dune fields of the Eastern Cape of South Africa. Their chapter also includes a brief overview of the sand dunes of South Africa as a whole.

1.5.8 Australia and New Zealand

Extensive sand dunes occur around the coastline of Australia. Bird (2010) includes numerous references to them. Those to the west and south have calcareous sand derived from shells deposited on the sea floor. Waves drive the sand onshore during periods of high sea level, whilst wind is the agent when sea level is low. On southeast,

east and northern coastlands, the sand comes from weathering of granites and sandstone. Their distribution follows a pattern that is similar throughout the world i.e. larger sand sheets, smaller pocket dunes and sand deposits blown over rising land including sea cliffs. A sandy coastline backed by 200,000 ha of Holocene dunes (Short 1988) bound more than half of the 14,500 km of southern Australia. Important sites include the aeolian barrier of the Younghusband peninsula, South Australia and East Gippsland, Victoria (Bird 1961). Queensland, in the northeast has a series of sand dunes around the coastline including those of Shoalwater Bay (Thompson et al. 1993). The number of mainland beaches in Western Australia is 2,051, with many backed by sand dunes. They include the D'Entrecasteaux National Park, which has a large (1,800 ha) mobile dune Yeagarup (Short 2005a). There are also extensive sand dunes on Fraser Island, Australia, which stretches over 123 km in length and 22 km at its widest point.

In the 1990s, an inventory of New Zealand's active coastal dunes was undertaken. This included maps derived from published topographic maps and other historical sources from the 1950s, 1970s and 1980s for each region. Maps of active dunes in the 1990s from the most recent aerial photographs held by local authorities showed they were widespread, occurring along about 1,100 km of the shore. They ranged in size from a few square metres to thousands of hectares in area, in total covering an estimated 39,000 ha (Hilton et al. 2000).

1.6 Sand Dune Vegetation Regional Variation

The process of succession leads to two broad vegetation types; one associated with temperate (cool to warm-temperate and boreal to subarctic climates), the other tropical (humid and arid climates) regions of the world. The two regions show major differences in the foredune vegetation with taller grasses, sedges and a variety of pioneer herbs dominating in the former; low-growing creepers dominate the latter (Hesp 2004). In temperate regions a wide variety of species colonise as the sand dune becomes stabilised creating a mosaic of plants communities, described below. The description of regional variation in European vegetation that follows is summarised from the European Sand Dune Inventory (Doody 2008). This summary includes some additional references.

1.6.1 Northwest and Western Europe

Most of the sand dunes described for northwest Europe and western Europe have a sequence of vegetation types, which includes all the more important successional communities from strandline (driftwalls) to yellow and grey dune, dune grassland, heath and scrub (Sect. 1.3). In areas where beach erosion occurs some of the early stages of succession may be absent with the sand dune forming a cliff above the beach. In inland areas, grazing favours dune grassland and heath, preventing the natural progression to scrub and woodland. Native woodlands are generally scarce. There are exceptions with secondary mixed scrub, broad-leaved woodland and occasionally succession to natural pine forest (with *Pinus sylvestris*) as for example in Poland (Zoladeski 1991). See also Ellenberg and Strutt (1988) for a general description of the vegetation of coastal sand dunes for Germany, Denmark and the southern Baltic states and chapters in van der Maarel (1993a).

Vegetation of the sand dunes in Portugal is particularly rich in species due to the climatic diversity of their location lying between the colder regions of the north and warmer southern shores. The communities include a number of species endemic to the area (Caldas and Honrado 2001).

1.6.2 Mediterranean Coast

Typically in the Mediterranean, open vegetation forms where Sea Daffodil *Pancratium maritimum* is a common species on the beach and in the foredune. Although successional development is less obvious, narrow zones of pioneer vegetation backed by inland scrub are present. Stands of *Pinus pinea* occur, and in the Camargue have an undergrowth of species such as Flax-leaved Daphne *Daphne gnidium*, Sage-leaved Cistus *Cistus solvifolius* and Rosemary *Rosmarinus officinalis* (Polunin and Walters 1985).

In Italy, Acosta et al. (2007), describe eight communities occurring in a zonation. These include annual strandline with *Cakile maritima* and Prickly Saltwort *Salsola kali*, low embryo dunes with *Elytrigea juncea*, through more stable forms with *Ammophila arenaria* subsp. *australis* and *Calystegia soldanella* to pioneer Juniper scrub (*Juniperus oxycedrus* subsp. *macrocarpa* and *Juniperus phoenicea*) and evergreen communities with False Olive *Phillyrea angustifolia* and Mastic Tree *Pistacia lentiscus*. Dune slacks also occur with Black Bog-rush *Schoenus nigricans* and Ravenna Grass *Erianthus ravennae* and another evergreen community with Holm Oak/Evergreen Oak *Quercus ilex*. The climax vegetation with *Quercus ilex*, can occur as a pure stand or with deciduous elements. In the mouths of some rivers, *Fraxinus* and *Populus* species are widespread. However, as elsewhere there are very few sand dunes with natural woodland.

The grassland and open scrub areas can be rich in species. In Greece, vegetation with the shrub Sea Grape *Ephedra distachya* includes the herb Sand Catchfly *Silene conica* ssp. *subconica*. Vegetation dominated by shrubs (dune heath) includes evergreen sclerophyllous shrubs such as *Juniperus phoenicea*, True Myrtle *Myrtus communis*, *Pistacia lentiscus*, Spanish Broom *Spartium junceum*, Strawberry Tree *Arbutus unedo*, Tree Heath *Erica arborea* and Kermes Oak *Quercus coccifera*. The vegetation is more similar to garigue[5] and maquis[6] typical of grazed inland

[5] An open dwarf, often spiny and aromatic shrub community about 60 cm and rarely >1 m high, also called phrygana.

[6] A dense mostly evergreen shrub community >1 m high.

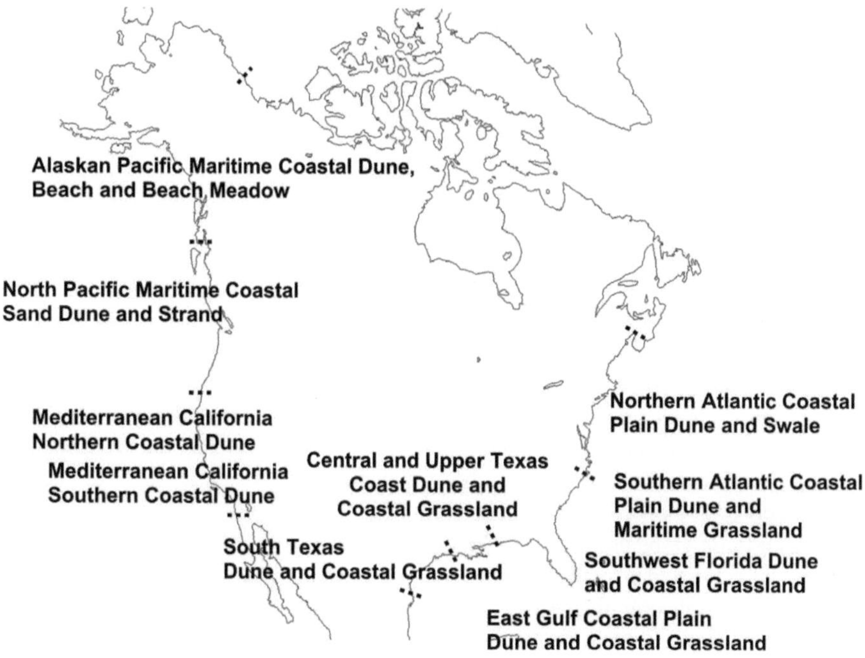

Fig. 1.14 The location of coastal sand dune community types in North America, as summarised, 1–5 above (NatureServe 2010 http://www.natureserve.org/explorer/index.htm)

areas (Polunin and Walters 1985). A recent study by Sýkora et al. (2003) identified one strandline and two sand-dune plant communities with 29 sub-associations.

In Albania (Fig. 1.9 above) and Greece, forests of Aleppo Pine *Pinus halepensis* are probably the best developed and most natural. In Bulgaria Common Reed *Phragmites australis*, *Schoenus nigricans* and Creeping Bent *Agrostis stolonifera* are common. At some sites in the south, the wetter dune sand is colonised by pine woodland, presumed to be natural, which stands out in contrast to the dry, sparse vegetation found elsewhere on the dune ridges.

1.6.3 North America

The geographical range of vegetation types is considerable but coastal dunes include strandline, foredunes, and active to stable dune grassland, heath or scrub (backdunes) and woodland. The typical zonation on the east Atlantic coast includes dunes dominated by *Ammophila breviligulata* in the north, with other species such as Seaside Goldenrod *Solidago sempervirens* and Beach Wormwood *Artemisia stelleriana*. To the south on the Gulf coast and in the southeast, Sea Oats *Uniola paniculata* is the most abundant species. Like the dunes of northwest Europe, more stable communities develop inland and woodland is the climax vegetation in the absence of grazing.

Maritime dune scrub is a frequent community dominated by Sand Heather *Hudsonia tomentosa*. In this zone, Poison Ivy *Rhus radicans* and Virginia Creeper *Parthenocissus quinquefolia* are common (Fig. 1.8 above).

There are five main coastal geographical regions of the United States of America with vegetated sand dune, summarised below:

1. "Alaskan Pacific Maritime Coastal Dune, Beach, and Beach Meadow" and "North Pacific Maritime Coastal Sand Dune and Strand";
2. "Northern Atlantic Coastal Plain Dune and Swale", includes the vegetation of barrier islands. It marks a northern element in the flora where *Ammophila breviligulata* replaces *Uniola paniculata*;
3. "Southern Atlantic Coastal Plain Dune and Maritime Grassland" and "Southwest Florida Dune and Coastal Grassland" includes *Uniola paniculata,* Coastal Panicgrass *Panicum amarum var. amarulum*, and Dune Marsh-elder *Iva imbricata*. Inkberry *Scaevola plumieri,* and Seagrape or Baygrape *Coccoloba uvifera*, which help distinguish this system from those to the north;
4. "Central and Upper Texas Coast Dune and Coastal Grassland", "East Gulf Coastal Plain Dune and Coastal Grassland" and "South Texas Dune and Coastal Grassland" includes plant communities of primary and secondary dunes and interdunal swales (equivalent to dunes slacks in European terminology);
5. "Mediterranean California Northern Coastal" and "Mediterranean California Southern Coastal Dune" include beaches, foredunes, sand spits, and active to stabilising backdunes and sandsheets derived from acidic sands (Fig. 1.14).

1.6.4 South America

Along the coast of Mexico, sand dunes have both vegetation types directly related to local dune physiography and regional patterns resulting mainly from soil and climate differences. In addition to the pioneer communities occurring along the coast with Sea Oats *Uniola paniculata* and their transition to inland types with thickets on fixed dunes, there is a regional variation. To the north are species in common with Texas and the southeast United States. To the south, thickets have species such as Crucillo Bush *Randia laetevirens*, *Coccoloba barbadensis* and Berlandier's Fiddlewood *Citharexylum berlandieri* in common with inland vegetation types (Moreno-Casasola and Espejel 1986; Castillo et al. 1991).

1.6.5 Australia and New Zealand

A distinct feature of the vegetation of Australian dunes is the difference between the northern regions with a wide-ranging tropical flora and the southern temperate regions, where more of the species are endemic. A second major division is between the dry west coast and the humid southeast, with foredunes having mainly inland species in the former and coastal species in the latter.

Within this wide-ranging climatic variation, major vegetation zones develop in sequence as elsewhere in the world and exemplified by northwest Europe (Sect. 1.3 above). However, here the transitions are to scrub rather than grassland or heath, which dominates temperate regions in northern Europe. Along the Queensland (northern tropical) coast, for example, there are three major dune vegetation zones (Clark 1994):

1. A pioneer zone with stabilising herbaceous plants, such as *Spinifex sericeus*;
2. A zone with stabilising stunted woodland trees or scrub species with a few associated herbs and grasses;
3. A forest (or coastal heath zone).

In southern temperate regions of Western and South Australia, well-developed foredune ridges have a wide variety of vegetation, including the native grasses Hairy Spinifex *Spinifex hirsutus* and Beach Fescue *Festuca littoralis*. The vegetation also includes *Cakile maritima* and *Salsola kali*, two species also present on sandy shores in northwest Europe. Behind this foredune ridge Coast She-oak *Casuarina equisetifolia* may dominate along the tropical and subtropical coastlines of north and northeast Australia (Clark 1994).

In New Zealand, an inventory of dynamic coastal dunes (Hilton et al. 2000) includes studies of individual sites, providing information on sand dunes that are relatively free from human interference. Beyond the high water mark, two native sand-binding plants Golden Sand Sedge *Desmoschoenus spiralis* and *Spinifex sericeus* colonise damp sands, which also include *Calystegia soldanella* in the open dune (Sykes and Wilson 1991).

With stabilisation, a variety of non-dune species, such as Matagouri *Discaria toumatou* invades. Scrub and sometimes forest develops in wetter climates, while in dry conditions sparse grassland often persists (Partridge 1992). Between the foredune and stabilised dunes, hollows close to the water table (dune slacks/swales) develop with wetland plants such as Jointed Wire Rush *Apodasmia similis*, New Zealand Flax *Phormium tenax* or Selliera *Selliera radicans* (Roxburgh et al. 1994).

1.7 Conclusion

This chapter introduced the origins of sand dunes, their location, physical formation and vegetation largely without reference to human interference. Their development within the temperate regions of the world depends on the availability of material eroded, moved and deposited at the end of the last glacial period. Although it is possible to look farther back in time, for the purposes of this book changes in the Holocene have most relevance. Human actions may have changed the landscapes in the Mediterranean many thousands of years ago and before the end of the last Ice Age. However, the last 2,000 years are particularly important to our understanding of the way climate variation and human activities combine to influence sand dune development. Chapter 2 provides information on the role humans have played in modifying coastal sand dunes, an essential prerequisite for determining their conservation, management and restoration needs.

References

Acosta A, Ercole S, Stanisci A, De Patta Pillar V, Blasi C (2007) Coastal vegetation zonation and dune morphology in some Mediterranean ecosystems. J Coast Res 23(6):1518–1524

Bagnold RA (1954) The physics of blown sand and desert dunes, 2nd edn. Methuen, London, 265 pp

Baldwin KA, Maun MA (1983) Microenvironment of Lake Huron sand dunes. Can J Bot 61:241–255

Bever JD, Schultz PA, Pringle A, Morton JB (2001) Arbuscular mycorrhizal fungi: more diverse than meets the eye, and the ecological tale of why. Bioscience 51(11):923–931

Bird ECF (1961) The coastal barriers of East Gippsland, Australia. Geogr J 127(4):460–468

Bird ECF (1996) Beach management. Wiley, Chichester, 281 pp

Bird E (2008) Coastal geomorphology: an introduction, 2nd edn. Wiley, Chichester, 320 pp

Bird ECF (ed) (2010) Encyclopaedia of the world's coastal landforms. Springer, Dordrecht/Heidelberg/London/New York, 1495 pp

Caldas FB, Honrado JJ (2001) Flora and vegetation of the Atlantic dunes of the northwest coast of Portugal. In: Houston JA, Edmondson SE, Rooney PJ (eds) Coastal dune management, shared experience of European conservation practice. Liverpool University Press, Liverpool, pp 337–342

Carter RWG (1989) Coastal environments. An introduction to the physical, ecological and cultural systems of coastlines. Academic, London, 617 pp

Castillo S, Popma J, Moreno-Casasola P (1991) Coastal sand dune vegetation of Tabasco and Campeche, Mexico. J Veg Sci 2(1):73–88

Clark PJ (1994) Coastal dune vegetation. In: Groves RH (ed) Australian vegetation, 2nd edn. Cambridge University Press, Cambridge, pp 501–522

Clarke ML, Rendell HM (2006) Effects of storminess, sand supply and the North Atlantic Oscillation on sand invasion and coastal dune accretion in western Portugal. Holocene 16(3):341–355

Clarke ML, Rendell HM (2010) Atlantic storminess and historical sand drift in Western Europe: implications for future management of coastal dunes. J Coast Conserv 15(1):227–236

Clarke M, Rendell H, Tastet J-P, Clave´ B, Masse´ L (2002) Late-Holocene sand invasion and north Atlantic storminess along the Aquitaine Coast, southwest France. Holocene 12:231–238

Clemmensen LB, Pedersen K, Murray A, Heinemeier J (2006) A 7000-year record of coastal evolution, Vejers, SW Jutland, Denmark. Bull Geol Soc Den 53:1–22

Cooper WS (1958) Coastal sand dunes of Oregon and Washington, vol 72, Memoir. The Geological Society of America, New York, 169 pp

Davy AJ, Grootjans AP, Hiscock K, Peterson J (2006) Development of eco-hydrological guidelines for dune habitats – phase 1. English Nature Research reports no 696, Peterborough. http://publications.naturalengland.org.uk/publication/61018. Accessed 14 Aug 2012

Dawson S, Smith DE, Jordan J, Dawson AG (2004) Late Holocene coastal sand movements in the Outer Hebrides, N.W. Scotland. Mar Geol 210(1–4):281–306

de la Vega Leinert AC, Keen DH, Jones RL, Wells JM, Smith DE (2000) Mid-Holocene environmental changes in the Bay of Skaill, mainland Orkney, Scotland: an integrated geomorphological, sedimentological and stratigraphical study. J Quat Sci 15(5):509–528

de Villiers M, Hirtle S (2006) Sable island: the strange origins and curious history of a dune adrift in the Atlantic. Walker and Company, New York, 276 pp

Delaney C, Devoy R (1995) Evidence from sites in western Ireland of late Holocene changes in coastal environments. Mar Geol 124:273–287

Dias JMA, Boski T, Rodrigues A, Magalhaes F (2000) Coastline evolution in Portugal since the Last Glacial Maximum until present – a synthesis. Mar Geol 170:177–186

Doing H (1995) Landscape ecology of the Dutch coast. J Coast Conserv 1:145–172
Doody JP (ed) (2008) Sand dune inventory of Europe, 2nd edn. National Coastal Consultants/ EUCC, the Coastal Union, in collaboration with the IGU Coastal Commission. Published on CDROM, Liverpool Hope University, Liverpool, 126 pp
Ellenberg H, Strutt GK (1988) Vegetation ecology of Central Europe, 4th edn. Cambridge University Press, Cambridge, 731 pp
European Commission (2007) Interpretation manual of European habitats. Natura 2000. European Commission, DG Environment, Nature and Biodiversity, Brussels. Source: http://ec.europa.eu/ environment/nature/legislation/habitatsdirective/docs/2007_07_im.pdf. Accessed 4 Apr 2011
Gay SPJnr (2005) Blowing sand and surface winds in the Pisco to Chala Area, Southern Peru. J Arid Environ 61(1):101–117
Gilbertson DD, Schwenninger JL, Kemp RA, Rhodes EJ (1999) Sand-drift and soil formation along an exposed North Atlantic coastline: 14,000 years of diverse geomorphological, climatic and human impacts. J Archaeol Sci 26:439–469
Grove AT, Rackham O (2001) The nature of the Mediterranean Europe: an ecological history. Yale University Press, New Haven/London, 384 pp
Haslett SK, Davies P, Curr RHF (2000) Geomorphologic and palaeoenvironmental development of Holocene perched coastal dune systems in Brittany, France. Geogr Ann Ser A Phys Geogr 82(1):79–88
Havholm KG, Ames DV, Whittecar GR, Wenell BA, Riggs SR, Jol HM, Berger GW, Holmes MA (2004) Stratigraphy of back-barrier coastal dunes, northern North Carolina and southern Virginia. J Coast Res 20(4):980–999
Hellemaa P (1998) The development of coastal dunes and their vegetation in Finland. Fennia 176(1):1–157. Source: http://ethesis.helsinki.fi/julkaisut/mat/maant/vk/hellemaa/sisalto.htm. Accessed 4 Apr 2011
Hesp PA (2004) Coastal dunes in the tropics and temperate regions: location, formation, morphology and vegetation processes. In: Martínez M, Psuty NP (eds) Coastal dunes: ecology and conservation, vol 171, Ecological studies. Springer, Berlin, pp 29–49
Hilton M, Macauley U, Henderson R (2000) Inventory of New Zealand's active dunelands. Sci Conserv 157:30. Source: http://www.doc.govt.nz/upload/documents/science-and-technical/ sfc157.pdf. Accessed 4 Jan 2011
Houghton JT, Ding Y, Griggs DJ, Noguer M, van der Linden PJ, Dai X, Maskell K, Johnson CA (eds) (2001) Climate change 2001: the scientific basis. Contribution of Working Group I to the third assessment report of the Intergovernmental Panel on Climate Change. Cambridge University Press, Cambridge/New York, 881 pp
Illenberger WK, Burkinshaw JR (2008) Coastal dunes and dunefields. In: Lewis CA (ed) Geomorphology of the Eastern Cape, South Africa. NISC, Grahamstown, pp 85–106
Issar AS (2003) Climate changes during the Holocene and their impact on Hydrological systems, International hydrology series. Cambridge University Press, Cambridge, 144 pp
Jones PS (2008) Aspects of the population biology of *Liparis loeselii* (L.) Rich. var. ovata Ridd. ex Godfery (Orchidaceae) in the dune slacks of South Wales, UK. Bot J Linn Soc 126(1–2):123–139
Ketner-Oostra R, Sýkora KV (2004) Decline of lichen-diversity in calcium-poor coastal dune vegetation since the 1970s, related to grass and moss encroachment. Phytocoenologia 34:521–549
Klijn JA (1990) The younger dunes in the Netherlands: chronology and causation. In: Bakker T, Jungerius P, Klijn JA (eds) Dunes of the European coasts, geomorphology-hydrology-soils, vol 18, Catena supplement., pp 89–100
Kutiel P (2001) Conservation and management of the mediterranean coastal sand dunes in Israel. J Coast Conserv 7(2):183–192
La Cock GD, Lubke RA, Wilken M (1992) Dune movement in the Kwaaihoek region of the Eastern Cape, South Africa, and its bearing on future developments of the region. J Coast Res 8(1):210–217
Lamb HH (1995) Climate history and the modern world, 2nd edn. Routledge, London, 433 pp
Lamb HH, Frydendahl K (1991) Historic storms of the North Sea, British Isles and Northwest Europe. Cambridge University Press, Cambridge, 228 pp

Lancaster N (1988) Development of linear dunes in the southwestern Kalahari, Southern Africa. J Arid Environ 14:233–244
Lubke RA (2004) Vegetation dynamics and succession on sand dunes of the eastern coasts of Africa. In: Martínez M, Psuty NP (eds) Coastal dunes: ecology and conservation, vol 171, Ecological studies. Springer, Berlin, pp 67–84
Madole RF, Romig JH, Aleinikoff JN, VanSistine DP, Yacob EY (2008) On the origin and age of the Great Sand Dunes, Colorado. Geomorphology 99:99–119
Martinez ML, Psuty NP (eds) (2004) Coastal dunes – ecology and conservation, vol 171, Ecological studies. Springer, Berlin, 386 pp
Maun MA (1998) Adaptations of plants to burial in coastal sand dunes. Can J Bot 76(5):713–738
Maun MA (2004) Burial of plants as a selective force in sand dunes. In: Martínez ML, Psuty NP (eds) Coastal dunes: ecology and conservation, vol 171, Ecological studies. Springer, Berlin, pp 119–135
Maun MA (2009) The biology of coastal sand dunes. Oxford University Press, Oxford, 265 pp
McLachlan A (1991) Ecology of coastal dune fauna. J Arid Environ 21:229–243
Moreno-Casasola P, Espejel I (1986) Classification and ordination of coastal sand dune vegetation along the Gulf and Caribbean Sea of Mexico. J Plant Ecol 66(3):147–182
Orford JD, Wilson P, Wintle AG, Knight J, Braley S (2000) Holocenecoastal dune initiation in Northumberland and Norfolk, eastern UK: climate and sea-level changes as possible forcing agents for dune initiation. In: Shennan I, Andrews J (eds) Holocene land-ocean interaction and environmental change around the North Sea, vol 166, Special publications. Geological Society, London, pp 197–217
Partridge TR (1992) The sand dune and beach vegetation inventory of New Zealand. I. North Island. DSIR Land Resources Scientific report no 15. DSIR Land Resources, Christchurch, p 250
Peterson CD, Stock E, Price DM, Hart R, Reckendorf F, Erlandson JM, Hostetler SW (2007) Ages, distributions, and origins of upland coastal dune sheets in Oregon, USA. Geomorphology 91:80–102
Pilkey OH, Pilkey-Jarvis L (2007) Useless arithmetic – why environmental scientists can't predict the future. Columbia University Press, New York, 230 pp
Polunin O, Walters M (1985) A guide to the vegetation of Britain and Europe. Oxford University Press, Oxford, 368 pp
Prøsch-Danielsen L, Selsing L (2011) Aeolian activity during the last 9200 calendar years BP along the southwestern coastal rim of Norway. Norsk Geologisk Forening. Source: http://www.geologi.no/data/f/0/20/44/0_2401_0/Sandfl_p_Jaeren_2011.pdf. Accessed 3 Apr 2012
Provoost S, Jones LML, Edmondson SE (2009) Changes in landscape and vegetation of coastal dunes in northwest Europe: a review. J Coast Conserv 15(1):207–226
Psuty NP (2004) The coastal foredune: a morphological basis for regional coastal dune development. In: Martínez M, Psuty NP (eds) Coastal dunes: ecology and conservation, vol 171, Ecological studies. Springer, Berlin, pp 11–27
Psuty NP, Silveira TM (2010) Global climate change: an opportunity for coastal dunes? J Coast Conserv 14(2):153–160
Pye K, Tsoar H (1990) Aeolian sand and sand dunes. Springer, Berlin/Heidelberg, 458 pp
Pye K, Saye S, Blott S (2007) Sand dune processes and management for flood and coastal defence. Part 2: sand dune processes and morphology. DEFRA R&D technical report FD1392/TR. Source: http://sciencesearch.defra.gov.uk/Document.aspx?Document=FD1302_5396_TRP.pdf. Accessed 4 Apr 2011
Ranwell DS (1959) Newborough Warren, Anglesey. 1. The dune system and dune slack habitat. J Ecol 47:571–601
Ranwell DS (1972) Ecology of salt marshes and sand dunes. Chapman and Hall, London, 258 pp
Ranwell DS, Boar R (1986) Coast dune management guide. Institute of Terrestrial Ecology, HMSO, London, 105 pp

Rhind P, Stevens D, Sanderson R (2006) A review and floristic analysis of lichen-rich grey dune vegetation in Britain. Biol Environ Proc R Acad 106B(3):301–310

Rodwell JS (ed) (2000) British plant communities. Volume 5, maritime communities and vegetation of open habitats. Cambridge University Press, Cambridge, 512 pp

Rosati JD (2005) Concepts in sediment budgets. J Coast Res 21(2):307–322

Ross S (1992) The Culbin sands – fact and fiction. Centre for Scottish Studies, University of Aberdeen, Aberdeen, 196 pp

Roxburgh SH, Wilson JB, Gitay H, King WMcG (1994) Dune slack vegetation in southern New Zealand. N Z J Ecol 18(1):51–64

Salisbury E (1952) Downs and dunes, their plant life and its environment. G. Bell and Sons, London, 328 pp

Seely M (1992) The Namib: natural history of an ancient desert, 2nd edn. Shell Namibia, Windhoek, 104 pp

Sherman DJ, Nordstrom KF (1994) Hazards of wind-blown sand and coastal drift sands: a review. In: Finkl CW (ed) Coastal hazards: perception susceptibility and mitigation. J Coast Res 12:263–275

Short AD (1988) Holocene coastal dune formation in southern Australia: a case study. Sediment Geol 55(1–2):121–142

Short AD (2005a) Beaches of the Western Australian coast: eucla to Roebuck Bay: a guide to their nature, characteristics, surf and safety. Sydney University Press, Sydney, 433 pp. (This is one of a seven book series entitled "*Beaches of the Australian Coast*" published by Sydney University Press. http://sydney.edu.au/sup/)

Short AD (2005b) Sandy coasts. In: Schwartz ML (ed) Encyclopaedia of coastal science. Springer, Dordrecht, pp 821–825

Smith PH (1999) The sands of time – an introduction to the sand dunes of the Sefton Coast. National Museum and Galleries on Merseyside, Liverpool, 196 pp

Steers JA (1969) The coastline of England and Wales, 2nd edn. Cambridge University Press, Cambridge, 762 pp

Sun J, Liu T (2006) The age of the Taklimakan desert. Science 312/5780:1621

Sykes MT, Wilson JB (1991) Vegetation of a coastal sand dune system in southern New Zealand. J Veg Sci 2(4):531–538

Sýkora KV, Babalonas D, Papastergiadou ES (2003) Strandline and sand dune vegetation of coasts of Greece and some other Aegean countries. Phytocoenologia 33(2–3):409–446

Sýkora KV, van den Bogert JCJM, Berendse F (2004) Changes in soil and vegetation during dune slack succession. J Veg Sci 15:209–218

Tansley AG (1949) The British islands and their vegetation, vol 2. Cambridge University Press, Cambridge, 930 pp

Thomas PJ, Murray AS, Granja HM, Jain M (2008) Optical dating of late Quaternary coastal deposits in northwestern Portugal. J Coast Res 24/2B:134–144

Thompson CH, Walker J, Drinnan JC, Clarke SF, Edwards JM, Ross DJ, Pedersen CH (1993) Coastal dunes at Shoalwater Bay, Queensland: data from a reconnaissance of vegetation, soils and landforms. Technical Memorandum, 93/10, Division of Water Resources, CSIRO, Australia. Source: http://www.clw.csiro.au/publications/divisionofwaterresources/techmemos/wr-tm-93-10.pdf. Accessed 30 Oct 2011

Tinley KL (1985) Coastal dunes of South Africa. South African National Scientific report no 109. Foundation for Research Development, Pretoria. Source: http://researchspace.csir.co.za/dspace/bitstream/10204/2353/1/SANSP%20109.pdf. Accessed 4 Apr 2011

van der Maarel E (ed) (1993a) Ecosystems of the world, 2A. Dry coastal ecosystems, polar regions and Europe. Elsevier, Amsterdam, 600 pp

van der Maarel E (ed) (1993b) Ecosystems of the world, 2B. Dry coastal ecosystems, Africa, America, Asia and Oceania. Elsevier, Amsterdam, 636 pp

Verwaest T, De Wolf P, Herrier J-L, Leten M (2005) Windows in the dunes – the creation of sea inlets in the nature reserve de Westhoek in De Panne. In: Herrier J-L, Mees J, Salman A, Seys J,

Van Nieuwenhuyse H, Dobbelaere I (eds) Proceedings 'Dunes and estuaries 2005' – international conference on nature restoration practices in European coastal habitats, Koksijde, Belgium, 19–23 Sept 2005, VLIZ special publication vol 19. xiv+p 685, pp 433–439
Wentworth CK (1922) A scale of grade and class terms for clastic sediments. J Geogr 30:377–393
Wilson P, Braley SM (1997) Development and age structure of Holocene coastal sand dunes at Horn Head, near Dunfanaghy, Co Donegal, Ireland. Holocene 7(2):187–197
Wilson P, Farrington O (1989) Radiocarbon dating of the Holocene evolution of Magilligan Foreland, Co. Londonderry. Proc R Ir Acad B Biol Geol Chem Sci 89B:1–23
Zoladeski CA (1991) Vegetation zonation in dune slacks on the Łeba Bar, Polish Baltic Sea coast. J Veg Sci 2(2):255–258

Chapter 2
Human Occupation, Use and Abuse

Abstract Occupation of sand dunes dates back many thousand years. This chapter provides a summary of the history of human occupation from Mesolithic Times to the present. It considers the way human activities have modified or destroyed the surface of the, sometimes extensive dune fields (Sects. 2.1 and 2.5). It describes the indirect impact these may have on the beach/foredune together with the implications of sediment depletion (Sect. 2.6). The cumulative nature of these developments helps identify a 'sand dune squeeze' (Sects. 2.7 and 2.8).

2.1 Early Human Occupation

The picture of early Humans is one of settlement near the coast because of the relatively open nature of the land and availability of an abundant food supply. The populations were small, and given their hunter-gatherer existence not rooted to a particular place. Thus, they moved in response to the changing coastal scene and appeared to have little impact on the environment. It is possible that there were Mesolithic people living on areas of coastal land, including sand dunes, overwhelmed by sea level rise in the late Pleistocene (Bailey and Milner 2002/2003).

Some of the earliest evidence of human occupation comes from the Mediterranean and may date from between 37,000 and 20,999 years ago, as for example on the island of Sicily. Following this further colonisation took place in stages, and by the early Neolithic 7,400 years ago agricultural development began. It is questionable if any Mediterranean islands had true wild landscapes even in the early Holocene, given the onset of warming and the presence of humans at the same time (Vogiatzakis et al. 2008).

J.P. Doody, *Sand Dune Conservation, Management and Restoration*,
Coastal Research Library 4, DOI 10.1007/978-94-007-4731-9_2,

Fig. 2.1 Midden exposed following erosion of the machair, western Ireland 1991

2.1.1 Middens

Many sand dune sites have examples of 'middens' with the remains of shellfish in them (Fig. 2.1). An analysis of flint tools suggests that settlers of Torrs Warren, a sand system in southwest Scotland, may have appeared between 5,500 and 7,000 years ago (Coles 1964). Evidence from Hayle to Gwithian Towans in Cornwall, southwest England dates back 6,500–4,500 years, with hunter-gatherer remains found on the site. Also discovered were later Bronze Age artefacts from about 3,000 years ago. Each of these cultures seems to have been subject to the vagaries of the shifting sands (Lewis 1992). A Late Neolithic village called Skara Brae in Orkney was also at the mercy of sand blow. It is located on the southern shore of Sandwick's Bay o' Skaill; occupied from about 5,200 to 4,200 years ago and abandoned following burial by sand. Exposed again in 1850 by a storm, the site provides one of the best examples of this type of settlement anywhere in Europe (Sigurd Towrie 2011 http://www.orkneyjar.com/history/skarabrae/).

However, the picture of ancient people living close to the sea, but having little influence on the development of coastal systems is not be all that it seems. It appears that human activity may have had a greater influence on the stability of sand dunes and over longer periods than previously thought. Even low-intensity human activity could have had significant impact on stability in the Neolithic/Bronze Age, as described for sand dunes in northwest Ireland (Knight and Burningham 2007).

2.2 Sand Drift

Sand drift is the movement of windblown sand, which can occur because of changes in climate, sea level or sediment availability. There are correlations between changes in climate including storms, and sand movement in the late Holocene (Sect. 1.1.5). It is likely that human use may have been equally if not more important. This section considers the main anthropogenic factors that have resulted in sand drift.

2.2.1 Deforestation

Forests covered much of Europe in historical times including sand dunes. In southwest Norway deforestation of mixed forest in the mid-Holocene spanned more than 3,600 years from 6,000 to 2,400 years ago. Beginning with Mesolithic people, forage from broad-leaved trees was important to the still dominant hunter-gatherer economy. Abrupt deforestation and its replacement by heathland coincide with the start of an agro-pastoral economy involving burning and grazing (Kvamme and Kaland 2009). Thereafter, Neolithic farming developed along two paths, the first towards heath and the second to grassland and permanent fields (Høgestøl and Prøsch-Danielsen 2006). Throughout this time, there were periods of sand drift, as for example along the coast of Norway (Prøsch-Danielsen and Selsing 2011).

In other areas, such as the Łeba Bar in Poland, oak woodland developed from c 3,000 years ago. From c 2,000 years ago, fire (presumably caused by human activity) created an open and highly mobile system with wandering dunes, moving at up to 10 m per year (Piotrowska 1989). On the Curonian Spit on the southern Baltic, deforestation and overgrazing in the sixteenth century, exacerbated by the Russians during the Seven Years War (1756–1763) resulted in massive sand movement. This sand movement overwhelmed entire villages (Povilanskas et al. 2009).

In the Mediterranean, fire probably caused the loss of most of the oak/ash forest. Two of the more common shrub communities, the garigue and the maquis, the products of this management, occur on dunes. In the Coto Doñana, southern Spain, the destruction of the original mixed forest also appears to have resulted from fire and shifting agriculture (Corona et al. 1988; Garcia Nova 1997). In the Danube Delta, forest dominated the coastal sand dunes. In historical times its removal, coupled with grazing created instability and a mosaic of shifting and fixed sands (Kelemen 1992). Today, a climax vegetation of natural dune forests is rare in Europe.

2.2.2 Overuse

In temperate regions of Europe, a variety of human activities can cause loss of vegetation, exposure of the sand surface, erosion and sand drift. These include overgrazing and the removal of plant material. The extent and importance of overuse on the instigation of sand drift is apparent from the enactment of special laws in some countries to reverse the incidence of blowing sand. There were two main approaches, firstly prevent removal of sand-binding vegetation and secondly promote planting of the same species.

In Iceland, the natural vegetation evolved largely in the absence of large herbivores and thus has little resistance to grazing by domestic stock. Erosion because of overgrazing has been a major problem for many years, with 50% of the original vegetation lost since 874 when settlement began, especially on sand dunes (Runólfsson 1987). These effects continued into the late twentieth century.

In Denmark, historical records indicate overgrazing by domestic stock and the removal of vegetation for fuel, animal bedding and thatch began nearly 1,000 years ago. Sand drift problems of were so acute by the sixteenth century that a Royal Decree in 1539 prohibited the removal of any vegetation (Skarregaard 1989). Again, in 1792 and 1867 there were special laws to prevent removal of plant material from the dunes. These laws also made provision for rehabilitating the dunes using *Ammophila arenaria* (Skarregaard 1989; Feilberg and Jensen 1992). Almost all conifer forests planted in the Dutch dunes around 1900 were a response to massive dune mobility brought about by overuse. In Belgium, sand drift may have been responsible for the abandonment of settlements in the village of Pulle in the Early Middle Ages (Deresea et al. 2010).

On the Sefton coast in northwest England, *Ammophila arenaria* was cut and sold for making mats, brushes and brooms. An Act of Parliament passed in 1742, "for the more effectual preventing of the cutting of 'Starr' or 'Bent' (*A. arenaria*)" showed how serious the problem had become. This Act made it an offence to cut the grass and sell or be in possession of it "within 5 miles of the sand hills". It was also apparent that the "best way to preserve the sand hills" was to plant the same species (Smith 1999).

Also on the Sefton coast photographs from the 1930s show extensive inland invasion of blown sand between Woodvale and Birkdale, northwest England (Pye et al. 2007). There are also pictures showing extensive planting with *Ammophila arenaria* (Jones et al. 1993, Fig. 1.6). In the early 1900s, Culbin Sands in Scotland was described as a "desert-like wilderness" before the massive afforestation took place (Ross 1992; Sect. 2.3.2). In the early 1950s, Braunton Burrows, Devon, England also had extensive areas of blowing sand resulting from wartime activities and rabbit grazing (Sect. 12.3.2). In Ireland the deflation of the machair, following the breakup of the vegetated surface continues today, exposing evidence of early occupation (Fig. 2.1).

A visit to the Mediterranean coast of Turkey in the 1980s revealed extensive sand movement brought about by overgrazing, especially by goats (Fig. 2.2). When combined with burning by shepherds, this resulted in enormous damage to vegetation, which in turn has led to erosion and deflation of the vegetated dune surface.

2.3 Vegetation Stabilisation and Afforestation

It is clear from the above that human activity has played an important role in destabilisation, in addition to that resulting from climate change. Faced with blowing sand overwhelming agricultural land, farms and villages, preventing further erosion

Fig. 2.2 A boy herding goats along the southern coastline of Turkey in 1985. At this time, the number of animals kept on many of the dunes visited was far in excess of their carrying capacity

and sand drift became a priority in many countries. Over the years, land managers everywhere continued to develop techniques for creating conditions for sand deposition and vegetation development. This included planting dune-building species such as *Ammophila arenaria* and *Leymus arenarius* as part of schemes of restoration management. The following sections describe some of these early approaches.

2.3.1 Planting Ammophila spp.

Early attempts to restore mobile sands involving planting with *Ammophila arenaria* to arrest sand drift were in use as early as 1423 in the Netherlands (Pilon 1988). At Newborough Warren, Anglesey, Wales planting was important enough for the enactment of an injunction from Queen Elizabeth 1st in 1561, such that anyone found cutting, uprooting or carrying away the planted 'Marram' should be punished (Owen 1953, reported in Ranwell 1959). By 1795, experimental methods included sowing seeds and Marram planting accompanied by physical structures designed to reduce wind speed and aid the stabilisation process, were well established, Fig. 2.3 (Carey and Oliver 1918).

More recently, *Ammophila arenaria* planting has taken place in the United States of America, South Africa, Australia and in New Zealand. In New Zealand, active sand dunes showed a marked decline and by the early 1990s, the area had fallen from 129,000 ha to about 39,000 ha, a 70% reduction (Hilton et al. 2000). This and many other species introduced to prevent sand blow have become invasive, which has not been without consequences for the native sand duneflora and fauna (Chap. 8).

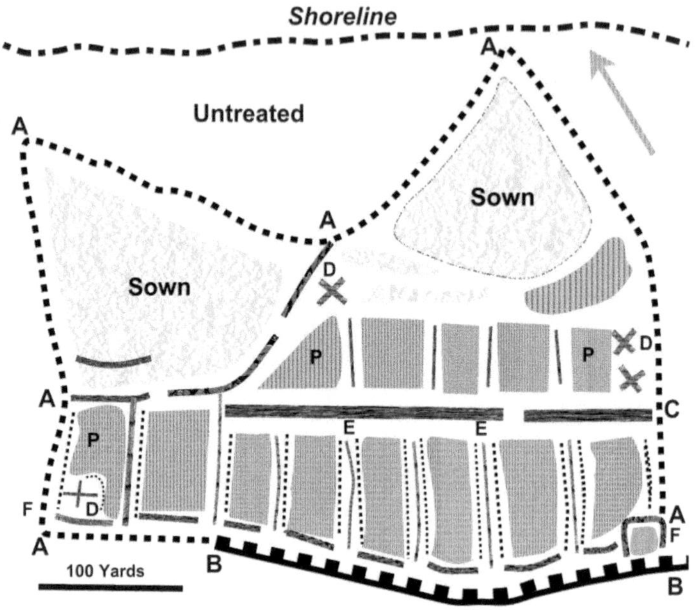

Fig. 2.3 Detailed chart of early experimental Marram planting by Sören Biörn (1795) redrawn from Carey and Oliver (1918). *A* brushwood fence; *B* planked paling fence; *C* hurdle of Alder; *D* depressions with brushwood fences placed in a cross; *E* living poplar with *Salix repens*; *F* low brushwood. The sown area (*mottled*) had seeds of marram and the rest (*stippled grey*), some marked *P*, had cuttings of Marram (After Gerhardt 1921)

2.3.2 *Afforestation*[1]

Early attempts at dune afforestation took place on a large scale in Denmark, where planting began in 1853. Although these were not altogether successful, after some experimentation and persistence over the next 150 years, plantations covered more than 10% of the 80,000 ha or so of sand dune (Skarregaard 1989; Feilberg and Jensen 1992).

In the Netherlands, some of the earliest sand drift problems occurred, and by the end of the sixteenth century over-exploitation had destroyed most woodland and other sand dune vegetation. Afforestation began in earnest in the first half of the twentieth century with large scale planting on coastal dunes (Laar 1995) and inland state-owned sand dunes, mostly with pines (Riksen et al. 2006; Koster 2009).

Large projects have also taken place on the west coast of France. The 230 km long dune system of Les Landes forests called *la forêt des landes de Gascogne*,

[1] Afforestation is a term usually used to describe tree plantations on land not formerly forested. As it is not always clear if the sand dunes were previously tree-covered or not, this term is used here for any planting designed to prevent or restrict sand mobility.

Fig. 2.4 Young pine plantation Benone dunes, Northern Ireland, United Kingdom 2011

literally "the forest of the moors/marshes of Gascogne", lies within the larger region of Aquitaine. In the nineteenth century, tree planting helped turn a vast area of mobile sand dunes and marshes into dense forest. Maritime pines (*Pinus maritima*, *P. pinaster* and *P. pinea*) under planted in places with oak, which included *Quercus ilex* and *Q. suber* were used (Favennec 1998).

In the United Kingdom, some of the earliest systematic attempts began in 1839 on Culbin Sands, northeast Scotland, one of the most mobile areas of dune at that time. The Forestry Commission became involved in 1922 when planting began in earnest with various species of pine, and by 1950 nearly two thirds of the dune was forested (Ovington 1950; Ross 1992). Extensive plantations took place from 1948 on Newborough Warren, Anglesey, Wales with more than 50% of its area planted predominantly with Corsican Pine *Pinus nigra* subsp. *laricio*. A similar fate overtook the Tentsmuir sand dune system, in eastern Scotland. By 1950, planting had taken place on a large part of the area with *Pinus laricio*, *P. contorta*, *P. sylvestris* and in wetter areas Larch *Picea abies* (Ovington 1951; Sect. 2.8.3). Planting continued although on a smaller scale in a few areas, such as on the coastal sand dunes near Benone Northern Ireland (Fig. 2.4).

Planting forests with *Pinus pinaster* (pinhals) has also been a traditional way of controlling coastal sand drift in Portugal (Martins 1989; Danielsen 2008). In much of the Mediterranean, afforestation did not begin until the 1940s (Fabbri 1997). In Turkey, projects partly designed to stabilise the dunes, began on a small scale in 1885. Later between 1961 and 1990, planting took place on 10,672 ha of coastal dune (approximately 30% of the Turkish coastal dunes). In many areas replacement of the natural vegetation was mostly with Wattle *Acacia cyanophylla* and Eucalyptus *Eucalyptus camaldulensis* (both exotic), while *Pinus pinea* was also commonly used. Maritime Pine *Pinus maritima*, Italian Cypress *Cupressus sempervirens*, Spanish Broom *Spartium junceum*, False Acacia *Robinia pseudoacacia* and

Tamarisk *Tamarix smyrnensis* were also planted (Uslu 1995). Towards the end of the twentieth century, the area of sand dune planted with non-native trees in Europe (mostly various species of pine) amounted to nearly 25% (Tekke and Salman 1995).

In 1903, the New Zealand Parliament debated a "Sand Drift Bill" designed to stop the "onward march" of "the evil" drifting sands. In the event nothing happened until the 1930s. In 1951 the New Zealand Forest Service took over responsibility for 9,000–10,000 ha of mobile dunes, planting them with *Ammophila arenaria* and *Lupinus arboreus*, and 3,800 ha of forest trees. In the 1960s and 1970s experiments took place to increase the stabilisation properties of Monterey Pine *Pinus radiata* by planting *Lupinus arboreus* first. This work continued into the 1980s when large areas had become commercial forests (Gadgil 1971; McKelvey 1999; Walrond 2007). This caused the decline in the area of active dunes from approximately 129,000 ha in the early 1900s to about 39,000 ha or 70% since the early 1950s (Hilton et al. 2000). Revised figures suggest the remaining open dune area is an over estimate and that only 21,000 ha (11.6%) survive (Ministry for the Environment 2007).

2.4 Agriculture and Aerial Pollution

Coastal sand dunes may have been amongst the first areas settled by Neolithic farmers. Grazing and other forms of agriculture, such as growing cereals probably took place on fixed dunes. No doubt, the light soils were amenable to ploughing. In some areas, substantial parts of the inland dune are in intensive agricultural use and little of the original dune topography remains. Grazing by domesticated stock is widespread especially in northwest Europe, and an important component on many coastal dunes. Aerial pollution (acid rain and nitrogen deposition) are the result of activities taking place at some distance from the sand dune. These can all have a significant influence on the dune vegetation.

2.4.1 Cultivation

Cultivation with cereals and potatoes continues on the extensive machair plains in northwest Scotland (Chap. 11). In the Netherlands, the sandy soils associated with 'older' inland dunes provide places for growing flower bulbs and onions. In Peru, prehistoric sunken fields in large pits dug in coastal dunes to reach the water table, provided opportunities for agriculture (Parsons and Psuty 1975). A practise continuing into the twentieth century, although with problems associated with salt-water intrusion (Psuty pers. com.). In Portugal excavation of the sand dune surface to accommodate potatoes and other crops in the low-lying areas near the water table, also provides opportunities for growing grapes on the sides (Fig. 2.5).

Fig. 2.5 Excavated sand dune growing potatoes and vines, southwest Portugal 1992

2.4.2 Grazing

Grazing by native herbivores, introduced rabbits and domesticated stock (cattle and sheep) is important in the history of development of sand dunes and their vegetation. The introduction of grazing animals as both stock and game, probably took place as early as 5,000 years ago in the Dutch dunes. Human settlements revealed remnants of sheep and goats suggesting these grazing animals were an important food resource (Klijn and Bakker 1992, reported in Baeyens and Martínez 2004).

At this point, it is sufficient to recognise that rabbits played an important role in the development of sand dunes and sand dune vegetation since their introduction to northwest Europe (probably from Spain by the Romans). This and subsequent introductions has had an impact on both stability and vegetation development worldwide. Chaps. 5 and 7 (Sects. 5.3.4 and 7.3) deal with these in more detail.

Studies of the sand dunes of Doñana National Park give an insight into the change from forest to mobile sand brought about by grazing and other uses. Up until 1636, the dunes had a forest dominated by Cork Oak *Quercus suber*. From 1636 to 1682, the introduction of cattle and pigs, the use of prescribed fire and timber and cork exploitation caused the destruction of the forest and mature scrub. Pioneer scrub became dominant (Corona et al. 1988). The continued presence of domesticated animals and the effects of the Little Ice Age around 1700 led to erosion. It was not until 1737, when *Pinus pinea* plantations began, that the sand dunes became stable again. However, even today they include large areas of mobile dune with relatively

Fig. 2.6 Mobile dunes Coto Doñana, Spain October 2006. The *pine forest* is visible beyond the bare sandy plain

little vegetation, together with scrub and pine woodland in damp hollows (Fig. 2.6), amounting to about 7,000 ha.

2.4.3 Acid Deposition and Nutrient Enrichment

Nitrogen oxides (NO_X) and ammonia (NH_3) occur naturally in the atmosphere but with the advent of industrial pollution, they have increased considerably mainly from burning fossil fuels. Emissions of sulphur dioxide (SO_2) and nitrogen oxides react with water molecules in the atmosphere resulting in sulphuric acid (H_2SO_4) and nitric acid (HNO_3). Deposition occurs when these acids and their derivatives reach the ground through precipitation. Falling as 'acid rain', they have two main effects:

1. Increasing the soil acidity causing the loss of nutrients from the soil;
2. Increasing the input of nitrogen, (nitrogen deposition) improving soil fertility and with it plant growth.

This can have a significant impact on the vegetation especially, but not limited to, dune heath (Sect. 7.7.2).

2.5 Infrastructure Development

There were settlements on sand dunes dating back several thousand years, as at Skara Brae (Sect. 2.1.1). Most of these have long-since disappeared through erosion or burial by sand. A few continuously inhabited villages built on coastal dunes still survive, as for example, the eleventh to thirteenth centuries on the sand dunes in Hungary on the Black Sea coast (Kiss et al. 2004). This is rare and most urbanisation is much more recent.

2.5.1 Urbanisation

Urbanisation in Belgium has resulted in the loss of nearly 50% of the dune landscape to buildings, gardens and roads (Provoost and van Landuyt 2001) a decrease from 6,000 ha at the beginning of the twentieth century to 3,800 ha in the 1980s (Bonte et al. 2002). This includes a decrease in area from 730 to 350 ha of rare grey dune along the western part of the Belgian coast, since the 1950s (Provoost and van Landuyt 2001). In addition to the overall losses, the dune areas are highly fragmented with only a narrow beach surviving along much of the coastal fringe (Fig. 2.7). Illegal huts in Portugal were built on dunes during the 1970s, such as those in the Leira district (Martins 1989) and in the Ria Formosa Natural Park near Faro. Together these further reduce the extent of open dune habitat and increase pressure on the remaining areas from other uses.

Barrier island urbanisations in the eastern United States support significant residential populations. These developments took place despite the regular occurrence of severe storms, which drive the barrier islands landward. There is an extensive literature on approaches to understanding and managing these inhabited zones, which is not the subject of this book. See Walker and Coleman (1987) for a general review of the Atlantic and Gulf Coastal areas and Psuty and Ofiara (2002) for a detailed appraisal of the situation on the New Jersey shoreline.

Other publications provide geographically wide-ranging information on loss of coastal areas from the USA (Stauble 1989) and worldwide (Schwartz 1973; Pilkey and Fraser 2003; Nordstrom 2004). The important point is that surviving undeveloped sand dunes are a diminished resource in many parts of the world.

2.5.2 Victorian Tourism

Dr. Richard Russell who came to Brighton, southern England in 1753 has the accolade of 'inventing' the seaside holiday (the forerunner of mass tourism). By 1822 when the famous Brighton Pavilion was complete, a resort was well established. The coming of the railway in 1841 stimulated further development and by 1900,

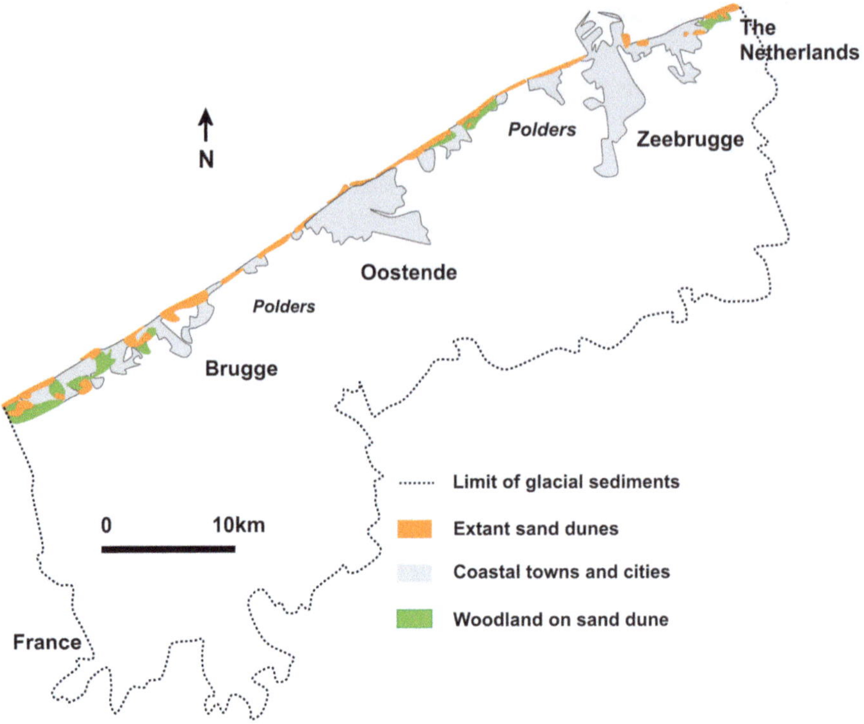

Fig. 2.7 Distribution of the surviving sand dunes along the Belgium coast. (Derived from Lanckneus et al. 2001)

there was 'ribbon' development along much of the south coast (Doody 1995). In 1783, Royal patronage encouraged the British aristocracy to visit other places like the Norfolk coast to take the seawater, thought to have medicinal properties similar to spa water.

Victorian workers usually had 1 week's holiday a year, often spent at the seaside. Subsequently the desire to live near the sea resulted in extensive urbanisation including the "traditional English Victorian seaside town" of Southport on the Sefton Coast. This together with housing development in the 1960s and 1970s obliterated significant areas of sand dune (Sect. 2.8.2). Similar losses occurred on the south Wales coastline, where Kenfig Burrows National Nature Reserve is one of the last remnants of huge dune systems that once stretched along the coast of Swansea Bay.

2.5.3 *Tourism in Europe*

The advent of tourism as we know it today began much later. With the increasing affluence of Europe's population and the arrival of cheap air travel in the latter half of the twentieth century, the holiday by the sea with sun and sand became an annual attraction. This manifests itself in the upsurge in package holidays with large numbers of tourists 'migrating' to the warm south (Bramwell 2004; Davenport and Davenport 2006). More permanent summer residences built by the local population, caused further losses.

Sandy shores and sand dunes provided sand for building, a location close to the sea, a beach for sun bathing and when left undeveloped, a recreational zone. The combined impact caused devastation of some of the finest coastal areas (including sand dunes) in southern Europe. Spanish urbanisations were amongst the first to affect the coast, and during the 1960s and 1970s developments were largely uncontrolled. Many areas were "totally altered and destroyed" (Gómez-Pina et al. 2002).

This pattern of development continued throughout the Mediterranean. Losses of sand dune habitat along the Spanish, French and mainland Italian coasts over 30 years amounted to 75–80% (van der Meulen and Salman 1996). In Greece from the 1960s, expansion of tourism transformed the socio-economic structure of the country but not without adverse social and ecological effects (Tzatzanis and Wrbka 2002). With each new development, there were fewer areas to exploit and increased competition led to the development of new 'low-cost' resorts proceeding eastwards to what was then Yugoslavia and on to Turkey. In North Africa, including the Nile Delta large-scale development also caused considerable loss of dune habitat (El Banna 2004). Examples from Bulgaria and Estonia, two countries recently joining the European Union in the early part of the twenty first century, show how tourism further extends the scale and geographical range of sand dune loss (Stancheva et al. 2011).

Many tourist urbanisations include a golf course as an essential part of the attraction for a "high quality" resort. A view of the dunes of Maspalomas in Gran Canaria in Google Earth show them surrounded by an extensive tourist "Time Share" urbanisation (Playa del Inglés) and an extensive golf course (Campo de Golf Maspalomas; Fig. 2.8). The remaining desert-like landscape includes natural 'wetlands' occurring as damp slacks with Canary Island Date Palm *Phoenix canariensis*.

There are no less than 27 golf courses in the Algarve, southern Portugal, some of which have affected dunes. The development of the greens, tees and fairways, coupled with the requirement for water, further destroys and degrades the natural environment. However, there are trade-offs associated with these developments especially when the management takes account of nature conservation values (Chap. 10).

2.5.4 Trends in World Tourism

Coastal areas around the world are also the focus for development. This has had a profound and adverse effect through the construction of coastal resorts and roads, marinas and jetties on sand dunes, causing habitat fragmentation and reduced biodiversity (Davenport and Davenport 2006). More specifically, it has resulted in severe environmental degradation when building hotels and houses directly on the dunes. Trampling by people further destroys vegetation causing increasing sand movement and dune erosion (Martinez et al. 2008). Comments in documents emanating from the UN Environmental Programme often refer to the growth and impact of world tourism on coastal systems. Mass tourism is, "together with urbanisation accused of

Fig. 2.8 Golf course adjacent to a timeshare development on Gran Canaria, Maspalomas dunes in the background 1992

being one of the main causes of the littoralisation phenomenon (ribbon development) and severe ecological losses in coastal areas".... "Land-grab and demand for resources are at the root of severe erosion phenomena, loss of valuable habitats (such as sand dunes coral reefs, wetlands and mangrove forests), the irreversible destruction of pristine areas and the loss of rare plant and animal species." (Markovic et al. 2009). A "Tourism Vision 2020" using 1995 as the base year prepared by the World Tourism Organisation predicts that by 2020 the global number of travellers will reach 1.6 billion (an increase of 4.1% per annum (Anon 2001)), so these problems can only increase.

In Newfoundland, Canada loss of fishing has had an unexpected consequence as the search for new commercial outlets has led to greater tourist development affecting sand dunes. The pressures generated have caused degradation of the dune systems. This has in turn, led to pressure to erect boardwalks and other infrastructure to cater for tourists and reduce their impact (Catto 2002).

2.6 Beach Sediment Depletion

The development of a sandy beach and foredune depends on the supply of sediment. Anything that reduces this supply will adversely affect its ability to deliver sand for the formation of new dunes. The removal of sand from the beach has an obvious

Fig. 2.9 Sand dune mining, the beach at Quendale, Shetland, September 1982

and direct impact. Offshore sand extraction and river damming, may deplete sediment availability. However, it may be difficult to establish direct cause and effect. The outcome of sediment loss from the beach can be all too obvious in areas with coastal development on or near the beach.

2.6.1 Onshore Sand Mining

Sand mining occurs on foreshores around the world. Most sandy beaches have at one time or another suffered from sand extraction ranging from a few wheelbarrow loads to industrial scale production (Pilkey et al. 2011). In Northern Ireland, for example, extraction of shell sand for agricultural use has a cumulative, if local impact (Carter et al. 1992). Small-scale local mining operations for agricultural and building purposes have taken place in Scotland (Fig. 2.9). There are many other examples in the Caribbean, the Azores, Africa especially northern Morocco where possibly the largest operation occurs, United States of America, Australia and New Zealand (Pilkey et al. 2011).

In Monterey Bay, California, extraction from the surf zone began in 1906, continuing up to 1990. The volume of material removed between 1940 and 1984 was approximately 50% of the total lost along the dune front due to erosion. As a result, sand mining exacerbated erosion trends along some parts of the coast

(Thornton et al. 2006). In Australia, extensive mining of the Cronulla sand dune system on the Kurnell Peninsula in Botany Bay, associated with the construction of Sydney, began in the 1930s. The removal of vast quantities of sand has severely weakened the system and from the 1950s resulted in campaigning for its conservation (National Trust of Australia http://www.nationaltrust.com.au/default.asp).

There are also other large-scale commercial operations, for mineral extraction. Examples include the Zululand coast, South Africa with the extraction of heavy minerals from the dunes since 1977 (Lubke et al. 1996). The extraction of valuable minerals also takes place on North Stradbroke Island's high sand dunes on the coast of Queensland, Australia. However, phasing out of this activity is due to take place by 2025 (Queensland Government 2011).

2.6.2 Offshore Extraction and River Damming

The extraction of sand and gravel from offshore deposits is a relatively recent phenomenon. Major uses are for the construction industry and the nourishment of eroding beaches. Great Britain and Japan produce two thirds of the world's marine aggregates from offshore sand and gravel mining. In the early 1920s, only a few million tonnes were landed in Great Britain, whereas today the landings from licensed extraction areas around England and Wales are regularly more than 20 million tonnes a year, representing 20% of the material used in England and Wales (Crown Estate http://www.thecrownestate.co.uk/marine/aggregates/). In the Netherlands, the extension of the Port of Rotterdam (Maasvlakte 2) will create 2,000 ha of additional port facilities. Much of the sand (365 million m^3) will come from deposits offshore extracted to a depth of 20 m to reduce the area of seabed affected (Port of Rotterdam Authority, "Sand extraction: Sand for land" http://www.maasvlakte2.com/kennisbank/sand_extraction_1.pdf). In the United States of America, offshore extraction is mainly for beach nourishment, where there are more than 200 nourished sites.

River dams affect many of the major rivers of the world and their construction increased considerably in the second half of the last century (Vörösmarty et al. 2003). Examples include the Aswan High Dam, on the Nile River, creating the world's third largest reservoir. In Spain, there are no less than 170 dams in the Ebro drainage system, (Sanchez-Arcilla et al. 1998; McManus 2002). Offshore sediment extraction together with river damming, which prevents sediment from reaching the coast, reduces the supply for building new sand dunes (Sects. 6.3.1 and 6.3.3 respectively).

2.6.3 Sea Defence

Beaches with a depleted sediment supply also have implications, especially for hotels and other infrastructure established near frontal dunes. As the beach losses sediment it becomes susceptible to erosion. The response is typically the construction

Fig. 2.10 Building on eroding shorelines not only destroys the sand dune habitat but also creates the need for additional unsightly 'protection', which may ultimately become economically unsustainable. An abandoned hotel on the New Jersey shoreline, United States of America in 1996

of concrete sea walls, revetments or other defences. These may help protect the buildings in the short term. However, they can exacerbate erosional trends by cutting off the supply of sediment to the backshore and preventing alongshore drift. In addition, they may reflect wave attack, forcing the sediment offshore and away from the beach. Reduced to a narrow zone, the beach and foredune no longer act as a buffer to waves and storms by absorbing wave energy. They are also increasingly vulnerable to erosion, overtopping and flooding. Eventually, despite the variety of artificial protective measures, the active coastal fringe disappears and the buildings become undermined (Fig. 2.10) and uninhabitable.

2.7 Sand Dune Loss – A European Perspective

Humans have influenced sand dunes for several thousand years. Prior to about 6,000 years ago, little evidence remains because of the rapid sea level rise that took place after the end of the last glaciation. It is possible to produce a trend line representing the intensity of development over the last 2,000 years or so. This conceptual approach suggests human activities have become more and more intense and directed towards deliberate intervention as opposed to accidental or incidental

Table 2.1 Estimated figures for loss of vegetated sand dunes through urbanisation and conversion to forestry or agriculture in Europe

Country	Afforested[a] (%)	'Natural' 2000[b] (%)	Main reason for loss/change
Norway	–	25	Urbanisation
Poland	–	50	Afforestation, tourism
Denmark	49	50	Afforestation, recreation
Germany	4	60	Urbanisation, recreation
Netherlands	11	60	Afforestation, recreation
Belgium	28	31	Tourism, urbanisation
Great Britain	14	58	Afforestation
Ireland	15	50	Recreation, agriculture
France	19	33	Afforestation, tourism
Spain	51	60	Afforestation, tourism & recreation
Portugal	35	40	Afforestation, agriculture & tourism
Italy	–	25	Tourism, urbanisation
Albania	–	33	Agriculture, tourism
Turkey	–	–	Afforestation, tourism

Percentage loss figures derived from afforestation (Tekke and Salman 1995[a]); surviving in a 'natural' state (Salman and Strating 1991, Internal EUCC Report, Leiden[b]), – no figure available. The percentage figures are included in one table for convenience; they are not directly comparable but give an indication of the prevalence of plantations of non-native trees over other forms of habitat loss

affects of earlier years. The curve rises rapidly from the nineteenth century as human population pressures increased (Nordstrom 2004).

Loss to urbanisation and other development is more or less permanent. These include major tourist, port and other infrastructure. Even quite small developments such as building individual summerhouses, can have a cumulative impact. This is in addition to the isolation associated with habitat fragmentation. Other uses, including afforestation and agriculture also have a major impact but are to some extent reversible.

2.7.1 A European Summary

The combined impact of the above human activities results in increased erosion, landward migration of the foredune and loss of vegetated inland sand dune. In Europe, there was a decline from an estimated area of approximately 707,000 ha in the 1900s to 530,000 ha in the year 2000. Of the 75% remaining dunes only 320,000 ha (45%) were considered to be in a 'natural' state. This figure dropped to 25% in the Mediterranean where between 1960 and 1995, 30 ha were estimated to have been lost every day (Salman and Strating 1991). The principal reasons for the losses are urbanisation, including European tourist development, afforestation and agriculture (Table 2.1).

2.7.2 *Water Abstraction*

Some sand dunes have an aquifer large enough to provide a supply of water abstraction, as in the Netherlands. Water abstraction here and in other locations for tourist development and agricultural adversely affected the sand dunes (e.g. Coto Doñana Section 7.7.1). More localised abstraction on 'links' golf courses can also change dune vegetation (Chap. 10).

2.8 Case Studies – The 'Sand Dune Squeeze'

It is important when considering conservation and management to recognise the cumulative nature of many human actions as they affect coastal sand dune systems. In this context, many coastal sand dunes are remnants of much larger systems. The following examples illustrate the way development and land use management, especially over the last 150 years, combine to alter the area of an individual system, its internal dynamics and surface topography.

2.8.1 *The Netherlands*

At a local level, the impact on individual habitats is considerable. In the Netherlands, for example, since 1850 sand mining, conversion to bulb fields, afforestation, urbanisations including road construction and industrial development have all taken their toll. The remaining dune areas have changed dramatically in other ways with the loss of dune slacks and scrub encroachment, partly due to drinking water abstraction (Table 2.2).

The figures included in the tables give an indication of the losses incurred. Two examples from the United Kingdom described next, illustrate cumulative losses to urbanisation, agriculture and forestry.

2.8.2 *The Sefton Coast, Northwest England*

The Sefton coast, northwest England has one of the largest continuous areas of blown sand in the United Kingdom. Human activities have influenced the area for several thousand years, and there may have been a well-developed farming system around the time of the 'Domesday Book', completed in 1086. Urbanisation of the Sefton coast did not begin until the late 1800s and early 1900s with the growth of Crosby (northern end of the Liverpool dock development), Formby and Southport. Golf course development (GC) causes loss of vegetated dune to greens, tees and fairways. In addition older inland dunes were probably already in use firstly as

Table 2.2 Change in the area of sand dune habitat in the Netherlands (Janssen 1992)

Date	1850 (ha)	1990 (ha)
Deciduous woodland	3,000	5,000
Coniferous forest	0	6,000
Thickets	5,000	9,000
Large-scale active dunes	4,000	300
Slufters[a]	4,000	1,500
Open water	500	700
Dune slacks	13,000	2,000
Dune heath	5,000	5,000
Other open dune habitat	6,500	8,500
Coastal village landscape	1,000	1,000
Total	42,000	39,000

Note: the dune area includes the inland dune field but excludes the beach, beach plain, foredunes and the wooded landward side of the dunes

[a]Slufters are breaches of the seaward line of dunes that permit seawater to penetrate the valley behind the dunes

rabbit warrens and later for agriculture (A). Other areas include an airfield and a military rifle range (MOD) (Fig. 2.11). Forestry planting also covers areas of mobile dune near Ainsdale (Jones et al. 1993). The figure does not show these areas, but see Sect. 9.7.2.

Each of these activities causes a cumulative loss of habitat. In areas where there is also loss of sediment and/or sea level is rising, there is a sand dune 'squeeze'. On the Sefton coast, some 40% of the sand dune has been lost to development. Restoration of this built environment (houses, roads and other infrastructure) to sand dune is largely unattainable. Of the remaining 15% or so, restoration of the diverse dune vegetation on the more intensively used golf course greens, tees and fairways is probably unlikely. They do however, offer opportunities for sympathetic management in parts of the playing area especially in the 'roughs' (Chap. 10), as do the afforested areas (Sect. 9.7.2). Today the surviving sand dunes outside these areas are largely in the ownership of conservation bodies and/or 'protected' by a variety of international and national nature conservation designations (Smith 1999).

2.8.3 *Tentsmuir, Southeast Scotland*

Tentsmuir in southeast Scotland provides an illustration of the way stabilisation can alter the dune surface, leaving only a small area for active dune processes to take place. The site has a long history of human occupation including use for agriculture (Whittington 1996). An extensive area of sand dune heath at the turn of the twentieth century, planting took place over a large proportion of the site with various

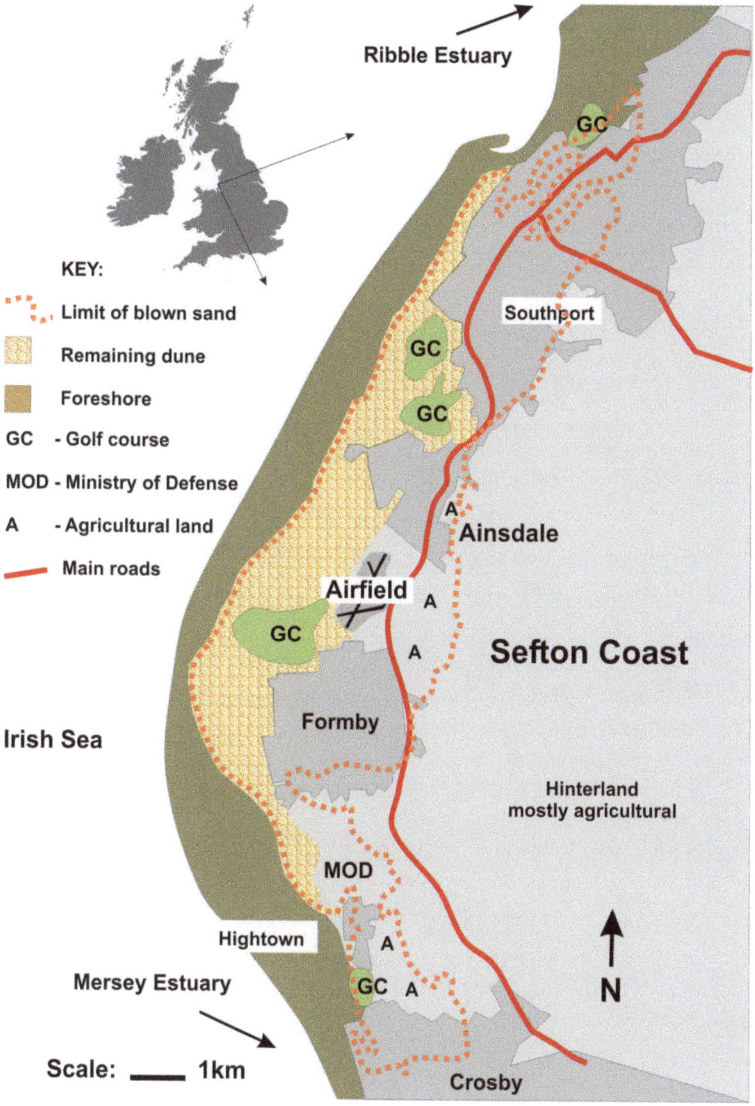

Fig. 2.11 Sketch map of the Sefton coast, northwest England showing the extant sand dunes in relation to the approximate known limits of blown sand. Information derived from a variety of sources, including maps and aerial photographs

species of exotic pines in the 1950s (Sect. 2.3.2). Today, the only dynamic part of the sand dune system is restricted to accretion at Tentsmuir Point, which forms part of the Tentsmuir National Nature Reserve (Fig. 2.12). This only exists because of the relatively abundant sediment coming from the River Tay to the north, which facilitates the growth of sand dunes at the point.

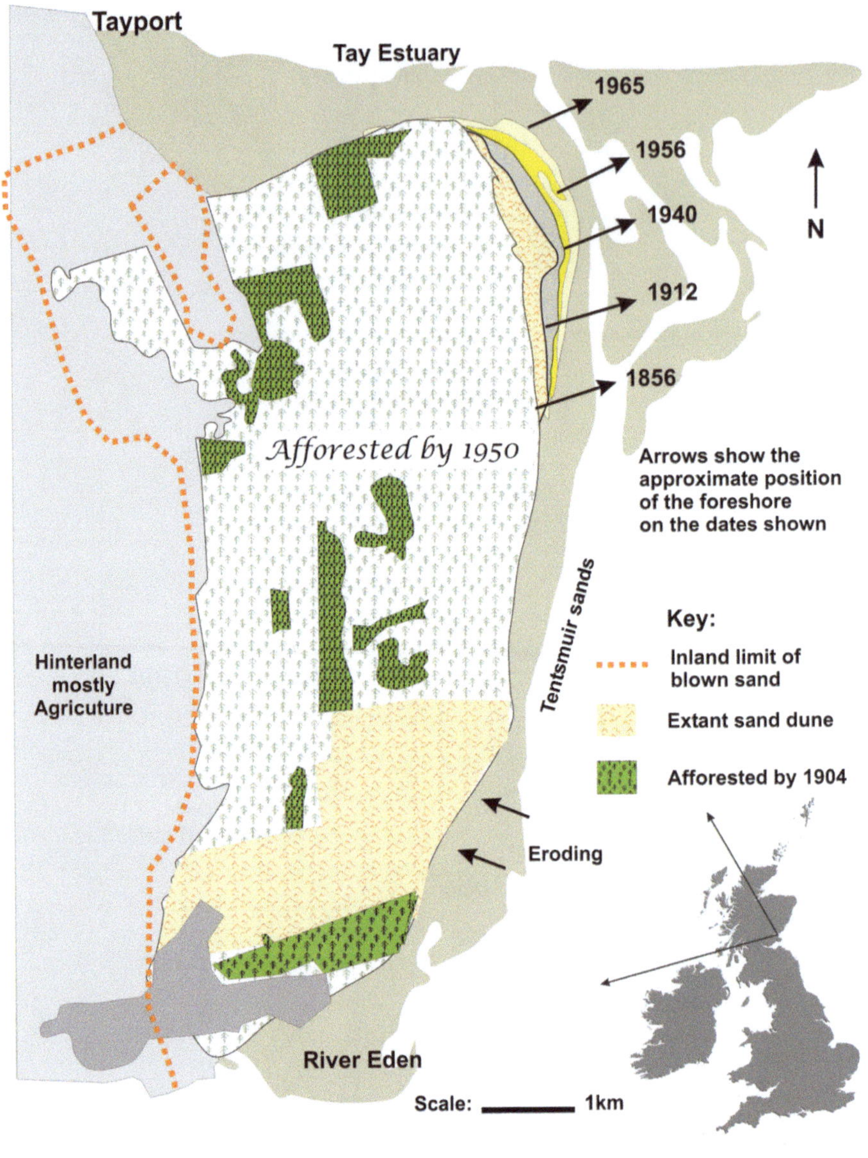

Fig. 2.12 Sketch map of Tentsmuir sand dune showing the approximate inland limit of blown sand, as shown by the Geology of Britain viewer (British Geological Survey http://mapapps.bgs.ac.uk/geologyofbritain/home.html). The area of afforestation and the remaining areas of open stable sand dune (Earlshall Muir in the south) and a series of accreting foredunes (Tentsmuir Point in the north east) are shown. Hansom (2003) provides an update of the position of the accreting mobile foredune up to 1990, together with an area of erosion in the south east

In the case of Tentsmuir only approximately 30% of the open stable sand dune remained free from forestry planting, with the majority lying within the southern section of the site. Removing the forest (deforestation) is physically possible but difficult to achieve because of economic forestry, recreational and political considerations (Sect. 9.7). Managing the limited area of accreting sand requires careful consideration of all the factors influencing the geomorphological status and biological diversity of the area, as does the larger area of the extant sand dune to the south. Both sites illustrate the extent to which destruction and modification leave a smaller area which is more sensitive to change.

2.9 Conclusion

This chapter has chronicled the losses of coastal sand dunes and changes to the habitat, over the last 2,000 years. The impacts have ranged from modification of the vegetation by introducing grazing by domesticated stock, to afforestation and destruction through infrastructure development. Infrastructure development and afforestation leave a smaller area of vegetated sand dune. When combined with rising sea levels or depleted sediment supply the beach/foredune sand system can migrate landwards creating a sand dune 'squeeze'. As the inland sand dunes become smaller, they are more susceptible to changing environmental conditions, especially from human activity. Recognition of the inherent natural values associated with sand dunes, has led to a range of legislation designed to protect these remaining areas from further damage (Chap. 3).

References

Anon (2001) Tourism 2020 vision – global forecast and profiles of market segments, vol 7. World Tourism Organisation, Madrid, 139 pp

Baeyens G, Martínez ML (2004) Dune animals: from exploitation to protection. In: Martinez ML, Psuty N (eds) Coastal dunes, ecology and conservation, vol 171, Ecological studies. Springer, Berlin/Heidelberg, pp 279–296

Bailey G, Milner N (2002/2003) Coastal hunter-gatherers and social evolution: marginal or central? Before Farm 4/1:1–15

Bonte D, Baert L, Maelfait J-P (2002) Spider assemblage structure and stability in a heterogenous coastal dune system (Belgium). J Arachnol 30(2):331–343

Bramwell B (ed) (2004) Coastal mass tourism: diversification and sustainable development in Southern Europe. Channel View Publications, Clevedon, 300 pp

Carey AE, Oliver FW (1918) Tidal lands, a study in shore problems. Blackie and Son, London, 284 pp

Carter RWG, Eastwood DA, Bradshaw P (1992) Small scale sediment removal from beaches in Northern Ireland: environmental impact, community perception and conservation management. Aquat Conserv: Mar FreshwEcosyst 2(1):95–113

Catto N (2002) Anthropogenic pressures on coastal dunes, Southwestern Newfoundland. Can Geogr 46:17–32

Coles JM (1964) New aspects of Mesolithic settlement of south-west Scotland. Trans Dumfries Galloway Nat Hist Antiqu Soc 12:67–98

Corona MG, Vicente AM, Garcia Novo F (1988) Long-term vegetation changes on the stabilized dunes of Doñana National Park (SW Spain). Plant Ecol 75(1–2):73–80
Danielsen R (2008) Palaeoecologial development of the Quiaios–Mira dunes, northern-central littoral Portugal. Rev Palaeobot Palynol 152(1–2):74–99
Davenport J, Davenport JL (2006) The impact of tourism and personal leisure transport on coastal environments: a review. Estuar, Coast Shelf Sci 67(1–2):280–292
Deresea C, Vandenberghea D, Eggermontb N, Bastiaensb J, Annaertb R, Van den haute P (2010) A medieval settlement caught in the sand: optical dating of sand-drifting at Pulle (N Belgium). Quat Geochronol 5(2–3):336–341
Doody JP (1995) Infrastructure development and other human influences on the coastline of Europe. In: Salman AHPM, Berends H, Bonazountas M (eds) *Coastal Management and Habitat Conservation*. Proceedings of the 4th EUCC Congress, Marathon, Greece, Vol. 1 EUCC, Leiden, Netherlands pp 133–152
El Banna MM (2004) Nature and human impact on Nile Delta coastal sand dunes, Egypt. Environ Geol 45(5):690–695
Fabbri P (1997) Tourism, recreation and coastal dunes of the Mediterranean. In: Drees JM (ed) Coastal dunes, recreation and planning. EUCC, Leiden, pp 18–20
Favennec J (1998) The dunes of the Atlantic coast of France, typology and management. In: Ovesen CH (ed) Coastal dunes, management, protection and research, report from a European seminar, Skagen, Denmark. National Forest and Nature Agency, Geological Survey of Denmark and Greenland, Copenhagen, pp 130–138
Feilberg A, Jensen F (1992) Management and conservation of sand dunes in Denmark. In: Carter RWG, Curtis TGF, Sheehy-Skeffington MJ (eds) Coastal dunes: geomorphology, ecology and management, proceedings of the third European Dune Congress, Galway, Ireland, 17–21 June 1992. A.A. Balkema, Rotterdam, pp 429–437
Gadgil RL (1971) The nutritional role of *Lupinus arboreus* in coastal sand dune forestry. Plant Soil 35:113–126
Garcia Novo F (1997) The ecosystems of Doñana National Park (southwest Spain). In: Garcia Novo F, Crawford RMM, Diaz Barradus CD (eds) The ecology and conservation of European dunes. Universidad de Sevilla, Savilla, pp 149–154
Gómez-Pina G, Muñoz-Pérez JJ, Ramírez JL, Ley C (2002) Sand dune management problems and techniques, Spain. J Coast Res, Special issue 36:325–332
Hansom JD (2003) Chapter 7: sandy beaches and dunes – GCR site reports Site: Tentsmuir (GCR ID: 1070). In: May VJ, Hansom JD (eds) Coastal geomorphology of Great Britain. GCR, 28, p 2731. Source: http://jncc.defra.gov.uk/pdf/gcrdb/GCRsiteaccount1070.pdf. Accessed 27 Nov 2011
Hilton M, Macauley U, Henderson R (2000) Inventory of New Zealand's active dunelands. Sciconserv, 157:30. Source: http://www.doc.govt.nz/upload/documents/science-and-technical/sfc157.pdf, which includes a series of 124 detailed distribution maps. Accessed 20 Apr 2011
Høgestøl M, Prøsch-Danielsen L (2006) Impulses of agro-pastoralism in the 4th and 3rd millennia BC on the south-western coastal rim of Norway. Environ Archaeol 11(1):19–34
Janssen MPJM (ed) (1992) Duinen voor de wind. Een toekomstvisie op het gebruik en het beheer van de Nederlandse duinen als natuurgebied van internationale betekenis. Stichting Duinbehoud. Leiden. Source: http://www.compendiumvoordeleefomgeving.nl/indicatoren/nl1122-Areaalverandering-van-duinbiotopen.html?i=4-26. Accessed 14 Aug 2012
Jones CR, Houston JA, Bateman D (1993) A history of human influence on the coastal landscape. In: Atkinson D, Houston J (eds) The sand dunes of the Sefton coast. Proceedings of the Sefton coast research seminar, Liverpool, 31st May 1991. National Museums and Galleries on Merseyside, Liverpool, pp 3–20
Kelemen J (1992) Conservation and management of sand dune areas in the valley of the River Danube. In: Carter RWG, Curtis TGF, Sheehy-Skeffington MJ (eds) Coastal dunes: geomorphology, ecology and management, proceedings of the third European Dune Congress, Galway, Ireland, 17–21 June 1992. A.A. Balkema, Rotterdam, pp 487–492

Kiss T, Sipos G, Bódis K, Barta K (2004) Selection of human influenced fixed sand dunes on the basis of morphometric measurements South Nyírség Region, Hungary. Int J Fieldwork Stud 2(1):1–23

Klijn J, Bakker Th (1992) Vijfduizend jaar dieren in de duinen. Duin 15:3–5

Knight J, Burningham H (2007) Coastal morphodynamics and prehistoric human occupation, County Donegal, NW Ireland. J Coast Res SI 50, Proceedings of the 9th International Coastal Symposium:104–108

Koster EA (2009) The "European aeolian sand belt": geoconservation of drift sand landscapes. Geoheritage 1(2–4):93–110

Kvamme M, Kaland PE (2009) Prescribed burning of coastal heathlands in western Norway: history and present day experiences. Int For Fire News 38:35–50

Laar J van (1995) Coastal dune afforestation and developments of dune forests during the ninetheenth and twentieth centuries in the Netherlands. In: Caring for the forest: research in a changing world. Abstract 20th IUFRO World Congress, Tampere, Finland, pp 431–432

Lanckneus J, Van Lancker V, Moerkerke G, Van Den Eynde D, Fettweis M, De Batist M, Jacobs P (2001) Investigation of natural sand transport on the Belgian continental shelf BUDGET (Beneficial usage of data and geo-environmental techniques). Final report, Federal Office for Scientific, Technical and Cultural Affairs (OSTC), p 104+p 87 Annex. Source: http://www.kustatlas.be/en/fysisch_milieu/een_dynamisch_en_broos_milieu/map/?id=3&lan=en. Accessed 20 Apr 2011

Lewis D (1992) The sands of time: Cornwall's Hayle to Gwithian Towans. In: Carter RWG, Curtis TGF, Sheehy-Skeffington MJ (eds) Coastal dunes: geomorphology, ecology and management, proceedings of the third European Dune Congress, Galway, Ireland, 17–21 June 1992. A.A. Balkema, Rotterdam, pp 463–473

Lubke RA, Avis AM, Moll JB (1996) Post-mining rehabilitation of coastal sand dunes in Zululand South Africa. Landsc Urban Plan 34(3–4):335–345

Markovic M, Satta A, Skaricic Z, Trumbic I (2009) Sustainable coastal tourism/an integrated planning and management approach. United Nations Environment Programme, Division of Technology, Industry and Economics (UNEP-DTIE) "Practical Manuals on Sustainable Tourism" publication series. Source: http://www.unep.org/pdf/DTIE_PDFS/DTIx1091xPA-SustainableCoastalTourism-Planning.pdf. Accessed 02 Nov 2011

Martins F (1989) Morphology and management of dunes at Leiria district (Portugal). In: van der Meulen F, Jungerius PD, Visser J (eds) Perspectives in coastal dune management. SPB Academic Publishing, The Hague, pp 81–90

Martinez ML, Psuty NP, Lubke RA (2008) A perspective on coastal dunes. In: Martinez ML, Psuty NP (eds) Coastal dunes, ecology and conservation. Ecological Studies Vol 171 Springer-Verlag Berlin Heidelberg, pp 3–10

McKelvey PJ (1999) Sand forests: a historical perspective of the stabilisation and afforestation of coastal sands in New Zealand. Canterbury University Press, Christchurch, 224 pp

McManus J (2002) Deltaic responses to changes in river regimes. Mar Chem 79:155–170

Ministry for the Environment (2007) Protecting our places: information about the statement of national priorities for protecting rare and threatened biodiversity on private land. Ministry for the Environment publication ME 805. Source: https://www.biodiversity.govt.nz/pdfs/protecting-our-places-brochure.pdf. Accessed 10 Dec 2010

Nordstrom KF (2004) Beaches and dunes of developed coasts. Cambridge University Press, Cambridge, 352 pp

Ovington JD (1950) The afforestation of the Culbin sands. J Ecol 38(2):303–319

Ovington JD (1951) The afforestation of Tentsmuir sands. J Ecol 39(2):363–375

Owen H (1953) Hanes Plwf Niwbwrch ym Mon. Caernarvon

Parsons JR, Psuty NP (1975) Sunken fields and prehispanic subsistence on the Peruvian coast. Am Antiq 40(3):259–282

Pilkey OH, Fraser MA (2003) A celebration of the world's barrier islands. Columbia University Press, New York, 400 pp

Pilkey OH, Neal WJ, Cooper JAG, Kelley JT (2011) The world's beaches: a global guide to the science of the shoreline. University of California Press, Berkeley, 283 pp
Pilon JJ (1988) Kustafslag op Voorne in het verleden. Waterschapsbelangen 73:255–259
Piotrowska H (1989) Natural and anthropogenic changes in sand dunes and their vegetation on the southern Baltic coast. In: van der Meulen F, Jungerius PD, Visser JH (eds) Perspectives in coastal dune management. SPB Academic Publishing, The Hague, pp 33–40
Povilanskas R, Baghdasarian H, Arakelyan S, Satkunas J, Taminskas J (2009) Secular morphodynamic trends of the holocene dune ridge on the Curonian spit (Lithuania/Russia). J Coast Res 25(1):209–215
Prøsch-Danielsen L, Selsing L (2011) Aeolian activity during the last 9200 calendar years BP along the southwestern coastal rim of Norway. Nor Geolog Foren. Source: http://www.geologi.no/data/f/0/20/44/0_2401_0/Sandfl_p_Jaeren_2011.pdf. Accessed 01 Nov 2011
Provoost S, van Landuyt W (2001) The flora of Flemish coastal dunes (Belgium) in a changing landscape. In: Houston JA, Edmondson SE, Rooney PJ (eds) Coastal dune management, shared experience of European conservation practice. Liverpool University Press, Liverpool, pp 381–401
Psuty NP, Ofiara DD (2002) Coastal hazard management – lessons and future directions from New Jersey. Rutgers University Press, New Brunswick, 448 pp
Pye K, Saye S, Blott S (2007) Sand dune processes and management for flood and coastal defence. Part 1, Project overview and recommendations. DEFRA, R&D technical report, FD1302/TR. Source: http://www.safecoast.org/editor/databank/File/sand%20dune%20processes%20and%20management%20%28overview%29.pdf. Accessed 17 Mar 2010
Queensland Government (2011) The North Stradbroke island protection and sustainability act, 14th April 2011. Source: http://www.legislation.qld.gov.au/LEGISLTN/ACTS/2011/11AC011.pdf. Accessed 14 Aug 2012
Ranwell DS (1959) Newborough Warren, Anglesey. 1. The dune system and dune slack habitat. J Ecol 47:571–601
Riksen M, Ketner-Oostra R, van Turnhout C, Nijssen M, Goossens D, Jungerius PD, Spaan W (2006) Will we lose the last active inland drift sands of western Europe? The origin and development of the inland drift-sand ecotype in the Netherlands? Landsc Ecol 21(3):431–447
Ross S (1992) The Culbin sands – fact and fiction. Centre for Scottish Studies, University of Aberdeen, Aberdeen, 196 pp
Runólfsson S (1987) Land reclamation in Iceland. Arct Alp Res 19(4):514–517
Salman AHPM, Strating C (1991) European coastal dunes and their decline since 1900. Internal Report EUCC, Leiden
Sanchez-Arcilla A, Jimenez JA, Valdemoro HI (1998) The Ebro delta: morphodynamics and vulnerability. J Coast Res 14:754–772
Schwartz ML (ed) (1973) Barrier islands. Dowden Hutchinson and Ross, Pennsylvania, 451 pp
Skarregaard P (1989) Stabilisation of coastal dunes in Denmark. In: van der Meulen F, Jungerius PD, Visser J (eds) Perspectives in coastal dune management. SPB Academic Publishing, The Hague, pp 161–181
Smith PH (1999) The sands of time – an introduction to the sand dunes of the Sefton coast. National Museum and Galleries on Merseyside, Liverpool, 196 pp
Stancheva M, Ratas U, Orviku K, Palazov A, Rivis R, Kont A, Peychev V, Tõnisson H, Stanchev H (2011) Sand dune destruction due to increased human impacts along the Bulgarian Black Sea and Estonian Baltic Sea coasts. J Coast Res 64:324–328
Stauble DK (ed) (1989) Barrier islands: process and management. American Society of Civil Engineers, New York, 324 pp
Tekke RMH, Salman AHPM (1995) Coastal woodlands, forestry and conservation along the Atlantic and North Sea shores. In: Salman AHPM, Berends H, Bonazountas M (eds) Coastal management and habitat conservation, proceedings of the 4th EUCC congress, Marathon, Greece, April, 1993. EUCC, Leiden, pp 395–409
Thornton EB, Sallenger A, Conforto Sesto J, Egley L, McGeee T, Parsons R (2006) Sand mining impacts on long-term dune erosion in southern Monterey Bay. Mar Geol 229(1–2):45–58

Tzatzanis M, Wrbka T (2002) Sun beds vs. sand dunes: a conservation – tourism conflict. In: Brebbia CA (ed) Coastal environment environmental problems in coastal regions IV, vol 8, Environmental studies. Boston WIT Press, Southampton, pp 25–34

Uslu T (1995) Coastal dune afforestation policy in Turkey: environmental aspects. In: Salman AHPM, Berends H, Bonazountas M (eds) Coastal management and habitat conservation, proceedings of the 4th EUCC congress, Marathon, Greece, April, 1993. EUCC, Leiden, pp 113–118

van der Meulen F, Salman AHPM (1996) Management of Mediterranean coastal dunes. Ocean Coast Manag 30(2–3):77–195

Vogiatzakis IN, Pungetti G, Mannion AM (eds) (2008) Mediterranean island landscapes. Natural and cultural approaches, vol 9, Landscape series. Springer, Dordrecht, 372 pp

Vörösmarty CV, Meybeck M, Fekete B, Sharma K, Green P, Syvitski J (2003) Anthropogenic sediment retention: major global impact from registered river impoundments. Glob Planet Change 39:169–190

Walker HJ, Coleman JM (1987) Atlantic and Gulf Coastal Province. In: Graf WL (ed) Geomorphic systems of North America. Geological Society of America, Boulder, pp 51–110

Walrond C (2007) Dune lands, Te Ara – the encyclopaedia of New Zealand, Updated 25 Sept 2011. Source: http://www.TeAra.govt.nz/TheBush/Landscapes/DuneLands/en. Accessed 16 Jan 2012

Whittington G (1996) Landuse of Tentsmuir: a history of diverse activities. In: Whittington G (ed) Fragile environments: the use and management of Tentsmuir NNR, Fife. Scottish Cultural Press, Edinburgh, pp 16–25

Chapter 3
Nature Conservation – Policy and Procedures

Abstract In historical times human occupation caused many sand dunes to become destabilised, and uninhabitable (Chap. 2). The enactment of laws to prevent this were amongst the first conservation measures.

Although some sand dunes became nature reserves early in the twentieth century, it was not until the second half of that century that they became places more widely valued for their wildlife interest and even later for their geomorphological value.

This chapter deals with mechanisms for the conservation of sand dunes at international and national levels. It introduces two examples of inventories designed to help identify important sites in Great Britain and Europe. Thereafter, it describes a number of approaches to habitat and site conservation through ownership of nature reserves, statutory sites designation and other legal mechanisms.

3.1 Coastal Sand Dune Inventories

A prerequisite for conserving any habitat is knowledge of the distribution, scale and status of individual sites. Section 1.5 provides a general review of sand dune distribution and types of vegetation in Europe, the United States of America and other parts of the world. These examples represent the first stage in developing a sound knowledge of the resource. The overview provided in Chap. 1 is a good starting point, but the identification and establishment of 'protected areas' requires more detailed information on individual sites. Examples of two 'inventories' are described next.

J.P. Doody, *Sand Dune Conservation, Management and Restoration*,
Coastal Research Library 4, DOI 10.1007/978-94-007-4731-9_3,

3.1.1 A Sand Dune Survey of Great Britain and the EU Habitats Directive

A comprehensive review of the sand dune resource in Great Britain took place in the late 1980s and early 1990s. The results, published in three reports were Vol. I, England (Radley 1994), Vol. II, Scotland (Dargie 1993) and Vol. III, Wales (Dargie 1995). The survey used a standard methodology for vegetation mapping and classification in accordance with the National Vegetation Classification (Rodwell 2000). Using the overall distribution map derived from existing surveys and Ordnance Survey maps, summarised in Fig. 1.13, a schedule of site surveys was undertaken. The survey included most of the sites in England and Wales but only 30% of the resource in Scotland. The survey was extended and a review of all the sand dunes in Scotland completed in 1999/2000 (Dargie 2002).

The survey of the English sites provided a national inventory (Radley 1992). This together with the other country reports helped form the basis for the selection of a Special Area of Conservation (SAC) under the European Union Habitats Directive (Hopkins and Radley 2001), identify 'Ecological Zones' and provide a basis for monitoring future habitat change. The results highlighted the enormous diversity of coastal sand dune vegetation, with more than 120 distinct vegetation types reflecting the considerable range of variation that exists between different geographical areas.

The English survey has a high level of detail but aggregation of the Sand Dune (SD) vegetation types provides quantitative data on ten key communities, including four 'Priority' habitats (Table 3.1).

Selected sites have:

1. Good representation of the types of vegetation and their associated animals;
2. A large proportion of the habitat within the Biogeographical region and/or the European Union as a whole;
3. Intact structure and function;
4. Representation at a regional level.

The UK is particularly important for Atlantic dune grasslands, and the SAC series reflects this. The main "Priority" dune types are: "Fixed dunes with herbaceous vegetation ('grey dunes')" EU Code 2130; "Decalcified fixed dunes with *Empetrum nigrum*" EU Code, 2140; "Atlantic decalcified fixed dunes (*Calluno-Ulicetea*)" EU Code, 2150; "Coastal dunes with *Juniperus* spp." EU Code, 2250.

The sites selected based on this classification, are the largest representative examples in the United Kingdom. Assessed as having good structure and function, they often form part of a wider coastal ecosystem. Transitions to landward vegetation are also important deciding factors in site selection. The background to selection includes a description of their ecological characteristics, UK and European status and distribution (McLeod et al. 2005).

Other surveys specifically targeted at sand dunes and sand dune vegetation include extensive work in Finland (Hellemaa 1998); strandline and sand dune vegetation of coasts of Greece and of some Aegean islands (Sýkora et al. 2003) and New Zealand's active sand dunes (Hilton et al. 2000) (see also Sect. 1.5).

Table 3.1 Coastal habitats of Community interest (sand dunes) forming the basis for the selection of Special Areas of Conservation (SAC), European Commission (2007)

National vegetation classification communities, relevant pages (Rodwell 2000) applicable to the UK only	Directive code and name – priority habitat[b]
SD 2 *Honkenya peloides-Cakile maritima* strandline community (pp 136–139)	1210 – Annual vegetation of drift lines (Communities of sandy/shingle shorelines)
SD 3 *Matricaria maritima-Galium aparine* strandline community (pp 140–143)	
SD 4 *Elymus farctus* ssp. *boreali-atlanticus* foredune community (pp 144–147)	2110 – Embryonic shifting dunes
SD 5 *Leymus arenarius* mobile dune community (pp 148–152)	
SD 6 *Ammophila arenaria* mobile dune community (pp 153–162)	2120 – Shifting dunes with *Ammophila arenaria* (white dunes)
SD 7 *Ammophila arenaria-Festuca rubra* semi-fixed dune community (pp 163–173)	2130 – Fixed dunes with herbaceous vegetation (grey dune)
SD 8 *Festuca rubra-Galium verum* fixed dune grassland (pp 174–187)	
SD 9 *Ammophila arenaria-Arrhenatherum elatius* dune (pp 188–193)	
SD 10 *Carex arenaria* dune community (pp 194–200)	
SD 11 *Carex arenaria-Cornicularia aculeata* dune community (pp 201–207)	
SD 12 *Carex arenaria-Festuca ovina capillaris* dune grassland (pp 208–213)	
H 11 *Calluna vulgaris-Carex arenaria* heath (Rodwell 1991; pp 484–491)	2140 – Decalcified dunes with *Empetrum nigrum*[b]
H 11 *Calluna vulgaris-Carex arenaria* heath (Rodwell 1991; pp 484–491)	2150 – Atlantic decalcified fixed dunes (*Calluno-Ulicetea*)[b]
SD 10 *Carex arenaria* dune community (*Festuca rubra* sub-community) includes *Corynephorus canescens* (pp 194–200)	
SD 18 *Hippophae rhamnoides* dune scrub (pp 242–246)	2160 – Dunes with *Hippophae rhamnoides*
SD 16 *Salix repens-Holcus lanatus* dune-slack community (pp 230–235)	2170 – Dunes with *Salix arenaria* ssp. *argentea* (*Salicion arenariea*)
SD 13 *Sagina nodosa-Bryum pseudotriquetrum* dune-slack community (pp 214–217)	2190 – Humid dune slacks
SD 14 *Salix repens-Camylium stellatum* dune-slack community (pp 219–223)	
SD 15 *Salix repens-Calliergon cuspidatum* dune-slack community (pp 224–229)	
SD 16 *Salix repens-Holcus lanatus* dune-slack community (pp 230–235)	
SD 17 *Potentilla anserine-Carex nigra* dune-slack community (pp 236–241)	
SD 8 *Festuca rubra-Galium verum* fixed dune grassland (pp 174–187)	21A0 – Machairs Priority habitat[b] in Ireland only

(continued)

Table 3.1 (continued)

National vegetation classification communities, relevant pages (Rodwell 2000) applicable to the UK only	Directive code and name – priority habitat[b]
No specific types but see note below[a]	2250 – Coastal dunes with *Juniperus* spp.
No specific types	2180 – Wooded dunes of the Atlantic, Continental and Boreal region

[a]The sand dune survey of Great Britain included a wide variety of vegetation types in addition to those typically associated with sand dunes. These ranged from neutral, calcareous and acid grassland, heathlands, wet heaths and mires, fens and swamps, scrub and woodland and transitions to other inland habitats. This served to emphasise the interrelationships between sand dunes and other coastal and inland terrestrial habitats, so important to the management and restoration of the dune itself

[b]Indicates Priority Habitat. Note SD 1 (not included) is a shingle beach community; SD 2 & 3 are also shingle beach communities but can be included here when there is a high proportion of sand. The equivalent National Vegetation Classification communities are shown alongside the equivalent Natura 2000 codes for sand dunes of the Atlantic, North Sea and Baltic

3.1.2 Sand Dune Inventory of Europe

Early descriptions of the sand dunes of Europe included a Council of Europe Report (Géhu 1985) and a wide-ranging description in a special *Catena* publication (Bakker et al. 1990). The Coastal & Marine Union (EUCC) in an effort to promote dune conservation in Europe proposed the production of a 'Sand Dune Inventory of Europe' (Doody 1991). The continuing adverse impact of human activity on the flora, fauna and physical characteristics of the coast provided the driving force for its preparation. The meeting 'Dunes and Estuaries' an International Conference on Nature Restoration Practices in European Coastal Habitats, held at Koksijde, Belgium, 19–23 September 2005 prompted the production of the revised inventory (Doody 2005).

The results gave a brief outline of the European sand dune resource for most countries, stretching from Iceland in the northwest to Turkey and Israel in the east. The inventory provides a brief description of the type of dune formation, size of the overall resource, vegetation, important sites, comment on conservation issues and a short list of references. This provides a first stage in the review of Europe's sand dunes, helping to establish the basis for the identification of important sites and policy issues affecting their conservation (Doody 2008).

The report shows that despite enormous losses, there are many surviving areas of important dunes. With the increasing recognition of their value, many of the sites identified in this report form part of the European-wide Natura 2000 network.

3.2 Nature Reserves and Other Protected Areas

The identification and management of nature reserves are cornerstones of nature conservation policy. The following sections provide an illustration of the type of reserves at different levels of jurisdiction. It is not exhaustive. Most of the information comes from the inspection of individual site reports on the Internet.

3.2.1 International

Some examples of World Heritage Sites with major coastal sand dune interest identified by the United Nations Educational, Scientific and Cultural Organization (UNESCO), include:

1. Fraser Island which lies just off the east coast of Australia, is 122 km long and reputed to be the largest sand island in the world;
2. Doñana National Park in Andalusia, Spain situated on the right bank of the Guadalquivir River, which empties into the Atlantic Ocean;
3. Sigatoka Sand Dunes, Fiji, are located directly west of the mouth of the Sigatoka River. The product of fluvial erosion in the coastal hinterland they cover an area of 650 ha;
4. Curonian Spit on the southern Baltic Sea coast is 98 km long and up to 4 km wide, an outstanding example of a sand dune landscape.

In 2012 there were 580 Biosphere Reserves (also defined by UNESCO) worldwide in 114 countries. They included 31 sites, partially or wholly recognised for their sand dune interest. Information on these and other individual sites is available from the World Conservation Monitoring Centre (WCMC) Cambridge, United Kingdom, World Database on Protected Areas. In addition to the World Heritage Sites and Biosphere Reserves, they include sites designated under the Ramsar Convention on Wetlands. This is an intergovernmental treaty providing the framework for the conservation and wise use of wetlands signed in Ramsar, Iran, in 1971. Several of these wetlands include coastal sand dunes within their boundaries.

3.2.2 Natura 2000 Designated Sites and Nature Reserves

Information collection at national level combined within a Europe wide classification helps to define a network of habitats and species sites of European importance. Called Natura 2000 it seeks to conserve Europe's natural heritage. There are two principal legal instruments:

1. The European Community Council Directive (92/43/EEC) on the Conservation of Natural Habitats and of Wild Fauna and Flora (Habitats Directive). It aims to assure the long-term survival of Europe's most valuable and threatened species and habitats. It requires member states to designate Special Areas of Conservation, which are sites of European importance for habitats listed on Annex I and for species listed on Annex II of the Directive;
2. The European Community Council Directive (2009/147/EC) on the conservation of wild birds (Birds Directive). Special Protection Areas (SPAs) are sites of European importance identified under the Birds Directive. The species are those classified under Article 4 of the Birds Directive and additionally include regularly occurring migratory species.

The European Environment Agency is the primary organisation providing environmental information for the European Union. The European Nature Information

Fig. 3.1 Extensive stable acidic sand dunes including grassland with *Corynephorus canescens* and heath on the island of Spiekeroog on the German coast, part of the Wadden Sea Natura 2000 site, August 2008

System (EUNIS) collates data collected and maintained by the European Topic Centre on Biological Diversity to facilitate access to information on species, habitats and sites across Europe. It is possible to obtain maps showing the presence of specific habitat types within Natura 2000 sites. A search for "Shifting dunes along the shoreline with *Ammophila arenaria* (white dunes)" for example, revealed 665 sites in the database with this habitat type distributed widely throughout the European Union. The more restricted "Atlantic decalcified fixed dunes (Calluno-Ulicetea)" was present in only 69 sites. For some sites, the habitat forms a relatively small part of the designated area. For others, it is a major component as for example on the Wadden Sea islands (Fig. 3.1).

3.2.3 National Nature Reserves – Examples from the United Kingdom

Most nature reserves exist within a framework of national legislation or via private organisations with a specific local interest. Owning and managing nature reserves usually offers the best form of protection from development and allows nature conservation management to take place. In the UK, for example there are three Biosphere Reserves with sand dune interest, Dyfi, in Wales and Braunton Burrows and North

Fig. 3.2 Blakeney Point, a shingle spit overlain by sand dunes and enclosing saltmarsh. The lifeboat station in the far distance is now a visitor centre, September 2008

Norfolk in southwest and eastern England respectively. Ynyslas sand dunes National Nature Reserve (NNR), owned and managed by the UK Government, through the Countryside Council for Wales, lies within the Dyfi Biosphere Reserve.

The North Norfolk coast includes land owned and managed by Natural England and several Non-Governmental Organisations (NGOs) including the National Trust, the Royal Society for the Protection of Birds and the Norfolk Wildlife Trust. The National Trust acquired Blakeney Point (Fig. 3.2) in 1912, part of the larger North Norfolk Coast Special Area of Conservation, and one of the oldest coastal nature reserves in the United Kingdom. It includes a significant area of dune overlying a shingle spit. They also acquired the nearby Scolt Head Island in 1923 for £500. The National Trust acquired many more sand dune areas including its first Neptune Coastline Campaign purchase in 1965, Whiteford Burrows a sand dune on the Gower Peninsula in southwest Wales. This campaign continued its policy of purchasing land for conservation and management, and by 2008 it owned 1,142 km of coastline.

There are a further ten National Nature Reserves in England; three in Scotland and five in Wales that are predominantly sand dune habitat. Other organisations such as Local Authorities, including County and District Councils, have the authority to designate and manage statutory Local Nature Reserves. Of course, not all sand dunes are nature reserves, most are not. Conservation and management often relies on the cooperation of others within the context of the legislation. Special Areas of Conservation, for example, is a high-level form of protection, and requires European Union Member States to maintain, or restore these areas to a favourable conservation condition. Protecting the sites from damaging plans or projects is also mandatory, and achieved through national legislation.

3.2.4 UK National Legislation

In the United Kingdom, in addition to those sites protected as nature reserves including National Nature Reserves, owned and managed by Government agencies, many have statutory protection as Sites of Special Scientific Interest (SSSI[1]) or Areas of Special Scientific Interest (ASSI) in Northern Ireland. These designations cover a large proportion of the sand dune resource. Of 121 sites in England surveyed between 1987 and 1990, (the majority of the surviving dunes) nearly 50% were wholly or almost wholly designated, and another 23 were partly designated. In a partial survey of Scottish dunes, 24 of the 34 sites and in Wales, 24 of the 49 sites surveyed were designated Sites of Special Scientific Interest. In Northern Ireland 10 of the 26 sites surveyed were designated as ASSIs with a further four sites proposed.

3.2.5 France

In France, the conservation of coastal habitats took its inspiration from the National Trust in the UK (Sect. 3.2.3). The government established the Conservatoire du Littoral in 1975, whose primary focus is the protection of coastal areas, through purchasing a significant proportion of the French coastline. In January 2003, the organisation owned about 125,000 ha involving 500 sites and approximately 10% of the length of the coast. The selection of the coastal areas relied on a series of inventories, collected over 20 years (Richard and Dauvin 1996).

3.3 European Ecological Networks and Biodiversity

Implicit in Natura 2000 is the concept of developing a network of sites. However, there are often no physical links between them and indeed most are discrete entities. This makes individual sites, especially the smaller ones, vulnerable to human and natural perturbations. Providing links between sites, corridors are a means of facilitating movement of species between core areas. This approach may also provide areas that act as buffer zones outside the core sites.

---

[1] Sites of Special Scientific Interest (SSSIs) give statutory protection to the best sites for wildlife and geology in England, Scotland and Wales. Designated by Natural England and the Countryside Council for Wales and protected under the Wildlife and Countryside Act 1981 (as amended). Scottish Natural Heritage designates SSSIs under the Nature Conservation (Scotland) Act 2004.

3.3.1 Ecological Networks – The Netherlands

All major sand dunes in the Netherlands are already nature reserves and included in Natura 2000 sites. These form part of the National Ecological Network (*Ecologische Hoofdstructuur, EHS*) a system of sites either already owned or managed as nature reserves, new nature reserves, or land under management agreements. The sand dunes front large sections of the mainland coast in the central part of the country and occur on the Islands of the Wadden Sea. For the Wadden Sea, there is a collaborative Trilateral Agreement with Denmark and Germany. This provides a political framework and covers the protection and conservation of the Wadden Sea including management, monitoring and research. The Common Wadden Sea Secretariat established in 1987 in Wilhelmshaven, Germany provides administration support. The Wadden Sea became a World Heritage Site in 2009.

There is a close relationship between nature conservation on sand dunes and their sea defence function, recreational interest and water catchment values. Because of this, wide-ranging policies cover most sand dunes. A legal framework guarantees the required safety-level of dunes and protects the dune environment as well. In practice, this means that beach nourishment, a key component of the coastal defence strategy, takes account of the Natura 2000 legislation. Nourishment takes place at a time of year and location on the beach that helps minimise damage to nature conservation interests.

3.3.2 European Policy – Biodiversity Action Plans (BAP)

Biodiversity Action Plans developed by contracting parties to the United Nations Environment Programme (UNEP) Convention on Biological Diversity (CBD http://www.cbd.int/) include both habitats and species. Recognising the importance of biodiversity, the European Union is committed to halting losses both on its territory and beyond (EU http://ec.europa.eu/environment/nature/biodiversity/comm2006/index_en.htm). In this context, Member States are contracting parties to the UN Convention on Biological Diversity. Part of the strategy includes the identification, designation and management of Natura 2000 sites (Sect. 3.2.2). In addition, member states are encouraged to adopt specific approaches for conserving these sites. Braunton Burrows Devon, England illustrates how the national and international designations overlap (Fig. 3.3), helping to support a comprehensive series of management opportunities.

In response to the Convention on Biological Diversity 1992, the UK Government produced a series of Biodiversity Action Plans (BAP) including one for sand dunes. This is under review but in 2008 included the following objectives and targets:

1. "There should be no further net loss of the resource, its distribution and range of habitats of about 56,500 ha (71,600 ha with Scottish Machair) from losses to anthropogenic factors, whether caused directly or indirectly (e.g. by flood risk

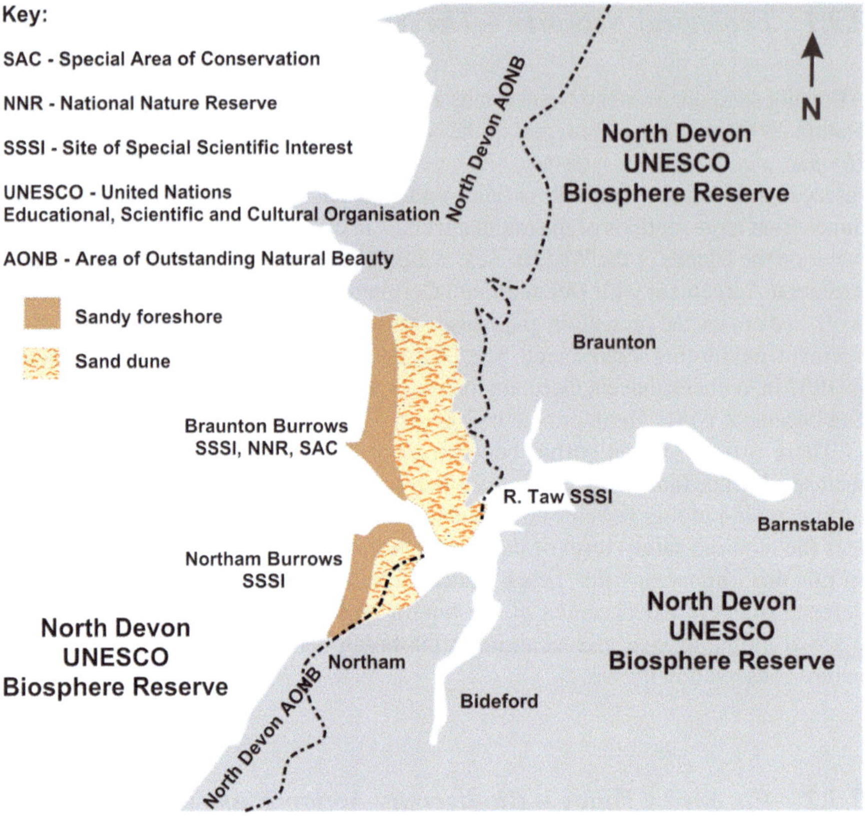

Fig. 3.3 Braunton Burrows sand dune forms the core of the much larger, United Nations Educational, Scientific and Cultural Organisation (UNESCO) Biosphere Reserve. Note the Biosphere Reserve covers an area much bigger than shown in the figure

management schemes affecting coastal processes)." This 'no net loss' target takes account of 'natural' losses due to the dynamic nature of sand dunes.

2. "Achieve favourable or recovering condition by appropriate management of XXha[2] of coastal sand dune systems currently in unfavourable condition by 2010. This should achieve the retention or enhancement of populations of BAP priority species associated with sand dunes";
3. "Control natural succession to scrub, woodland, bracken and other invasive non-native plants. A target value of 200 ha by 2010";
4. "Re-establish Atlantic dune woodland habitat to two sites by 2015 (applicable to Wales only)";
5. "Restore sand dune habitat lost or severely degraded as a result of afforestation, agriculture and infrastructure. A target figure of 1,000 ha (minimum) to

[2] At the time of writing, no specific target had been set.

be reinstated to dune habitat by 2010 (to be reviewed as a result of the inventory development)."

The original action plan had links with specific priority species, some of which have significant populations on sand dunes. These include a range of rare and endangered species included in Sect. 5.4. Coastal sand dunes (including machair) form part of the Supralittoral Sediments within Terrestrial & Freshwater Ecosystems.

The Biodiversity Action Plan is an important part of the process of assessing progress in halting the loss of biodiversity. At the United Kingdom level Biodiversity Action Reporting System, only one of the five targets showed "some progress (on schedule)". The others were either "progress unknown" or "unknown".

The original Biodiversity Action Plan (BAP) spawned 30 Local Area BAPs, some of which have specific targets for coastal sand dune vegetation and their associated species (Department of Environment, Food and Rural Affairs http://ukbars.defra.gov.uk/default.asp). At a local level, although it has provided some benefits there are limitations to its effectiveness not least because of a lack of clarity in the relationship between the United Kingdom national and local plans. Limitations on resources and the time consuming nature of the plan process, including generation of copious amounts of paper are also limiting factors (Cowley 2008).

3.4 Coastal Geomorphology

Recognition of the need to conserve geomorphological systems is relatively new. Over many decades, coastal defence to protect people and assets from erosion and flooding has often taken priority over biological conservation interests (Hooke 1998). Except in a relatively few protected areas, erosion of dynamic (including cliffs and sedimentary) landforms was something to be prevented not encouraged. The protection of dynamic landforms, such as sand dunes, came within legislation concerned largely with biological sites. By their nature, site boundaries depicted as lines on maps, determined by legislative 'rules' and burdened by bureaucratic processes, cannot easily accommodate the changing position of mobile systems such as sand dunes. Even in nature reserves, preventing the movement of mobile sand was often part of the management strategy. However, this changed with the recognition of the importance of dynamic processes as part of a functioning ecosystem. This section summarises some of the approaches to geological conservation (including geomorphological considerations), taking place at international and national levels.

3.4.1 International

There are several international organisations promoting geological conservation. These include the United Nations Educational, Scientific and Cultural Organisation

(UNESCO), whose aims include the encouragement of the development of a global series of geological sites, the 'Global Geoparks Network'. The International Geographical Union (IGU http://www.igu-ccs.org/) Commission on Coastal Systems encourages the study of coastal systems throughout the world. The International Association of Geomorphologists (IAG/AIG http://www.geomorph.org/) develops and promotes geomorphology as a science. These organisations help to raise the profile of geological conservation with the ultimate aim of influencing policy and promoting the establishment of geological sites and encouraging a wider understanding of landforms and processes.

Other international approaches, which might encompass geological conservation include, the Council of Europe, Pan-European Biological and Landscape Diversity Strategy (PEBLDS). Under Action Theme 4 "Conservation of Landscapes", it intends to "establish guidelines to address policies, programmes and legislation for mutually supportive protection of biodiversity, cultural and geological heritage" (Weighell and Torfason 2002).

Geosites is a project initiated by the International Union of Geological Sciences (IUGS) designed to identify geological areas (sites) of international importance (Wimbledon 1996). It set out to provide a comprehensive global inventory of geological and geomorphological sites selected and documented by regional groupings of geoscientists. The creation of Geoparks (based on the identified Geosites) in cooperation with UNESCO is active in Europe (Eder and Patzak 2004) and promoted by the European Association for the Conservation of the Geological Heritage (ProGEO). A trawl of the Internet suggests there are few if any of these sites specifically identified so far for their coastal geomorphological interest.

A review of UNESCO World Heritage Sites identifies 71 properties on the World Heritage List for outstanding universal value for the earth sciences. These include 18 coastal systems (Dingwall et al. 2005). One of these sites, Fraser Island, Australia includes a major active sand dune system. No doubt, detailed inspection of other individual site accounts would establish several more examples within larger sites. Despite this, conserving earth science interests is still the poor relation of biological conservation (Gray 2004).

3.4.2 National, United Kingdom, Ireland and United States of America

The United Kingdom Geological Conservation Review is a comprehensive national geological site inventory (Ellis et al. 1996). This provides the basis for identifying and conserving sites of geological interest. The statutory countryside agencies are responsible for designating them as Sites of Special Scientific Interest (SSSIs) notifying landowners and local planning authorities of their existence. These have legal protection in the same way as biological sites and some are included within National Nature Reserves (NNR).

There are over 3,000 sites, which represent individual geological sites, components of site networks or parts of the wider landscape. Within the 42 volume series, one volume covers coastal geomorphology (May and Hansom 2003). In England, in order to widen the context for conserving nature, the country has been divided into landscape units called 'Natural Areas'. Although this is successful as far as biological interests are concerned, it is much less so in respect of geomorphological conservation (Gray 2001).

In the United Kingdom geology has joined biodiversity in the action planning process, with the production of local Geodiversity Action Plans (Haffey 2008) As part of this process there are a number of initiatives set up to identify Regionally Important Geological/Geomorphological Sites (RIGS). In Scotland, there were no fewer than 204 potential geodiversity sites identified for West Lothian, Scotland (Barron et al. 2006). Several other counties also have Local Geodiversity Action Plans for sites based on these surveys, such as Norfolk, England, which includes the North Norfolk coast – "an outstanding assemblage of dynamic coastal landforms, including the shingle spit at Blakeney Point, the offshore barrier island at Scolt Head and the dunes at Holkham and Wells-Next-The-Sea" (Holt-Wilson 2010). In Devon, both the living and non-living environments combine to become the Devon Biodiversity and Geodiversity Action Plan (Devon County Council http://www.devon.gov.uk/devon_biodiversity_action_plan). Braunton Burrows National Nature Reserve and Dawlish Warren are two sand dunes identified both for their biological interest and as dynamic geomorphological features.

The Geological Survey of Ireland (GSI) Irish Geological Heritage Programme undertakes to identify, document, protect and promote the wealth of geological heritage in the Republic of Ireland. It is responsible for the identification of important sites that fit the criteria for designation as Natural Heritage Areas. Within 17 themes, sand dunes come within "Coasts, Rivers, Lakes" (Parkes and Morris 2001).

In the United States, the Bureau of Land Management (BLM) issued a statement in 1998, which defined the important aspects of geological heritage and the anthropogenic and natural threats to this heritage. One hundred and eighty parks have significant geological interest, with 97 having "dynamic shoreline geology" (National Park Service http://www.nature.nps.gov/geology/geoheritage/nps_sites.cfm). The National Park Service has a Coastal Geology Group, which provides coastal resource information and management throughout the country. The group is also beginning an inventory of all the significant geologic features within National Parks. It is working with State and local governments, as well as the U.S. Geological Survey and the geological community to evaluate and build upon the existing geology heritage programmes.

3.5 Protecting Coastal Dunes from Development

Development, especially tourist development, has often focused on sand dunes inland from the beach (Chap. 2). There is no comprehensive description of the scale of loss on a world basis. However, individual accounts give an indication of the devastation and in some cases the policy response. Sand dunes are not isolated from the rest of the coast and laws to protect them including those concerned with Integrated Coastal Zone Management are relevant.

There is a wide range of information on the web, which gives a more general introduction and links to the various approaches to coastal legislation throughout the world. A good starting point is a web site hosted by the Institute of Marine and Environmental Law, University of Cape Town, South Africa (Gibson "The UK Coastal Law Web Site" http://www.coastlaw.uct.ac.za/iczm/home.htm). This contains detailed legal material for the UK with links to sources of national, European Union and international law on coastal management. The following sections provide examples of legislative measures specifically targeted at coastal sand dunes.

3.5.1 North America – Spatial Zonation

Developing policies for controlling spatial activities has for many years revolved around the concept of Integrated Coastal Zone Management (ICZM). This sets a framework for coastal policy development within a horizontal (spatial) context and across sectors (Beatley et al. 2002). Defining the coastal zone is therefore an important part of policy development. For example, in New Brunswick, Canada coastal policy seeks to protect the coastal resource including sand dunes by defining three zones:

1. Zone A is the most sensitive and includes 'natural' habitats such as beaches and dunes, between the Lower Low Water Large Tide (LLWLT). This extends beyond the Higher High Water Large Tide (HHWLT). Development activities are restricted;
2. Zone B is an area 30 m landward from the inland edge of Zone A. Although Zone B is sensitive to sea level rise and storm damage, Zone A provides protection. A greater range of activities is acceptable;
3. Zone C lies landward of Zone B. Topography, elevation and erodability determines the distance inland.

In New Jersey, as early as 1930 there was recognition of the value of beaches and foredunes for coastal defence. This value was only formally recognised at the Federal level in 1972, with the enactment of the United States of America Coastal Zone Management Act. This provided the first national approach to the promotion of foredune restoration and maintenance (Psuty and Ofiara 2002). Development on foredunes in New Jersey is specifically prohibited except when there is no "practicable or feasible alternative in an area other than a dune, and that will not cause significant adverse long-term impacts on the natural functioning of the beach and dune system" (State of New Jersey 2011).

3.5.2 Other Laws in the USA

The Coastal Barrier Resources Act (CBRA) was an attempt by the Congress of the USA for a public policy to limit development on certain coastal barriers, as well as protecting the US treasury from paying for recovery when natural disasters occur.

However, research suggests that the Act, by itself, will not prevent development from occurring in the designated coastal areas (Salvesen 2005).

In Maine, a Natural Resources Protection Act (NRPA) recognising the value of 'natural' mobility of sand dunes includes a prohibition against construction on the frontal dune. Chapter 355 "Sand Dune Rules", a key component of the State's coastal management programme, recognises sea level rise as a by-product of global warming. Within the "Rules" on page 8 "A project may not be permitted if, within 100 years, the property may reasonably be expected to be eroded as a result of changes in the shoreline such that the project is likely to be severely damaged after allowing for a two foot rise in sea level over 100 years. Beach nourishment and dune restoration projects are excluded from this requirement." There are also limitations on rebuilding storm-damaged property, which must take place away from the dune front to take account of threats from erosion (Maine Department of Environmental Protection 2012 http://www.maine.gov/dep/land/nrpa/index.html#rule). In "An introduction to coastal management", Beatley et al. (2002) includes Maine as one of several "State Case Studies".

In the State of Virginia, the Marine Resources Commission added coastal primary sand dunes[3] and beaches to marine habitat protection legislation. Following principles already established for wetlands protection, the Virginia Marine Resources Commission (1980) requires permits to ensure that there is a reasonable balance between development and protection of coastal features. It recognises that inappropriate development, which destroys vegetation, can compromise the effectiveness of the storm-protecting features of sand dunes. Increased erosion can lead to flooding and property damage, destruction of wildlife habitat and an increased need for the expenditure of public funds. In 2008, the enactment of legislation expanded the area covered by the "Coastal Primary Sand Dunes and Beaches Act" from its original nine localities to the entire Virginia coastal zone (Mason 2009).

3.5.3 Spain

Before 1988, Spanish coastal dunes were unprotected. The 1988 Spanish Shore Act ("Ley de Costas") protects all coastal dunes within 100 m of the coast. It precludes their destruction by sand mining and many other (although not all) forms of development such as building hotels and apartments too close to the beach. Despite this losses continue. A review of the underlying causes suggests that short-term social and economic benefits and hence political influence militate against such control (Suárez de Vivero and Rodríguez Mateos 2005).

This is borne out by the reversal of the policy of destroying buildings built illegally in the 1950s on Tenerife. Initially the law prevented these properties from being bought and sold; some were even demolished. However, in 2009, because of pressure from local inhabitants, an amendment to the law allowed buying and selling beachside properties built before its introduction.

[3] "Primary sand dune" refers to the foredune.

Fig. 3.4 Part of Menie links, Scotland, converted to a golf course and future housing development with a loss of much of the mobile and fixed dune grassland. The picture, taken in March 1987 shows the unvegetated sand sheet, over 600 m long and over 400 m wide. This striking geomorphological feature remained active for most of the twentieth century. Before the golf course development, this formed part of the Foveran Links SSSI dune complex, recognised as a Geological Conservation Review site in its own right (Hansom 2007)

3.5.4 'Trumped' by Donald Trump

Despite the extensive legislation designed to protect features of nature conservation interest from development, modification and destruction of sand dunes continues. A relatively untouched sand dune in eastern Scotland and including some of the very few maintaining examples of mobile dune (Fig. 3.4) was the subject of a proposal by the United States tycoon Donald Trump to create the "greatest golf course in the world". Aberdeenshire Council originally rejected the proposed Trump International Golf Links, Aberdeen, Scotland in 2007 partly because of the damage to the Site of Special Scientific Interest. However, pressure from local businesses resulted in it being 'called in' for determination by the Scottish Government. The best efforts of the statutory agencies, in this case Scottish Natural Heritage and Scottish Environmental Protection Agency, the Royal Society for the Protection of Birds and the Scottish Wildlife Trust amongst others, failed to prevent the development, which was given approval following the Menie Estate Public Inquiry.

The £1 billion resort got permission because of the "significant economic and social benefit" of the project according to the Scottish Government. Once a haven for wildlife and a precious open access recreational area, it will eventually have two golf courses, a 450-bedroom hotel and housing as well as holiday apartments and golf villas (Milne 2008). This decision follows a tradition in Scotland, and affecting sand dunes throughout the world, to create the ultimate in 'links' golf course design (Sect. 10.6.1). Work began on the construction of the course in July 2010. Playing the new 18 hole course began on the 15th July 2012.

It is unlikely this will be the last major loss of sand dunes to development here or elsewhere in the world. Pressures for development will undoubtedly continue. A key wider policy issue, in the future, will relate to the value of sandy beaches and sand dunes in providing a sea defence function in the face of rising sea levels (Sect. 4.3.2). However, as far as the development of new links golf courses is concerned, it is possible for their construction to take place in an environmentally sensitive way, as shown by the Machrihanish development in western Scotland (Sect. 10.6.2).

3.6 A Reservoir of Sand

The European Union EUROSION study set out to provide the European Commission with quantified evidence on coastal erosion in Europe, the problems associated with it and the effectiveness and desirability of mitigation measures. The EUROSION report (Salman et al. 2004) provides a series of recommendations as part of an effort to bring coastal erosion into the mainstream of coastal management policy at the European, national, regional and local levels. General conclusions from the study were that the extent of urbanisation of the coast had turned a natural phenomenon (sand mobility) into a problem (erosion). Put another way "Beaches are indestructible, except when humans get involved." (Neal et al. 2007).

Erosion has a serious impact on some 20 % of the coastline of Europe, as assessed in 2004. Mitigation measures (coastal defence) are costly and most frequently undertaken by national and regional administrations. The owners of assets at risk or by the parties responsible for causing or exacerbating coastal erosion (such as developers building hotels on eroding cliffs or sand dunes) rarely bear any of the cost.

Two key findings from the study were recognition of the values associated with:

1. Having sufficient sediment to maintain 'natural' coastal defences such as sand dunes;
2. 'Space' for geomorphological processes to operate.

The report "Sediment and Space for Sustainability" (Doody et al. 2004) identifies the importance of 'coastal resilience'[4] and the ability of the coast to continue to offer goods and ecological services to the people who live, work and enjoy the coast.

[4] 'Coastal resilience' is defined as "the inherent ability of the coast to accommodate changes induced by sea level rise, extreme events and occasional human impacts, whilst maintaining the functions fulfilled by the coastal system in the longer term. The concept of resilience is particularly important in the light of the predictions for global climate change."

Larger sand dunes provide a major reservoir of sand and thus are potentially significant contributors to 'coastal resilience'. Chapter 6 provides a more detailed discussion of this concept, and of sand mobility as a positive force in dune conservation management.

3.7 Conclusions

This Chapter has shown that in many parts of the world there is increasing recognition of the value of sand dunes as features of high nature conservation and recreational value. Because of this, sand dune systems have a relatively high proportion of sites managed as nature reserves or designated under international and national legislation. However, proposals for development that will destroy these nature conservation values continue. In some parts of the world, there is legislation that seeks to retain the sea defence value of the beach/foredune. This often largely ignores inland dune areas. The next two chapters describe the States and Values of the dune/beach interface and of vegetated sand dunes, respectively.

References

Barron HF, Browne MAE, Finlayson A (2006) West Lothian geodiversity. British Geological Survey Commissioned report, CR/06/008N8N, p 190. Source: http://nora.nerc.ac.uk/7429/. Accessed 06 Nov 2011

Beatley T, Brower DJ, Schwab AK (2002) An introduction to coastal zone management, 2nd edn. Island Press, Washington, DC, 324 pp

Cowley DI (2008) Stakeholder's evaluation of local biodiversity action plans (LBAP). M.Phil., Bangor University. Source: http://www.wmbp.org/assets/userfiles/LBAPreportOpt.pdf. Accessed 08 Nov 2011

Dargie TCD (1993) Sand dune vegetation survey of Great Britain, part II – Scotland. Joint Nature Conservation Committee, Peterborough, 113 pp

Dargie TCD (1995) Sand dune vegetation survey of Great Britain, part III – Wales. Joint Nature Conservation Committee, Peterborough, 153 pp

Dargie TCD (2002) Sand dune vegetation survey of Scotland: national report, Survey and monitoring report series. Scottish Natural Heritage, Battleby

Dingwall P, Weighell T, Badman T (2005) Geological world heritage: a global framework. A Contribution to the Global Theme Study of World Heritage Natural Sites. Protected Area Programme, IUCN, World Commission on Protected Areas. Source: http://www.geoconservation.com/EHWH/Docs/whframework.pdf. Accessed 04 Nov 2011

Doody JP (1991) Sand dune inventory of Europe. Joint Nature Conservation Committee/The European Union for Coastal Conservation, Peterborough/Leiden, 80 pp

Doody JP (2005) Sand dune inventory of Europe. In: Herrier J-L, Mees J, Salman A, Seys J, Van Nieuwenhuyse H, Dobbelaere I (eds) Proceedings 'Dunes and Estuaries 2005' – international conference on nature restoration practices in European coastal habitats, Koksijde, Belgium, 19–23 September 2005, vol 19, VLIZ special publication., pp 45–54

Doody JP (ed) (2008) Sand dune inventory of Europe, 2nd edn. National Coastal Consultants/EUCC, the Coastal Union, in collaboration with the IGU Coastal Commission. Published on CDROM, Liverpool Hope University, Liverpool, 126 pp

Doody JP, Ferreira M, Lombardo S, Lucius I, Misdorp R, Niesing H, Salman A, Smallegange M (eds) (2004) Living with coastal erosion in Europe – sediment and space for sustainability. Results from the EUROSION study, European Commission, Office for Official Publications of the European Communities. Source: http://www.eurosion.org/project/eurosion_en.pdf. Accessed 08 Nov 2011

Eder W, Patzak M (2004) Geoparks – geological attractions: a tool for public education, recreation and sustainable economic development. Episodes 27(3):162–164

Ellis NV, Bowen DQ, Campbell S, Knill JL, McKirdy AP, Prosser CD, Vincent MA, Wilson RCL (1996) An introduction to the geological conservation review, vol 1, GCR series. Joint Nature Conservation Committee, Peterborough, 131 pp

European Commission (2007) Interpretation manual of European habitats. Natura 2000. European Commission, DG Environment, Nature and Biodiversity, Brussels. Source: http://ec.europa.eu/environment/nature/legislation/habitatsdirective/docs/2007_07_im.pdf. Accessed 05 Mar 2012

Géhu JM (1985) European dune and shoreline vegetation, vol 32, Nature and environment series. European Committee for the Conservation of Natural Resources, Strasbourg, 68 pp

Gray JM (2001) Geomorphological conservation and public policy in England: a geomorphological critique of English Nature's 'Natural Areas' approach. Earth Surf Process Landf 26:1009–1023

Gray JM (2004) Geodiversity: valuing and conserving abiotic nature. Wiley, Chichester, 434 pp

Haffey D (2008). Local geodiversity plans: a review of progress in England. Natural England Research report, 027. Source: http://publications.naturalengland.org.uk/publication/36005?category=30050. Accessed 06 Mar 2012

Hansom JD (2007) Assessment of the geomorphological interests at Foveran links SSSI. Scottish Natural Heritage Commissioned report no 232 (ROAME No. F05AC701). Source: http://www.snh.org.uk/pdfs/publications/commissioned_reports/232.pdf. Accessed 16 Nov 2011

Hellemaa P (1998) The development of coastal dunes and their vegetation in Finland. Fennia 176(1):1–157

Hilton M, Macauley U, Henderson R (2000) Inventory of New Zealand's active dunelands. Sci conserv, 157:p 30 + 124 detailed distribution maps. Source: http://www.conservation.net.nz/upload/documents/science-and-technical/Sfc157.pdf. Accessed 17 Mar 2010

Holt-Wilson T (2010) Norfolk's earth heritage, valuing our geodiversity. Norfolk Geodiversity Partnership,p74.Source:http://www.nbis.org.uk/sites/default/files/documents/Norfolk%27s%20Earth%20Heritage_Screen%201-17.pdf. Accessed 06 Nov 2011

Hooke J (ed) (1998) Coastal defence and earth science conservation. The Geology Society, Bath, 273 pp

Hopkins J, Radley G (2001) Sand dunes and the habitats directive: preparation of the UK national list. In: Houston JA, Edmondson SE, Rooney PJ (eds) Coastal dune management, shared experience of European conservation practice. Liverpool University Press, Liverpool, pp 283–301

Mason P (2009) Sand dunes and beaches in Virginia: science and management. Recommendations for guidance. Final report to Virginia Coastal Programme, Department of Environmental Quality. Source: http://ccrm.vims.edu/publications/pubs/beaches_dunes_rec_final.pdf. Accessed 16 Nov 2011

May VJ, Hansom JD (2003) Coastal geomorphology of Great Britain, vol 28, Geological conservation review series. Joint Nature Conservation Committee, Peterborough, 754 pp

McLeod CR, Yeo M, Brown AE, Burn AJ, Hopkins JJ, Way SF (eds) (2005) The habitats directive: selection of special areas of conservation in the UK. 2nd edn. Joint Nature Conservation Committee, Peterborough. www.jncc.gov.uk/SACselection

Milne D (2008) Blinded by the bling?? MilHouse Publishing, 124 pp

Neal WJ, Pilkey OH, Kelley JT (2007) Atlantic coast beaches. A guide to ripples, dunes, and other natural features of the seashore. Mountain Press Publishing Company, Missoula, 250 pp

Parkes MA, Morris JH (2001) Earth science conservation in Ireland: the Irish Geological Heritage Programme. Ir J Earth Sci 19:79–90

Psuty NP, Ofiara DD (2002) Coastal hazard management – lessons and future directions from New Jersey. Rutgers University Press, New Brunswick, 429 pp

Radley GP (1992) The dunes of England, an example of a national inventory. In: Carter RWG, Curtis TGF, Sheehy-Skeffington MJ (eds) Coastal dunes: geomorphology, ecology and management for conservation, proceedings of the 3rd European Dune Congress Galway, Ireland, 17–21 June 1991. A.A. Balkema, Rotterdam, pp 439–453
Radley GP (1994) Sand dune vegetation survey of Great Britain, part I – England. Joint Nature Conservation Committee, Peterborough, 126 pp
Richard D, Dauvin J-C (1996) Conservation strategies for French coastal areas. Aquat Conserv: Mar Freshw Ecosyst 6(4):205–214
Rodwell JS (ed) (1991) British plant communities, vol 2, Mires and heaths. Cambridge University Press, Cambridge, 638 pp
Rodwell JS (ed) (2000) British plant communities, vol 5, Maritime communities and vegetation of open habitats. Cambridge University Press, Cambridge, 512 pp
Salman A, Lombardo S, Doody JP (2004) Living with coastal erosion in Europe: sediment and space for sustainability. Part I — major findings and policy recommendations. Directorate General Environment, European Commission. Source: http://www.eurosion.org/reports-online/part1.pdf. Accessed 20 Mar 2011
Salvesen D (2005) The Coastal Barrier Resources Act: has it discouraged coastal development? Coast Manag 33(2):181–195
State of New Jersey (2011) Coastal zone management rules title 7 of the New Jersey Administrative Code N.J.A.C. 7:7E. Last amended March 2011. Source: http://www.state.nj.us/dep/landuse/--7e.pdf. Accessed 08 Nov 2011
Suárez de Vivero JL, Rodríguez Mateos JC (2005) Coastal crisis: the failure of coastal management in the Spanish Mediterranean region. Coast Manag 33:197–214
Sýkora KV, Babalonas D, Papastergiadou ES (2003) Strandline and sand dune vegetation of coasts of Greece and some other Aegean countries. Phytocoenologia 33(2–3):409–446
Th W Bakker, Jungerius PD, Klijn JA (1990) Dunes of the European coasts – geomorphology, hydrology and soils, vol 18, Catena supplement. Cremlingen, Holland, 223 pp
Virginia Marine Resources Commission (1980) Coastal primary sand dunes/beaches guidelines. Reprinted September 1993. Source: http://www.mrc.state.va.us/Regulations/dune_guidelines.pdf. Accessed 07 Nov 2011
Weighell T, Torfason H (2002) Geological and geomorphological conservation: a review of global, European and national programmes. Committee for the activities of the Council of Europe in the field of biological and landscape diversity, Working Group on the Geological Heritage, p 17. Source: http://www.progeo.pt/pdfs/Doc3.pdf. Accessed 04 Nov 2011
Wimbledon WAP (1996) Geosites – a new conservation initiative. Episodes 19(3):87–88

Chapter 4
Physical States and Values – Beach/Foredune

Abstract Sand dunes develop through geomorphological processes that rely on deposition of sediment on a beach. Waves, then wind drive the particles onto the upper shore. In temperate regions, vegetation plays an important part in the accretion of sediment and the development of strandline and foredune vegetation (Chap. 1). Human actions interfere with these processes causing a change to the system (Chap. 2). This chapter describes the physical characteristics of coastal sand dunes (Sect. 1.2) and vegetation along a continuum from the beach to the foredune (Sects. 1.3.1 and 1.3.2 respectively). It considers the forces that drive the system and the way these change the state and the ecosystem services (values) each provides.

4.1 Driving Forces, Pressures, States, Impacts and Response (DPSIR) Affecting the Beach/Foredune

This chapter uses an extension of the Organisation for Economic Development (OECD) Pressure – State – Response model to identify and analyse the values of sand dunes in different states. This model provides a causal framework for describing the interactions between society and the environment. It relates the natural forces and human activities, **Driving forces** that lead to **Pressure** on the environment. These help determine the **State** of the resource, in this case the beach/foredune. The different states have a positive or negative **Impact** on the values associated with each state. Society's **Response** includes changes in policy, raising awareness, management and restoration (Fig. 4.1).

The driving forces associated with population growth show up in the extent of losses suffered because of human actions, such as sediment depletion (Sect. 2.6). These pressures act to produce a particular State for the beach/foredune in its relationship with the inland sand dune. This involves a movement landward or seaward depending on the balance between the driving forces. In order to help determine the

J.P. Doody, *Sand Dune Conservation, Management and Restoration*,
Coastal Research Library 4, DOI 10.1007/978-94-007-4731-9_4,

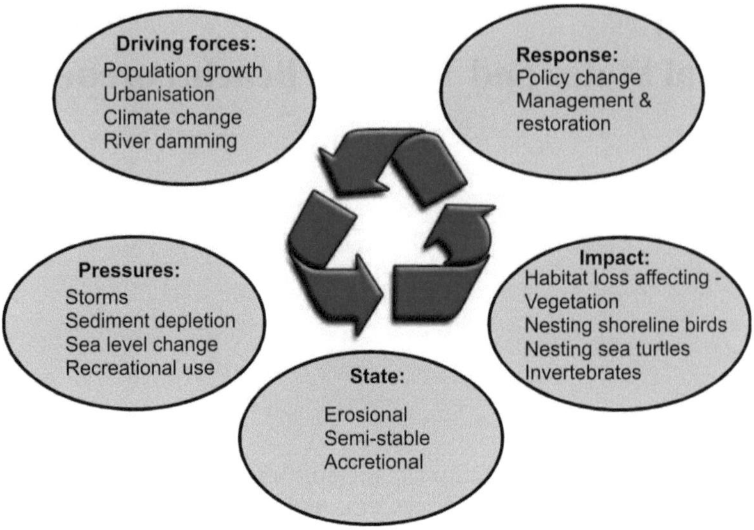

Fig. 4.1 Examples of the various stages in the DPSIR cycle relevant to mobile foredunes

most appropriate form of management it is important to know what the pressures are and how they influence the value of each state. This depends on a variety of factors including the processes originally helping to form the ecosystem, including sediment movement and vegetation succession. The underlying geology and physical location are important, as are plant and animal interactions and human interference.

The result is a series of values including those associated with ecological systems services values (Sect. 4.3) This elicits a response by way of management action, including habitat restoration, re-creation or creation. This response will depend largely on the perception and relative priority attached to each state by the 'viewer'. For example, in coastal defence terms a mobile sand dune could be a threat to human assets, requiring remedial action to prevent sand blow and landward retreat. To a nature conservationist this represents an opportunity for the formation of vegetation on the foreshore, together with its associated specialist and rare species. In this example, the needs appear diametrically opposed. However, as we will see, this need not be the case. This Chapter seeks to identify and describe the different values associated with each state.

4.2 Physical States – Description

The sand dune is classifiable into three distinct zones of development, namely:

1. Primary dune – this is the zone in which there is an active exchange of sediment between the beach and the first sand dune ridge;

2. Secondary dune – this involves modification of the primary dune and/or migration inland. This zone includes active blowouts and stable forms, which are separated from the active beach/dune interface;
3. Sand sheet/washover – are highly active areas where large quantities of material move inland (Psuty 2004).

The focus of this chapter is the primary dune (beach/foredune). It is essentially a 'sand-sharing system', having a variety of complex forms, where there is an active interchange of material across the beach. The secondary dune and sand sheet/washover areas are, in the context of this book, components of inland sand dunes no longer in direct contact with the shore (dealt with in Chaps. 5 and 7 respectively).

Early stages in the development of the primary dune include mobile sand with a strandline and foredune. The position of the accreting sand dune front depends on numerous factors. The first of these derives from the presence of a supply of suitable beach material. As the movement of sand along and onto the upper beach takes place through the action of waves and wind, obstacles on the beach impede its movement and deposition occurs. Hummocks of sand, sometimes associated with pioneer vegetation or debris tossed high on the beach profile coalesce to form a narrow strandline (Fig. 1.3). Additional sediment aids the process and a mobile foredune develops. Under favourable conditions, this may remain in situ and provide the basis for the development of more stable dunes behind, so long as a new strandline forms to seaward. On the other hand, with a limited sediment supply, or where the factors promoting erosion and loss of the dune front are dominant, a foredune may not be present at all.

A classification of Dutch foredunes into *regressive*, *stable* and *progressive* depends on whether there is morphological evidence of a landward or seaward movement or vertical accretion over time (Arens and Wiersma 1994). This translates into three states:

- **State 1 –** Eroding;
- **State 2** – Dynamic equilibrium or 'Semi-stable';
- **State 3** – Accreting.

Note that this classification relates to sand dunes in close proximity to the sea, where there is interplay between the beach and early forms of sand dune development. The three physical states can exist on the same beach. Sediment availability, changes in storm frequency (timing and intensity), in sea level and wind direction all help determine whether the dune front is moving landward (erosional) or seaward (accretional). The following sections describe each of the three states.

4.2.1 State 1 – Eroding

Eroding beaches are those where the proportion of bare sand is high and where sand movement is such that there is little or no vegetation. If a vegetated strandline or

Fig. 4.2 An example of a beach with a negative sediment budget and eroding dune face, Great Island Trail, Cape Cod, eastern United States of America, September 1991. Note material (debris) left by successive high tides. Waves and currents undermine the base of the dune. (See also Fig. 11.7)

incipient foredune develops this only survives until the next storm, waves, currents or wind remove it from the beach. Overall there is a sediment budget deficit with more sand removed from the beach than is gained. With a negative sediment budget there is often nothing to prevent loss of the inland dunes. Typically, the dune front has a steep or cliffed face (Fig. 4.2).

The causes of beach erosion (mechanisms for change Sect. 6.3) are many. These include direct effects brought about by human activities, such as sediment removal from the beach and coastal stabilisation structures locking up or deflecting sediment away from the shoreline. Depletion of the reservoir of offshore sand (including glacial material) through extraction, combined with the action of the natural forces such as storms and tidal movement.

Other effects include river dams, which reduce the rate of delivery of sediments to the sea at river mouths. Stabilisation of eroding land in the catchment because of reduced grazing and/or afforestation may have a similar outcome, although cause and effect are difficult to prove. In all Bird (1996) identifies 20 causes of beach erosion. Between 1976 and 1984, an estimated 70 % of the world's beaches were eroding (Bird 1985). These include major deltas, such as those in the Mediterranean, most of which have retreating shorelines (Grove and Rackham 2001, Chap. 18). In summary, there are nine key driving forces helping to initiate erosion along the beach/foredune interface (Table 4.1; Sect. 6.3).

Table 4.1 Driving forces and pressures leading to beach/foredune erosion

Driving forces	Pressures
1. Natural depletion of offshore glacial material	Reduced sediment supply
2. Offshore sediment extraction	Reduced sediment supply
3. Foreshore sand mining	Reduced sediment supply
4. River damming	Reduced sediment supply
5. Afforestation/reduction in grazing in the hinterland	Reduced sediment supply
6. Increased recreational use	Trampling, fires, excavating 'sun-traps' in the foredune, beach 'cleaning'
7. Vehicle traffic	Damage to the beach, loss of vegetation
8. Sea defence	Reduced sediment supply resulting from protection of eroding sea cliffs
9. Climate and sea level change	Increased storm activity, balance between isostacy[a] and eustasy[b] leading to rising sea levels

[a]Isostacy, changes in land levels
[b]eustacy, changes in sea level

4.2.2 State 2 – Dynamic Equilibrium or 'Semi-Stable'

The beach/foredune sediments may appear to be in a 'natural balance', sometimes described as being in a 'dynamic equilibrium' (Lennon et al. 1996). However describing early stages of sand dune formation in this way or as 'semi-stable' is something of a misnomer. In fact, beaches and foredunes are highly mobile moving seaward or landward in response to changing weather conditions, especially during storms and on high tides (Chap. 1). There is considerable variation and strandlines such as those shown in Fig. 1.3 can be ephemeral. During winter storms the beach becomes flatter, with the movement of sand from the foredune to lower beach levels. In summer months, a more gradual build up of sand occurs on the upper beach.

A study on the Oregon coast also showed marked seasonal variation with accretion in the summer especially between July and August. Winter storms stripped the sand from the beach returning it as bars between storms (Fox and Davis 1978). In Northern France, studies show the nature of change on a macro-tidal coast with sand movement taking place throughout the year. The most energetic period was during winter and spring. Accumulation only occurred during summer months, for the rest of the year the beach eroded. This was greatest during storm events and high waters levels, when complete removal of summer accumulations took place (Ruz and Meur-Ferec 2004).

Because changes take place over different timescales, it is not always easy to identify whether erosion or accretion is dominant. They frequently occur together along the beachfront creating an 'internal dynamic'. However, the system rarely remains in the same place for long. The important point from a nature conservation perspective is that the dynamic sand-sharing system includes bare sand, a strandline and vegetated foredune (Fig. 4.3) whether it is migrating landward (eroding) or seaward (accreting).

Fig. 4.3 Parker River, State of Massachusetts, United States of America. A healthy foredune and shallow shelving beach, September 1991, indicative of a dynamic beach/foredune interface. The beach and foredune are slowly shifting inland with the barrier island upon which they occur

In summary, four key driving forces influence 'natural' change (Table 4.2). Note any of the driving forces relating to human activity and causing erosion or accretion (Tables 4.1 and 4.3) may alter this 'balanced' position.

4.2.3 State 3 – Accreting

Accreting beaches generally occur when there is an abundance of sediment. As the beach accumulates and vegetation develops, the strandline and foredune may shift seawards. This can result in a series of ridges that are essentially abandoned foredunes, which may continue to accrete sediment creating inland dunes. Exceptionally, the foredune may expand seaward creating one large, high and wide dune.

Rapid progradation occurs in locations with abundant sediment derived from erosion in the hinterland and delivered to the coast via a river mouth into a micro-tidal sea (Fig. 4.4). Easily eroded sea cliffs or offshore deposits, including glacial material, are also important sources of sediment. With sufficient suitable material, accretion can occur on shorelines even during periods of relative sea level rise (Bird 1996).

The early stages of growth include typical strandline and dune-forming species. These vary geographically (Sect. 1.6) but *Ammophila* spp. is ubiquitous in temperate regions of the world. Occasionally other species may dominate, such as on the coast of Albania (Fig. 4.4) where Prickly Parsnip *Echinophora spinosa* is the main colonising species, along with Sea Daffodil *Pancratium maritimum*.

Table 4.2 Driving forces and pressure influencing the 'natural dynamic' in a beach/foredune system

Driving forces	Pressures
1. Weather (wind and rain)	Winter versus summer
2. Waves (change in strength and duration)	Winter versus summer
3. Tides and tidal range	Spring versus neap tides
4. Sea level change	Rising versus falling

Table 4.3 Driving forces and pressures leading to foredune accretion

Driving forces	Pressures
1. Land levels (isostacy) rise faster than sea level (eustacy)	Displacement of the location of accretion seaward
2. Sea cliff erosion	Increased sediment supply
3. Increased precipitation and river flows	Increased sediment supply
4. Deforestation in the hinterland	Increased sediment supply
5. Sea defence structures (fencing and other sediment trapping structures)	Sediment deposition. These can cause sediment depletion elsewhere
6. *Ammophila* invasion/planting	Sediment deposition and plant colonisation
7. Alongshore drift	Increased sediment supply
8. Artificial beach/foreshore nourishment	Increased sediment supply

Fig. 4.4 Accreting foredune, northern Albania, August 1993. The abundant sediment derives from extensive deforestation and erosion in the uplands, delivered to the coast via the many rivers flowing into the Mediterranean Sea

4.3 Ecosystem Services (Values)

The United Nations Millennium Ecosystem Assessment (2005) defines ecosystem services simply as "the benefits people derive from Ecosystems". Within this assessment, the only reference to dune systems is in the section dealing with flood and storm regulation. This is not the only benefit provided by sand dunes, as we will see. This section deals with the ecosystem service values associated with the beach/foredune. This takes into account their contribution to human wellbeing according to their supporting services (Sect. 4.3.1); regulating services (Sect. 4.3.2), provisioning services (Sect. 4.3.3) and cultural services (Sect. 4.3.4). Chapter 5 describes those that specifically relate to the vegetated inland dunes. Chapter 6 analyses the trends and trade-offs between the different degrees of erosion or accretion (mobility).

Note there is a growing literature that attempts to put economic values on these services, i.e. as a measure of the benefits people derive from them. In global terms, an estimate of the average value of these services was US$33 trillion annually. Of this, coastal systems provided US$10.6 trillion (Costanzal et al. 1997). Further analysis appears in a special issue of "Ecological Economics" on the "The Dynamics and Value of Ecosystem Services: Integrating Economic and Ecological Perspectives". A paper by Farber et al. (2002) provides some history and background for applying economic values to ecological systems as a context for other papers in the special issue. This and subsequent chapters in this book do not attempt to assign monetary values, but use the concept of ecosystem services to help understand the trade-offs associated with moving between different states.

4.3.1 Supporting Services[1]

Sand dunes are particularly valuable in supporting coastal processes at the margins between marine and terrestrial habitats. The most important are sediment storage and exchange between the beach and the foredune. These also facilitate the development of more stable inland dune forms, described earlier and in Chap. 5. The sediment store contributes directly to the regulating service of sea defence (Sect. 4.3.2).

The interchange between the beach and the foredune also help form the basis for other values, including those associated with nature conservation (Sect. 4.3.4). Beaches provide a number of other ecological services, including fresh water and

[1] Supporting services: are those ecosystem services that provide the basis for the other values associated with the habitat.

nutrients flowing from the foredune to the sea helping to sustain the nearshore food chain (McLachlan and Brown 2006). In turn, these may also contribute to supporting some coastal fisheries (McLachlan et al. 1996). Other components include insects and litter carried on the wind from the inland dune to the shore, intertidal invertebrates and carrion, which are prey for other animals (McLachlan 1991; Fig. 1.11).

4.3.2 *Regulating Services*[2] *– Sea Defence*

Sand dunes often form a significant element within a wider coastal ecosystem. In estuaries and deltas they help provide shelter, aiding the formation of tidal flats and saltmarshes. Beaches provide a sea defence directly by dissipating or attenuating wave energy especially during storms. As the dune erodes material moves from the foredune onto the beach, helping replenish and improve the resistance to further wave attack. A detailed study of beach profiles along the Lincolnshire coast, soon after the 1953 North Sea storm surge, showed that there was a close correlation between height and width of the beach and damage to the coast. The availability of marine sediment for beach replenishment was important, as was the reserve of sand in the foredune. This was 'drawn down' helping to raise and cushion the upper beach. By contrast, the presence of inflexible sea walls was detrimental to the ability of the beach to withstand wave attack (Barnes and King 1953).

Studies of dunes in Ireland clearly show the relationship between sediment trapped in foredunes and the release of material to 'feed' the beach during periods of storms. As material returns to the upper beach, it helps restore dunes when the storm has passed (Carter 1989). This may be particularly important where there is a landward progression of the beach/foredune resulting from sea level rise or lack of sediment supply. The key to this value lies an ability to absorb the energy by shifting location in response to these major environmental perturbations.

The economic value of the beach/foredune has a particular significance when it protects high-density residential or industrial developments, or agricultural land growing valuable commercial crops. Low-lying areas may be prone to flooding, especially in areas where sea level is rising relative to the land. Through the ability to buffer storm waves, the system helps prevent or reduce flooding in these areas. In the face of global warming and as sea levels rise, this service is likely to become more important. In India, sand dunes are amongst the most effective natural defences against storms, cyclones and tsunamis (Mascarenhas and Jayakumar 2008). Behind the foredune the inland dune provides a sediment store, which under extreme conditions also helps replenish the beach. This is scale-dependent and whilst a single foredune ridge can provide sea defence, it has more resilience if it has a larger dune inland.

[2] Regulating services are those that help sustain ecosystem processes. For example, sea defence, regulation of climate and water regulation.

4.3.3 Provisioning Services[3]

Beaches and foredunes have a relatively limited set of provisioning services. They are readily accessible and provide a source of material for building purposes, especially when these are associated with coastal tourist developments. As a result, there are many hotels, villas and associated infrastructure on sandy beaches throughout the world (Sect. 2.5). This is however, detrimental to many of the other values attached to system.

Close proximity to the sea also make the soft substratum of the beach/foredune (and inland sand dunes) ideal locations for pipeline construction and landfall sites North Sea oil and gas pipelines come ashore at St Fergus and Cruden Bay eastern Scotland and Theddlethorpe dunes National Nature Reserve, eastern England. Talacre Warren in north Wales receives gas via a pipeline from Liverpool Bay.

In some parts of the world, the beach provides a surface, which is stable enough to land aircraft. Examples include the recreational Copalis Beach State Airport, located on the beach near Grays Harbor County, Washington, United States of America, Sable Island off the coast of Nova Scotia and Fraser Island, on Australia's east coast, Queensland. On the island of Barra in the Outer Hebrides, Scotland a beach airport landing strip is the only one anywhere in the world with a scheduled airline service. Historically, land speed record attempts also used flat coastal sands. Following abortive attempts at Fanöe beach in Denmark, it was on Pendine Sands in south Wales on September 25th 1924, that Malcolm Campbell set a world land speed record in his car Bluebird. From there he moved to Daytona Beach, Florida, United States of America (Hough 1960).

4.3.4 Cultural Services[4]

Cultural services, within the context of this book, include those that relate particularly to nature conservation values. This is dependent on the characteristic natural dynamic of these early stages in sand dune succession. Other services relate to the aesthetic appeal. In combination with inland sand dunes, it is also important for research and teaching.

European plant communities – Dune plant communities help form the basis for identifying habitats of Community Interest under the European Union Habitats Directive (Sect. 3.2.2). Along the Atlantic, North Sea and Baltic coasts these include, "Annual vegetation of drift lines (Communities of sandy/shingle shorelines)" Code 1210, "Embryonic shifting dunes" Code 2110 and "Shifting dunes along the

[3] Provisioning services include the products obtained from ecosystems such as food, fibre & fresh water.

[4] Cultural services are the non-material benefits people obtain from ecosystems through spiritual enrichment, recreation, aesthetic experience etc.

shoreline with *Ammophila arenaria* (white dunes)" Code 2120. On the Mediterranean coast several communities of 'fixed' dune including "*Crucianellion maritimae* fixed beach dunes" Code 2210, "Dunes with *Euphorbia terracina*" Code 2220 and "*Malcolmietalia dune grasslands*" Code 2230 (Table 5.1) are present along the beach front.

Biodiversity, examples of invertebrates – Strandlines supply organic material through the decomposition of seaweed (wrack) on the upper beach, providing habitat for a variety of invertebrates including amphipods, isopods and insects. On the island of Malta in the Mediterranean large deposits of Neptune Grass *Posidonia oceanic*, along with other marine debris, are also important. This habitat includes 14 main taxa of which gastropods and isopods are the most numerous (Deidun et al. 2009). Similar deposits occur elsewhere on the micro-tidal Mediterranean coast, which in addition to their nature conservation value also help combat erosion of the upper beach (Fig. 4.5). Unfortunately, this material is often 'cleaned' from the shore for amenity reasons (Sect. 6.3.4) removing both these values. Driftwood and other material embedded on broad sandy beaches can also shelter both rare and abundant species.

The early stages in sand dune development include a variety of slope, aspect, sparse vegetation and bare ground. These provide habitat for a range of rare invertebrates, particularly suited to warmth-loving species, which bask on the dry warm slopes in more northern latitudes. The number of different species can be large. Of six invertebrate groups present in the *Ammophila arenaria* transition zone on sand dunes in Scotland, there were 655 species recorded (Welch 1989). Individual colonising plants may have specialist species associated with them such as *Cakile maritima*, a typical strandline species, which in the British Isles has a host-specific flea beetle *Psylliodes marcida*. Its mining larvae damage leaves, stems and fruits whilst adults nibble green parts of the plant (Cox 2007). The beetle can damage the plants, especially in high-density populations of *C. maritima*. The beetles are long-lived, very mobile, and isolated plants can have several resident individuals (Davy et al. 2006).

Biodiversity, examples of birds – The invertebrates of the beach and foredune are food for foraging shorebirds. In southern California, these include the Piping Plover *Charadrius melodus* (Dugan et al. 2003) and many other species of shorebirds (Hubbard and Dugan 2003). In the United Kingdom waders such as Oystercatcher *Haematopus ostralegus,* Turnstone *Arenaria interpres and* Curlew *Numenius arquata* feed on the strandline, along with passerine birds such as Rock Pipit *Anthus petrosus,* Linnet *Carduelis cannabina,* Twite *Carduelis flavirostris and* Chough *Pyrrhocorax pyrrhocorax.* On the east coast of England, along the shore of the north Norfolk sand dune complex, 130 of 164 (79 %) records of Snow Bunting *Plectrophenax nivalis* were from the strandline (Brown and Atkinson 1996).

The upper beach also provides suitable habitat for some ground nesting birds such as Ringed Plover *Charadrius hiaticula* (Europe), Piping Plover *Charadrius melodus* and Wilson's Plover *Charadrius wilsonia* (USA) and the Hooded Plover *Thinornis rubricollis* and Red-capped Plover *Charadrius ruficapillus* (Tasmania). The species nest just above the high tide mark on bare sand or among drift line

Fig. 4.5 Newly deposited *Posidonia* on a shoreline in southern Ibiza, May 2008

debris and in open vegetation (Fig. 4.6). Terns such as Sandwich Tern *Sterna sandvicensis*, nest on open coastal sandy beaches in Europe.

Biodiversity, example of sea turtles – Sea turtles rely on traditional sites to dig nest chambers in which to lay their eggs. Loggerhead Turtles *Caretta caretta* return to the same stretch of coastline every breeding year, and from June to September lay between 80 and 100 eggs. Hauling themselves onto a sandy beach at high tide, they deposit their eggs above high water (Fig. 4.7). Strictly speaking the location where the females dig their nesting chambers lies in the zone just below the first colonising vegetation, where salinity levels are high and the sand is just wet enough to facilitate excavation (Karavas et al. 2005).

There is some evidence to suggest that sea turtles introduce nutrients into beach ecosystems and thus help maintain the stable foredunes that are critical to their reproductive success (Bouchard and Bjorndal 2000).

Recreation – The beach provides the preferred place for many coastal recreational activities, such as lying in the sun and playing games or as access to the sea. The foredunes also provide places to lie out of the wind, sometimes in excavated sun-traps, lighting fires for picnics and 'sand sliding' from higher dune ridges. These uses all have the potential to create instability, exacerbating erosional trends the impact of which varies with scale and intensity of use (Sect. 6.3.4).

Landscape – The beachscape can contribute significantly to the visual experience of low-lying coastal areas. Natural geomorphological processes and associated vegetation development that characterise the habitat help create scenes of great aesthetic appeal. The sweep of the dune with a seascape backdrop appears in many

Fig. 4.6 Fire Island, New York State, United States of America, habitat for the Piping Plover, September 1991

Fig. 4.7 Zakynthos Beach, Zakynthos, Greece, October 2003 showing the location of a Turtle nest (marked with *red-topped canes*) at the base of colonising vegetation in the *middle left* of the picture. Beach furniture, including parasols and people can damage nesting sites

Fig. 4.8 Beachscape, foredunes along the west coast of Ireland September 1991, after Kroyer P.S. (1851–1909) "Summer Evening on Skagen's Southern Beach with Anna Ancher and Marie Krøyer"

paintings and photographs. They are highly valued by the public as a place for rest and relaxation, including walking and bird watching. Painters, photographers and poets amongst other creative artists alike eulogise these areas (Fig. 4.8).

Research and teaching – Sand dunes, particularly at the beach/foredune interface, provide opportunities for studying 'natural' systems. Some of the earliest studies of vegetation succession were on sand dunes, which helped develop our understanding of plant ecology. Classic works include Chapman (1934) working on Scolt Head Island, eastern England and the book on "Salt Marshes and Sand Dunes" (Ranwell 1972) is still a valuable reference.

In the United Kingdom, sand dunes form part of the General Certificate of Education at AS and A Level. In Scotland the Geography National Qualifications, have a resource on vegetation succession for sand dunes.

4.4 Conclusion

The beach/foredune interface is a significant component of any sand dune present in a coastal location. The system has its own characteristics that are different from those of inland sand dunes. It supports a range of values including a rich variety of specialist plants and animals, often dependent on the dynamic nature of the system. State 1 eroding beaches are the least valuable and are indicative of a negative sediment budget and/or sea level rise. State 2 'dynamic' beaches and foredunes occur when there is a stable or mildly negative sediment budget. They retain their values so long as there is space for the sequence of forms to move in response to changing environmental conditions. State 3 accreting beaches and foredunes are much less common but are places with a positive sediment budget where new sand dunes form. As such, they can aid the development and protection of more stable inland dunes. Chapter 6 deals with the trends and trade-offs associated with these three states. Chapter 5 describes the states and values associated with vegetated sand dunes that lie behind the foredune.

References

Arens SM, Wiersma J (1994) The Dutch foredunes: inventory and classification. J Coast Res 10(1):189–202

Barnes FA, King AM (1953) The storm floods of 1 February 1953 II the Lincolnshire coastline and the 1953 storm flood. Geography 38:141–160

Bird ECF (1985) Coastline changes: a global review. Wiley, Chichester, 219 pp

Bird ECF (1996) Beach management. Wiley, Chichester, 281 pp

Bouchard SS, Bjorndal KA (2000) Sea turtles as biological transporters of nutrients and energy from marine and terrestrial ecosystems. Ecology 81(8):2305–2313

Brown AF, Atkinson PW (1996) Habitat associations of coastal wintering passerines. Bird Study 43/ 2:188–200

Carter RWG (1989) Coastal environments. An introduction to the physical, ecological and cultural systems of coastlines. Academic, London, 617 pp

Chapman VJ (1934) The ecology of Scolt Head Island. In: Steers JA (ed) Scolt Head Island. W. Heffer and Sons Ltd., Cambridge, pp 77–145

Costanzal R et al (1997) The value of the World's ecosystem services and natural capital. Nature 387(15):253–260

Cox ML (2007) Atlas of the seed and leaf beetles of Britain and Ireland. Pisces Publications, Newbury, 344 pp

Davy AJ, Scott R, Cordazzo CV (2006) Biological flora of the British Isles: *Cakile maritima* Scop. J Ecol 94(3):695–711

Deidun A, Saliba S, Schembri PJ (2009) Considerations on the ecological role of wrack accumulation on sandy beaches in the Maltese islands and recommendations for their conservation management. J Coast Res, Special Issue 56, In: Proceedings of the 10th international coastal symposium, Lisbon, Portugal, pp 410–414

Dugan JE, Hubbard DM, McCrary MD, Pierson MO (2003) The response of macrofauna communities and shorebirds to macrophyte wrack subsidies on exposed sandy beaches of southern California. Estuar, Coast Shelf Sci 58S:25–40

Farber SC, Costanza R, Wilson MA (2002) Economic and ecological concepts for valuing ecosystem services. Ecol Econ 41:375–392

Fox WT, Davis RA (1978) Seasonal variation in beach erosion and sedimentation on the Oregon coast. Geol Soc Am, Bull 89(10):1541–1549

Grove AT, Rackham O (2001) The nature of the Mediterranean Europe: an ecological history. Yale University Press, New Haven/London, 384 pp

Hough R (1960) Book of the racing Campbells. Stanley Paul, London, 127 pp

Hubbard DM, Dugan JE (2003) Shorebird use of an exposed sandy beach in southern California. Estuar, Coast Shelf Sci 58(1):41–54

Karavas N, Georghiou K, Arianoutsou M, Dimopoulos D (2005) Vegetation and sand characteristics influencing nesting activity of *Caretta caretta* on Sekania beach. Biol Conserv 121:177–188

Lennon G, Bush DM, Neal WJ, Pilkey OH, Bullock J (1996) Living with the South Carolina coast. Duke University Press, Durham, 264 pp

Mascarenhas A, Jayakumar S (2008) An environmental perspective of the post-tsunami scenario along the coast of Tamil Nadu, India: role of sand dunes and forests. J Environ Manage 89(1):24–34

McLachlan A (1991) Ecology of coastal dune fauna. J Arid Environ 21:229–243

McLachlan A, Brown AC (2006) The ecology of sandy shores, 2nd edn. Academic, New York, 373 pp

McLachlan A, Dugan JE, Defeo O, Ansell AD, Hubbard DM, Jaramillo E, Penchaszadeh PE (1996) Beach clam fisheries. Oceanogr Mar Biol Annu Rev 34:163–232

Psuty NP (2004) The coastal foredune: a morphological basis for regional coastal dune development. In: Psuty NP, Martínez M (eds) Coastal dunes: ecology and conservation. Springer, Berlin, pp 11–27

Ranwell DS (1972) Ecology of salt marshes and sand dunes. Chapman and Hall, London, 258 pp

Ruz MH, Meur-Ferec C (2004) Influence of high water levels on aeolian sand transport: upper beach/dune evolution on a macrotidal coast, Wissant Bay, northern France. Geomorphology 60(1–2):73–87

United Nations Millennium Ecosystem Assessment (2005) Ecosystems and human wellbeing, a synthesis. United Nations. Source: http://www.maweb.org/en/Synthesis.aspx. Accessed 16 Nov 2011

Welch RC (1989) Invertebrates of Scottish sand dunes. Proc R Soc Edinb 96B:267–287

Chapter 5
Vegetated States and Values – Inland Dune

Abstract This chapter describes the vegetation of sand dunes as determined by systematic analysis of plant species composition. It includes the stable forms of sand dune inland of the beach and foredune, where there is one. Vegetation usually covers the surface, as described in Sects. 1.3.3, 1.3.4, 1.3.5 and 1.3.6. States 1–4 relate to different grazing pressures, State 5 to afforestation and State 6 the presence of native woodland. The chapter provides the basis for assessing the trends and trade-offs associated with changes in the extent and intensity of grazing, the degree of scrub and woodland development (also dealt with in Chap. 8 in relation to the invasion of alien plants) and the presence of native woodland. The following sections, using examples taken mostly from Europe, describe the vegetated inland sand dune states and their ecosystem services values in more detail.

5.1 Driving Forces, Pressures, States, Impacts and Response (DPSIR) Affecting Vegetated Sand Dune

This chapter uses the same approach as for Chap. 4 i.e. the Driving Forces, Pressure, State, Impact, Response model. In this case, the key driving force is the presence or absence of grazing animals and the extent to which the sand dune supports open dune grassland, heath, scrub or woodland. The key factors are changes in farming and forestry practice associated with both economic and social conditions. Other forces such as climate change, especially storm events, can override the effects of grazing. Nitrogen deposition also influences the vegetation, but to a lesser extent. These pressures lead to changes in the state of the sand dune in relation to the vegetation cover, which can cause deterioration in the habitat and its nature conservation value (Fig. 5.1).

As discussed in Sect. 4.1, reviewing the pressures and the values associated with each state helps identify the impact on the habitat. Loss of value may require management action. This response will depend largely on the perception and relative priority

J.P. Doody, *Sand Dune Conservation, Management and Restoration*,
Coastal Research Library 4, DOI 10.1007/978-94-007-4731-9_5,

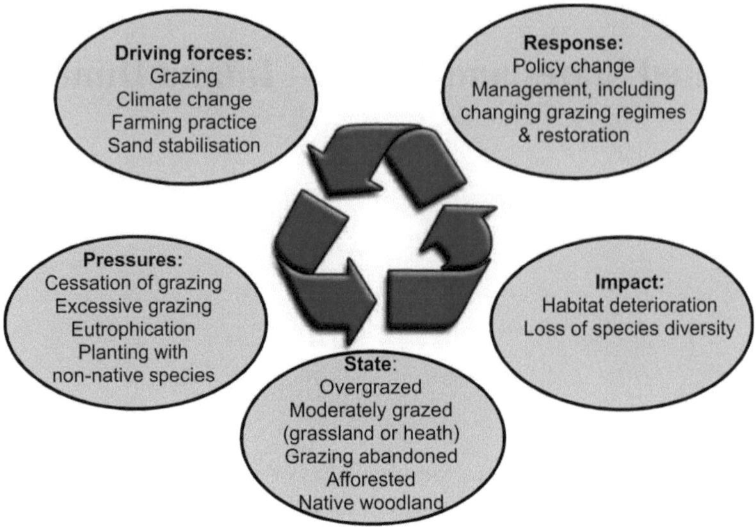

Fig. 5.1 Examples of the various stages in the DPSIR cycle relevant to stable (fixed), and semi-stable vegetated sand dunes

attached to the value of each state by the 'viewer'. For example, a farmer might consider the grazing potential paramount, preferring a 'fixed' dune with close-cropped vegetation; the conservationist might tend towards a less intensive regime, producing more structurally diverse vegetation with bare sand. As with concerns about the earlier stages in sand dune development at the beach/foredune interface, (Chap. 4) these interests can complement each other and it may be easy to agree a move from the existing state to a new one. On the other hand, it may not!

5.2 Vegetated States – Description

With the establishment of the pioneer stages of sand dune vegetation, more stable forms develop. The succession described in Chap. 1 (Sect. 1.3) is typical of the sequence leading to ever more stable dunes, and in the absence of human interference development of grassland, heathland, scrub and woodland. Modification of this sequence may take place because of natural factors such as destabilisation by storms or fire. However, human intervention has had a profound impact in many temperate regions of the world. Overuse resulted in sand drift, and measures designed to restore vegetation cover were common. Grazing and burrowing by a variety of native (and introduced) animals helped modify the vegetation resulting in the development of dune grassland or heathland, particularly in Europe.

The level and intensity of grazing helps to define four states for vegetated sand dunes:

- **State 1** – Heavily grazed;
- **State 2** – Moderately grazed, dune grassland;
- **State 3** – Moderately grazed dune heath;
- **State 4** – Abandoned formerly grazed (scrub dominated).

Two additional states, not directly related to grazing pressures are:

- **State 5** – Afforested dune;
- **State 6** – Native dune woodland.

The discussion relates mainly to sand dune management during the last 50 years or so. The descriptions refer to the main body of the sand dune behind the first dune ridge and the changes occurring within it. These exist as a mosaic within which inland mobile dunes and dune slacks form important components.

5.2.1 State 1 – Heavily Grazed

Heavily grazed dune vegetation occurs in areas where animal densities are high. Grazing may be by domestic stock – mainly sheep and/or cattle and ponies, or native animals such as hares and introduced rabbits. Domestic stock, especially if accompanied by rabbit grazing can create grass swards with little or no structural diversity, an increase in grazing tolerant grasses and a loss of herbs. Although some herbs survive, they often fail to flower or lose flowering spikes at an early stage of growth. This gives the close-cropped turf the appearance of having an impoverished flora (Fig. 5.2). The lack of structural diversity leads to a loss of invertebrates, breeding birds and other animals that rely on dune plants for food, nesting sites or shelter.

As grazing pressure increases, especially when burrowing rabbits are amongst the herbivores, patches of bare sand may appear. At its most extreme the vegetation may break up, exposing the underlying sand to the wind and rain and leading to extensive sand movement, loss of vegetation and dune surface deflation. An extreme form occurs in the machairs of western Ireland (Fig. 5.3) and several sites in north-west Scotland. Once the surface vegetation is lost, it can remain so for several years if not decades, particularly if the destabilising agent remains.

5.2.2 State 2 – Moderately Grazed Dune Grassland and State 3 – Dune Heath

Moderately grazed dune grassland and heath occur where grazing and/or 'natural' processes help to sustain transitions to mature grassland (Fig. 1.5) and/or heathland (Fig. 1.6), behind the foredune. In Europe this stabilised vegetation includes

Fig. 5.2 Heavily grazed dune grassland, Northam Burrows, Devon September 2005. The exercise of historical grazing rights results in a close-cropped grass sward. The only structural diversity in the vegetation occurs where less palatable plants such as the spiky *Juncus acutus* grow

Fig. 5.3 Extensive erosion of overgrazed machair on the west coast of Ireland. At many locations on these exposed coasts, there is no foredune. The interface with the beach has a steep eroding cliff (Fig. 11.8)

Fig. 5.4 Moderately grazed dune grassland and slack, Sandscale Haws, Cumbria, England, December 1988

Ammophila arenaria, which becomes less frequent in the sward as succession takes place. This seems to come about naturally, as the species is most vigorous when continually buried in sand (Sect. 1.3.2). In the absence of new sand, plant growth diminishes and new shoots are less frequent with the individual culms becoming more sparsely dispersed (Sect. 1.3.3).

The process of succession to typical calcareous dune grassland takes place under grazing regimes on sand dunes where the original sand grains have high calcium carbonate content. This often results in species-rich plant communities, which have affinities with inland chalk and limestone grasslands. Amongst a wide representation of species, orchids are often conspicuous components of the flora. The continuing presence of grazing animals, including domesticated stock and rabbits, helps to sustain the characteristics and value of this vegetation (Fig. 5.4).

In Europe, acid dune heathland is a priority habitat under the EU Habitats and Species Directive (Sect. 3.2.2). This plant community occurs on sand dunes, either where the original calcium carbonate content of the soils is low, or in older dunes where leaching has removed the carbonate ions. Mature heath is relatively unpalatable to rabbits and stock and often has a closed sward. This has a different but no less valuable suite of species, based on the presence of plants such as *Calluna vulgaris*, *Empetrum nigrum* on drier areas and *Erica tetralix* in wetter hollows. Grazing animals may be present, but usually in low numbers. Destabilisation is less pronounced and succession to scrub (State 4) more frequently encountered.

Where grazing is absent or reduced then instability becomes a much more important factor in maintaining the range of plant communities. Included within this are sites where erosion occurs in cycles, for example as 'blow-outs.' These act as precursors to dune slacks, which occur in vegetated States 2 and 3 within the inland dune where the water table is at or near the surface. Described in Sect. 1.3.4, they are a manifestation of the dynamic nature of the sand dune. The proportion of bare sand can vary in both time and space, but does not approach the levels depicted in Fig. 5.3.

5.2.3 State 4 – Abandoned, Formerly Grazed, Scrub Dominated

This state is characterised by sand dunes where the reduction or removal of grazing, results in the growth of coarse grasses and scrub. As the sand dune becomes overgrown, there is a loss of nature conservation values as species disappear from the open grassland or heath. Atmospheric nitrogen deposition can exacerbate this effect. For example, relatively tall graminoids[1], such as *Ammophila arenaria* and *Carex arenaria* replaced lichen-rich dune grasslands in parts of the Netherlands (Ketner-Oostra and Sýkora 2004). *Hippophae rhamnoides*, often introduced to control erosion, can also become highly invasive creating dense stands obliterating most other species of open sand dune (Chap. 8).

In warmer regions such as the Mediterranean, areas of scrub 'maquis' (Sect. 1.6.2) with an understory of dwarf shrubs or 'garrigue' forms the dominant vegetation (Groves and Rackham 2001). The vegetation is drought and grazing tolerant, and replaces the evergreen oak forest that probably covered most of the area historically. Here, open-range grazing by herds of goats and sheep has little obvious influence on the vegetation. This also appears to be the situation in the United States of America. Section 1.3.5 describes these vegetation types in more detail.

5.2.4 State 5 – Afforested Dune

Unlike the much rarer 'natural' dune woodlands, afforested sand dunes, usually planted with non-native tree species to control erosion, are ubiquitous. As described in Sect. 2.3.2 this has caused major loss of open sand dune habitat in many parts of the world. Pines are the most frequently used tree species, whilst shrubs, such as *Hippophae rhamnoides* (in Europe) and *Lupinus arboreus* (in New Zealand) may become part of the stabilisation process. These areas develop a closed canopy and have a significant negative impact on the native species of vegetated sand dunes. The build up of inert pine needles tends to dominate the woodland floor (Fig. 5.5), although not exclusively (Sect. 7.5.1).

---

[1] Grass or grass-like.

Fig. 5.5 Typical afforested dune, pine woodland floor dominated by inert needles, West Frisian island of Terschelling the Netherlands, 1987

5.2.5 State 6 – Native Dune Woodland

The destruction of most native dune woodland in Europe occurred in historical times (Sect. 2.2.1). Despite this, a few mature natural forests have survived. These include pine woodland remnants on the coast of Albania (Fig. 1.9), and in Italy. Elsewhere, woodland composed of native species is usually 'secondary'[2] in origin.

Secondary deciduous dune woodlands occur in the Netherlands, as for example in the Meijendel dunes. Here *Crataegus-Betula* woodlands appear to have developed between 1950 and 1980. Lack of grazing and an increase in water levels due to infiltration of the dunes from the River Rhine for drinking water, may have been responsible (van der Meulen and Wanders 1985). Scots Pine *Pinus sylvestris* was probably more extensive before human exploitation caused its extinction in the eighteenth century. Today Oak *Quercus robur* forests occur in Western Jutland, Denmark. The older (86–104 year) plantations have developed into a typical forest community, which could be unique both in Denmark and on an international scale (Lawesson and Wind 2002).

Secondary woodland with *Betula pendula* also occurs on a few sites in Scotland, *Pinus sylvestris* in Finland and *Quercus* spp. with Rowan *Sorbus aucuparia* on the

[2] 'Secondary' describes woodlands that have regenerated abandoned or neglected ground previously used for agriculture, grazing or other uses.

Fig. 5.6 Overgrown dune with developing native secondary woodland *Quercus spp.* and Rowan *Sorbus aucuparia*, Spiekeroog, Germany, August 2008

German East Frisian Island of Spiekeroog (Fig. 5.6). Although these woodlands are 'secondary', they have some of the characteristics of 'ancient'[3] woodlands. As we shall see in Chap. 7, the question arises as to the extent to which these habitats should be left to develop, re-establishing native woodland on State 4 ungrazed dunes or cut down to restore State 2 or 3 dune grassland or heath respectively.

Native woodland occurs on sand dunes in the United States of America. Nags Head Woods on North Carolina's Outer Banks is an extensive ecological preserve that includes a range of unique habitats, including forested dunes. In Virginia, maritime dunes include deciduous, coniferous and broadleaf evergreen woodlands occurring on inland dunes protected from regular salt spray. Similar communities are present along the Atlantic and Gulf coasts from New Jersey to Texas. They are relatively localised in the back dunes on Fire Island, New York and on Sandy Hook, New Jersey. Together they support the largest American Holly *Ilex opaca* forest in the United States of America (Psuty pers. comm. 2012).

[3] 'Ancient' describes woodlands in Great Britain, continuously wooded since the 1600s in England and Wales, or 1750 in Scotland.

The sub-tropical woodlands of Fraser Island, in Australia described as having "majestic remnants of tall rainforest growing on sand" in the United Nations Educational, Scientific and Cultural Organisation (UNESCO) World Heritage List is another example. However, these are the exception rather than the rule.

5.3 Provisioning Services

Sand dunes are the source of a wide variety of products (provisioning services defined in Sect. 4.4.3). Most of these derive from uses that represent modified rather than natural values. The following sections describe these values, which range from direct material benefits such as timber production to animals that live in or on the vegetation. Chapter 7 deals with the trade-offs in relation to nature conservation and cultural values (Sects. 5.4; 5.5 below) associated with each state.

5.3.1 Timber

Timber production from planted forests can provide commercial quantities of wood. Culbin Forest in northeast Scotland, for example, produced timber used for telegraph poles and wood pulp (Ross 1992). In the region of Aquitaine in France, the extensive area of forest in the French Département of "Les Landes" provides a significant source of timber products. Depending on the quality, larger logs make building timbers, plywood, parquet flooring, panelling and furniture. Being resinous, the pines are particularly valuable for areas that become periodically wet such as window frames. Poorer quality material is used by the pulping industry, providing raw material for boards or paper.

In New Zealand today, afforestation to arrest sand dune erosion, salt spray and wind effects using a variety of non-native shrubs and trees, also provides a successful commercial forestry operation (Berg 2006).

5.3.2 Hippophaë rhamnoides

Hippophaë rhamnoides has been used as a stabilising agent in many areas outside its native distribution. Although it is an alien invader in many parts of the world (Sect. 8.3), it has a number of values. It is especially important in China, where for at least twelve centuries it has been valued for its medicinal properties. Amongst its values are the following:

- Nutritious food;
- Medicine;
- Soil enhancer;
- Pollution control;

- Source of firewood;
- Landscape management tool;
- Wildlife enhancement (Li and Schroeder 1996; Small et al. 2002).

It is also valuable to wildlife, for example in the UK this includes:

- 28 species of invertebrates recorded as feeding on the genus;
- Flowers and fruits, which are valued for nectar and food;
- Forms thick dense cover for birds to nest and roost, especially at migration times;
- Its fruits are also much valued as a source of food for birds and mammals (Bacon 2003).

5.3.3 Domesticated Livestock

Domesticated livestock grazing sand dunes produce significant quantities of meat and wool, and in some parts of the world are integral to the local agricultural economy. Although this use is becoming less prevalent, it continues in some areas as part of a farming enterprise with established dune grasslands. For example, the machair of Scotland and Ireland both have domesticated stock (mostly sheep and some cattle). The former is also used for crop cultivation (Chap. 11).

It is possible to produce good quality beef from some breeds of cattle on what are impoverished grazing areas. These include:

- **Upland beef** such as Highland, Galloway, Welsh Black, Beef Shorthorn and Vaynol (Welsh), which are slow growing and late maturing. They may require supplementary feeding, or a period on improved grazing to reach marketable weight. They produce a good quality carcass often sold to a local or niche market;
- **Lowland Beef** such as Hereford, Aberdeen Angus, Sussex, South Devon and Lincoln Red (continental examples include Limousin and Charolais), which are moderately hardy. All have good quality meat and are thus highly marketable, although they require rather better quality grazing.

Sheep also produce saleable lamb on sand dune grassland. Hardy 'hill' sheep breeds including Swaledale, Cheviot, Welsh Mountain, Scottish Blackface and Herdwick are amongst the most frequently encountered (GAP/FACT 2009 http://www.grazinganimalsproject.org.uk/breed_profiles_handbook.html).

5.3.4 European Rabbit Oryctolagus cuniculus as Game

Oryctolagus cuniculus is native to southwest Europe[4] and a keystone species in the Iberian ecosystem, where it prefers a mixed habitat of Mediterranean oak or

[4] Note in its native range on the Iberian Peninsula it has declined to such an extent that it is now on the IUCN Red List of Threatened Species (International Union for Conservation of Nature and Natural Resources, http://www.iucnredlist.org/details/41291/0).

scrub-forest. In this native range it is an important prey item for the rare Iberian Lynx *Lynx pardinus*, probably the most endangered of the world's 36 cats. It occurs in some protected sand dune areas, including Doñana National Park in Spain (Delibes et al. 2000; Ward 2005). Outside this native range, it now occurs as an introduced species throughout most of western and central Europe, as far north as southern Sweden. It is also present on the islands of Corsica, Sardinia and Sicily and in parts of North Africa. Its export to Australia and New Zealand and the resulting environmental and economic consequences are well known. It also reached plague proportions in South America (Matthews and Brandt 2006).

Rabbits as a food resource were important in historical times. Their rapid rate of reproduction meant they were an ideal source of protein. Probably introduced by the Normans, the first records in England appear in 1176, when there were rabbits in the Scilly Isles. At sometime between 1183 and 1219 they were recorded on Lundy Island and in the early thirteenth century on the Isle of Wight. Rabbits seem to have been very expensive in the late thirteenth and fourteenth centuries (Medieval period) costing four or five times as much as chickens, suggesting their relative scarcity (Veale 1957; Bailey 1988).

In England in the medieval period, expansive areas of sand dunes were sheep-grazed. However, rabbit cultivation became more lucrative after the Black Death in the mid fourteenth century, and by the seventeenth century they were present on many large estates. Due to the relative ease with which rabbits can excavate burrows, sand dunes and sandy areas inland (such as the Brecklands in East Anglia) became favoured locations for their cultivation (Bailey 1988). By the eighteenth century, at sites such as on the Gwithian Towans in Cornwall, sheep grazing gave way to rabbit cultivation in 'warrens'. Sand dunes bearing the name 'warren' or 'burrows' are relatively common, testimony to the extent to which they were key locations for this use.

Rabbits, introduced to the Netherlands as early as 1280, were also cultivated for food in "waredes"[5] (Wallage-Drees 1988). In Ireland, they were also a source of meat and fur from the early nineteenth century until the early twentieth century. In the war years (1939–1945) they provided protein, and as demand for food increased the rabbit was nearly eliminated (Binggeli 1994).

5.3.5 Other Services

Sand dunes provide an important sediment source for infrastructure development. They also hold significant volumes of water and can provide a major source of drinking water. In the Netherlands, dune water extraction for domestic supply began in 1853, with the sand dunes effectively becoming water catchment areas. The main areas are located in the provinces of North and South Holland and supply about 800,000

[5] 'Warrens' – in Great Britain,

customers in the city of Amsterdam and surrounding areas. It is also important in parts of Israel where there is an underground water supply fed by rainwater, which provides potable water for settlements in the region of the Haifa-Acco sand dunes, Israel (Goldschmidt and Jacobs 1956).

Sand dunes also provide water for irrigation. Those with underground aquifers provide a ready local supply to water greens, tees and fairways on golf courses (Chap. 10).

The Ishikawa Agriculture Research Centre Sand Dune Experiment Station in Japan was specifically set up to promote cultivation on sand dunes. Also in Japan, horticultural enterprises use crops adapted to the properties of sand (Yoshichika 1999). In some parts of both temperate and tropical regions, beach and sand dune plants provide a variety of provisioning services such as food and fodder, and biochemical/pharmaceutical uses (Sridhar and Bhagya 2007). The machair of western Scotland has crops of cereal and potatoes (Chap. 11).

5.4 Cultural Services – Nature Conservation

Each of the vegetated states described above has a range of values; this section describes those associated with the nature conservation of areas with a reasonable cover of vegetation. By virtue of the mosaic of habitat, they support a wide range of plants and animals. Grazing management and/or periods of instability followed by stabilisation often sustain these values.

5.4.1 Vegetation

Many of the pioneer plants colonising sand dunes are restricted to this habitat. As the dunes stabilise the vegetation includes species with wider tolerances as described above (State 2 and State 3 dunes). Vegetation classifications provide the first stage in the identification of important nature conservation areas. These relate to the successional sequence from the beach/foredune interface to stabilised dunes. Geographical variation provides a second parameter. In Europe, there are two principal regions within which the vegetation is classified, the Atlantic, North Sea and Baltic coasts and the Mediterranean (Table 5.1). This provides the basis for selection of Special Areas of Conservation across northwest Europe and on the Mediterranean and Black Sea coasts.

Plant species diversity can be high on inland sand dunes. A detailed study of plants found on sand dunes in Wales recorded nearly 1,000 species including 439 vascular plants, 289 fungi, 171 bryophytes and 66 terricolous lichens. Of these a significant number were only present on sand dunes. Thirty two species of vascular plants (7%) were either wholly dependent or strongly associated with sand dunes; 22 species (nearly 13%) of the bryophytes also fell within this category.

Table 5.1 Coastal sand dunes of Community Interest forming the basis for the selection of Special Areas of Conservation (SACs). The table includes the Natura 2000 codes and description taken from the Interpretation Manual of European Union Habitats (European Commission 2007). NB. Plant communities with Codes 2110 and 2120 form part of the beach/foredune interface (Physical States 1–3, Chap. 4) and those with Codes 2210, 2220 and 2230 may also occur there (Sect. 4.3.4)

Codes	Directive name	Description
Sea dunes of the Atlantic, North Sea and Baltic coasts		
2110	Embryonic shifting dunes	"Formations of the coast representing the first stages of dune construction constituted by ripples or raised sand surfaces of the upper beach or by a seaward fringe at the foot of the tall dunes."
2120	Shifting dunes along the shoreline with *Ammophila arenaria* (white dunes)	"Mobile dunes forming the seaward cordon or cordons of dune systems of the coasts."
2130	[a]Fixed coastal dunes with herbaceous vegetation (grey dunes)	"Fixed dunes, stabilised and colonised by more or less closed perennial grass-lands and abundant carpets of lichens and mosses, from the Atlantic coasts (and the English Channel) between the Straits of Gibraltar and Cap Blanc Nez, and the shores of the North Sea and the Baltic."
2140	[a]Decalcified fixed dunes with *Empetrum nigrum*	"Decalcified dunes colonised by *Empetrum nigrum* heaths of the coasts."
2150	[a]Atlantic decalcified fixed dunes (*Calluno-Ulicetea*)	"Decalcified dunes of France, Belgium and Britain, colonised by heaths"
2160	Dunes with *Hippophae rhamnoides*	"Sea-buckthorn formations of forest colonisation in both dry and humid dune depressions."
2170	Dunes with *Salix repens* ssp. *argentea* (*Salicion arenariea)*	"*Salix repens* communities (*Salicion arenariae*), colonising wet dune slacks."
2180	Wooded dunes of the Atlantic, Continental and Boreal region	"Natural or semi-natural forests (long established) of the Atlantic, Continental and Boreal region coastal dunes with a well developed woodland structure and an assemblage of characteristic woodland species."
2190	Humid dune slacks	"Humid depressions of dunal systems. Humid dune-slacks are extremely rich and specialised habitats very threatened by the lowering of water tables."
21A0	Machairs ([a]in Ireland)	"Complex habitat comprised of a sandy coastal plain resulting partially from grazing and/or rotational cultivation, in an oceanic location with a cool, moist climate. The wind blown sand has a significant percentage of shell derived material, forming a lime rich soil with pH values normally greater than 7. Vegetation is herbaceous, with a low frequency of sand-binding species."

(continued)

Table 5.1 (continued)

Codes	Directive name	Description
Sea dunes of the Mediterranean coast		
2210	*Crucianellion maritimae* fixed beach dunes	"Fixed dunes of the western and central Mediterranean, of the Adriatic, of the Ionian Sea and North Africa with *Crucianella maritima*, *Pancratium maritimum*."
2220	Dunes with *Euphorbia terracina*	"Coastal dune grassland communities."
2230	*Malcolmietalia* dune grasslands	"Associations with many small annuals and often abundant ephemeral spring bloom."
2250	[a]Coastal dunes with *Juniperus* spp.	"Juniper formations." Also occurs in the north.
2260	*Cisto-Lavenduletalia* dune sclerophyllous scrubs	"Sclerophyllous or lauriphyllus scrubs established on dunes of the Mediterranean and Warm-Temperate Humid regions."
2270	[a]Wooded dunes with *Pinus pinea* and/or *Pinus pinaster*	"Coastal dunes colonised by Mediterranean and Atlantic thermophilous pines."

[a]Indicates a priority habitat. Table taken from the Interpretation Manual of European Union Habitats – EUR27 (European Commission 2007)

Of the fungi, only ten species (4%) appear to be restricted to sand dunes, and none of the lichen species (Rhind and Jones 1999).

Sand dunes can also support a significant proportion of the flora of some countries. In Israel, 173 plant species occurring on sand represent 8.2% of the total flora of the country and include many endemic species, mostly annuals. These make up 26% of all endemic species in Israel (Kutiel 2001).

5.4.2 Rare Plants

Sand dunes have a rich and varied flora including a number of specialised rare species. In Wales over 9% (91 species) of the total flora found on sand dunes are rare, scarce or endangered, and about 8% of the vascular plant species are considered endemic or near endemic to Europe (Rhind and Jones 1999). Amongst these dune slacks are especially important. Dune gentian *Gentianella uliginosa* and *Liparis loeselii* are two such rare plants in the UK.

Dune slacks also support rare mosses and liverworts. Petalwort *Petalophyllum ralfsii*, a bright green liverwort survives in Wales and at a few other scattered locations. Blunt Bryum *Bryum calophyllum*, a small upright moss colonising bare sand in dune slacks, is thought to be extinct in England but survives at one locality in Wales (on Anglesey) and at three sites in Scotland. Long-leaved Thread-moss *Bryum neodamense* is the only surviving population in England occurring on the Sefton Coast of Merseyside. Warne's Thread-moss *Bryum warneum* once known from as many as 35 localities around the British coast, now survives at only 12. Other rare and threatened bryophytes of coastal dune slacks include Knowlton's Thread-moss *Bryum knowltonii* and Sand Thread-moss *Bryum mamillatum*, now thought to be extinct (Plantlife International 2003).

There are several specialist dune fungi, which include species considered endangered, vulnerable or rare. Of some 502 taxa recorded in a study of Welsh sand dunes, at least 34 were included in this category (Rotheroe 1993). At Torrs Warren, southwest Scotland two rare species are Dune Brittlestem *Psathyrella ammophila* and a Bird's-nest Fungus *Cyathus stercoreus* recorded on rabbit dung amongst *Ammophila arenaria*.

Sand dunes may also provide habitats for small relict populations of other rare species. For example in Canada, the Contorted-pod Evening-primrose *Camissonia contorta* is restricted to seven semi-stable sandy flats on dunes no more than 15 m above sea. It is rare but more widespread further south in California where it occurs at elevations between 2,800 and 5,000 ft (COSEWIC 2006). *Viola rupestris* a species of mountains in Europe including Upper Teesdale, northern England (the origin of one of its common names, Teesdale Violet) occurs in sand dunes at sea level in at least one location in the Netherlands (Weeda 1992). This appears to be a postglacial relict population, similar to other species such as Spring Gentian *Gentiana verna* (Fig. 5.7), which also occurs in Upper Teesdale (Bradshaw and Doody 1978). These species survive in a few locations at sea level and at relatively low latitude.

Fig. 5.7 *Gentiana verna* an arctic-alpine growing in the sand dunes of the Burren, western Ireland

This suggests sand dune mobility and impoverished soils helped maintain the necessary open conditions for the species survival, throughout the postglacial period, especially when tree cover was at a maximum.

5.4.3 Invertebrates

Sand dunes are particularly important for invertebrates. As described in the previous chapter a mixture of plant communities and bare sand provide ideal conditions for a wide range of species. In all there are more than 300 scarce invertebrates restricted to, or mostly associated with sand dunes in the United Kingdom. These include those restricted to the beach/foredune (Sect. 4.3.4), others of inland dunes and some that range across both habitats.

Many invertebrate dune specialists require bare sand for their survival. Blowouts in the inland dunes provide suitable habitat for burrowing by invertebrates when nesting. In Wales, bare and sparsely vegetated sand dune held 172 species, the highest overall number for an individual vegetation type and nearly 40% of the total number of bee, wasp and ant species recorded (Howe et al. 2010). Species may include some that are restricted in their distribution due to the need for a soft substratum to excavate a nest. For example, Hymenoptera (especially wild bees, digger wasps,

Fig. 5.8 Hairy legged mining bee – *Dasypoda hirtipes* one of a large number of burrowing invertebrates that find sand dunes an especially suitable habitat

spider wasps and other solitary wasps) build nests mainly in sandy locations (Haeseler 1989, 1992). Although not restricted to sand dunes the Hairy Legged Mining Bee *Dasypoda hirtipes* is one such (Fig. 5.8). Others include *Colletes fodiens* present at its only Scottish site in the sand dunes of Torrs Warren and the nationally rare *Colletes cunicularius*, present in sand dunes on the coast of Wales and northwest England. Spiders are also important and include many rare and threatened species, especially in Northern (Almquist 1973) and Western Europe (Duffey 1968; Bell et al. 1998).

Invertebrate species not only rely on specific habitats but also on the mosaic of habitats. In the United Kingdom, the richest zones are those where clumps of stabilised vegetation occur in close proximity to areas of unstable open sand. This combination allows species nesting in open soils to feed from the flowers of tall vegetation. Calcareous dunes tend to be richer in invertebrates than acid ones (Kirby 1992).

Dune succession affects invertebrate species diversity, distribution and abundance. Species of Diptera (flies) increased from dry to humid habitat in Dolichopodids (small, metallic looking with large, prominent eyes), and from open to canopied sites in Empidids (dance flies). On the other hand, canopied sites appeared to function as home bases from which certain Empidid species colonised more sun-exposed areas (Pollet and Grootaert 1996). As the habitat becomes more stable, other species occur and these may show completely different assemblages. For spiders, the largest number of species, 117 occurred in dune slacks and 131 in ungrazed dune meadow, with only 65 in dune heath. This change in diversity appeared less related to plant species diversity and more to the complexity of habitat structure (Duffey 1968).

Mosaics of different individual patch sizes and vegetation types are also important. Larger patches tended to have less variation in the species complement between them than smaller ones. This occurs through edge effects, which can alter the spider assemblage considerably because of invasion by species from other vegetation types (Bonte et al. 2002). Even small sand dune remnants provide refuges for scarce invertebrate species. One species in the USA, the Oso Flaco flightless moth *Areniscythris brachypteris* occurs in one dune area only (the Oso Flaco Dunes) in California (Powell 1976, 1981). Another rare insect that inhabits dunes in the same area is the Smith's Blue Butterfly *Euphilotes enoptes smithi*, an endangered species (Black and Vaughan 2005).

5.4.4 Birds

Inland dunes support a wide range of avifauna. In addition to the birds frequenting the upper beach and foredunes (Sect. 4.3.4) there are many others, which take advantage of the range of habitats present in the inland dune zone. These include coastal species as well as those more typical of inland non-coastal habitats.

Vegetated dunes in the northern Europe have nesting gulls, including the Black-headed Gull *Larus ridibundus* and Eider duck *Somateria mollissima.* Breeding passerines include Skylark *Alauda arvenis*, Meadow Pipit *Anthus pratensis*, Linnet *Carduelis cannabina* and Stonechat *Saxicola torquata*, which can be relatively abundant when compared with adjacent farmland. Birds of prey including Short-eared Owl *Asio flammeus* and Merlin *Falco columbarius* hunt the dunes and dune slacks. The open cultivated and grazed machairs of the west of Scotland have high densities of nesting waders such as Dunlin *Calidris alpine* (Chap. 11). In the winter, Fieldfare *Turdus pilaris* and Redwing *Turdis iliacus* feed on the berries of Hawthorn *Crataegus monogyna* and *Hippophae rhamnoides*.

The Laughing Gull *Larus atricilla*, Caspian Tern *Sterna caspia* and the Gull-billed Tern *Sterna nilotica* nest in dunes in the eastern shores of the USA preferring areas with more dense vegetation. Several birds of prey, such as harriers *Circus* spp. (*Circus aeruginosus*, *C. pygargus* and *C. cyaneus*) and owls *Asio* spp. (*Asio flammeus* and *A. otus*) may hunt and breed in larger dunes. At a few sites in the Mediterranean, tree-nesting species such as the Squacco Heron *Ardeola ralloides* and Purple Heron *Ardea purpurea* are present in dense low scrub. Mutton bird or Short-tailed Shearwater *Puffinus tenuirostris* rookeries are also found in dune scrub. Little Penguin *Eudyptula minor* nest in Tasmania and South Australia and their burrows are often established in sand dune vegetation, including scrub.

5.4.5 Mammals

Sand dunes do not have a particularly significant mammal fauna. They provide habitat especially suited to the European Rabbit *Oryctolagus cuniculus* (Sect. 5.3.4). They are also frequented by predatory species such as the Fox *Vulpes vulpes* and Stoat *Mustela*

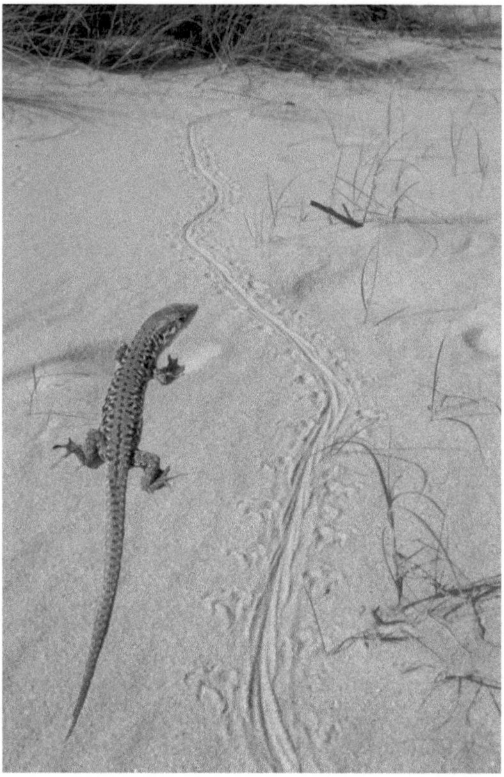

Fig. 5.9 Lizard tracks in a Spanish sand dune. Tracks such as these can often be seen in sand dunes free from human disturbance. Inset Sand Lizard

erminea. Paradoxically there are significant populations of the rare (in the UK) Red Squirrel *Sciurus vulgaris* in a number of forestry plantations on the Sefton Coast and Newborough Warren, Anglesey, Wales (Sect. 7.5.1). Tentsmuir Forest, south-east Scotland has Polecat *Mustela putorius*, Weasel *M. nivalis*, several species of bats, voles, mice and shrews occurring in most of the areas.

5.4.6 Reptiles and Amphibians

Reptiles are common inhabitants of sand dunes (Fig. 5.9). The Sand Lizard *Lacerta agilis*, widespread in Europe in sandy heaths and sand dunes has a patchy distribution. It is the rarest lizard in the UK, occurring in pockets of heathland and a few coastal sand dunes in northwest England (Moulton and Corbett 1999). The Argentine Sand Dune Lizard *Liolaemus multimaculatus* is an endemic species now almost confined to coastal sand dunes, where its colouring and ability to swim in the loose sands make it particularly well adapted (Kacoliris 2007).

In the USA snakes also live and feed in dune systems. Amongst these are Eastern Diamondback Rattlesnakes *Crotalus adamanteus*, which may use Gopher Tortoise *Gopherus polyphemus* burrows to seek refuge. Eastern Coachwhip snakes

Masticophis flagellum, Florida Rough Green snakes *Opheodrys aestivus carinatus*, and Coastal Dunes Crowned Snake *Tantilla relicta pamlica* (not found outside Florida) all live in grassy dunes or more stable woody dunes (University of Florida, Florida Museum of Natural History http://www.flmnh.ufl.edu/herpetology/fl-guide/fl-snakelist.htm).

Sand dunes are not known to be particularly rich in amphibians but for the Natterjack Toad *Epidalea calamita* it is an important habitat. Water bodies suitable for breeding must be temporary, shallow and exposed to the sun. Optimally these areas dry up in July–August helping the species to avoid competition from other amphibian species and predation from fish and insect larvae. Dry sandy areas provide habitat in which to hide and hibernate. It is on the International Union for the Conservation of Nature Red List of Threatened Species.

5.5 Cultural Services – Recreation and Research

Sand dunes provide a significant recreational resource in addition to their nature conservation values. As part of the coastal experience, they provide a place for rest and relaxation and enjoyment of scenery. Many activities including sunbathing, boating and swimming are associated with the beach/foredune (Sect. 4.3.4). Vegetated sand dunes inland are not however, just places that provide access to the sea. They have a range of other recreational values such as walking, bird watching as well as more active pursuits.

5.5.1 Recreation

Some sand dunes are large enough to accommodate many types of recreational activity, including hiking, photography, fishing, canoeing, horseback riding and camping. Coastal areas generally generate substantial economic benefits, especially in more remote areas, through the provision of opportunities for outdoor activities.

The Dune of Pilat, situated at the entrance of the Bay of Arcachon is probably the biggest in Europe. It is over 100 m high, 2,700 m long and 500 m wide and is one of the most visited coastal sites in the area. At 40 m high, the Råbjerg Mile near Skagen in northern Denmark and the Łeba sandbar, within the Slowinski National Park Poland are also popular local attractions. Their aesthetic appeal includes providing viewpoints over an otherwise low-lying landscape (Sect. 12.4.3).

They can also be exciting places, and in some areas off-road vehicles take people on rides across the dunes. In the United States of America, they include several coastal areas including sites within The Oregon Dunes National Recreation Area and in California (Crowley Offroad 2010 "Sand Dune Guide" http://www.duneguide.com/). 'Sand boarding' and 'sand-sliding' also take place both legally and illegally. The economic benefits of these attractions can be significant, but as we will see (Chap. 7) this is not without environmental consequences.

5.5.2 *Aviation*

In addition to the occasional use of beaches as landing sites for aeroplanes (Sect. 4.3.3) sand dunes provided the place for the first attempts at manned flight. The exposed northern Outer Banks, North Carolina are characterised by a series of bare sand hills just over 30 m high, known as the Kill Devil Hills. They provided the necessary elevation in an area remote from habitation with a steady onshore wind; the ideal conditions for the first experimental glider flights. These later led to the first powered flight made from the base of the same dunes. Today the site attracts many visitors and the dunes are lost to thousands of rental homes, restaurants, sports activities and shopping. The hill where the flights took place is now the Wright Brothers National Memorial and is stabilised.

5.5.3 *Research and Teaching*

Sand dunes provide a great place for research and teaching the fundamentals of ecology. Amongst the early studies, the growth habit of the plants themselves, including the root system, soil water and mineral nutrient relationships, provided one of the first classic studies on sand dunes (Salisbury 1952). This work helped to define the adaptations of individual species in overcoming environmental constraints. Their apparently 'natural' succession made them ideal for studying the processes associated with vegetation development and as an educational or research resource (see for example Ranwell 1972; Packham and Willis 2000; Martinez and Psuty 2004; Maun 2009)

There have also been several symposia and conferences devoted specifically to sand dunes. In 1989 the Royal Society of Edinburgh organised a conference on sand dunes (Gimmingham et al. 1989). Three early conferences organised by the European Union for Dune Conservation (now EUCC – The Coastal & Marine Union), were largely concerned with sand dunes (van der Meulen et al. 1989; Carter et al. 1992; García Nova et al. 1997). Subsequent EUCC conferences covered coastlines more generally, but relevant information on sand dunes is still included in the following volumes resulting from meetings in Greece (Salman et al. 1995, 1996) and Wales (Healy and Doody 1995; Jones et al. 1996). Subsequent conferences organised by EUCC – The Coastal & Marine Union and Eurocoast also included papers on sand dunes (Velosa Gomes et al. 2002; Green 2004).

There have also been several specialist meetings. Tourism, recreation and planning and the effect on sand dune conservation was a major concern at a meeting held in Holland in 1995 (Drees 1997). Two further meetings in Denmark considered the special case of the extensive dune landscapes of Denmark (Ovesen and Vestergaard 1992) and a wider geographic review of protection and management (Ovesen 1998). The implications of sand dune mobility, its geomorphological context and the importance of human actions in sand dune development were prominent issues at a seminar in Southport, northwest England, at the culmination of the European Union,

Sefton Coast Life Project (Houston et al. 2001). Two further international conferences devoted to sand dunes covered "Dunes and Estuaries" (Herrier et al. 2005) and "Changing Perspectives in Coastal Dune Management" held from the 31st March–3rd April 2008, in Liverpool, UK. The conference was organised under the auspices of the Sand Dune and Shingle Network, hosted by Liverpool Hope University and included a special issue of the Journal of Coastal Conservation (Rooney 2010).

A quick search of the Internet revealed one further conference specifically devoted to coastal sand dunes in Brazil (Klein and Maia 2008). No doubt, there are more, but there is already a considerably legacy from which to continue to develop our ecological understanding.

5.6 Conclusions

The inland sand dune zone depends on there being an active beach (or one having been present in the past) for the establishment of the primary dune (foredune) and the subsequent evolution into the broader secondary dune complex. Once established, grazing has a key influence on the nature of vegetation. This may have considerable nature conservation values (described under Cultural Services). These values change with the intensity of grazing. Historically moderately grazed sand dunes, tend to have the best representation of dune grassland on calcareous soils or heathland on acid soils. Increased stabilisation, often associated with reduced grazing pressure leads to the growth of scrub and woodland. This changes the sand dune and its associated plants and animals, eventually leading to the loss of nature conservation values.

Provisioning services, notably timber production from plantations of alien conifers results in the loss of most species of sand dune grassland or heath. Secondary woodland will also cause a significant change in species composition. However, because primary ancient woodland on sand dunes is rare, this change is likely to be more acceptable when colonising species are native.

The trends and trade-offs associated with these different states of vegetated sand dunes are the subject of Chap. 7.

References

Almquist S (1973) Spider associations in coastal sand dunes. Oikos 24:444–457

Bacon J (ed) (2003) The scrub management handbook: guidance on the management of scrub on nature conservation sites. A handbook produced by FACT (the Forum for the Application of Conservation Techniques). Natural England. Source: http://publications.naturalengland.org.uk/publication/72031. Accessed 16 Aug 2012

Bailey M (1988) The rabbit and the medieval East Anglian economy. Agric Hist Rev 36(1):1–20

Bell JR, Haughton AJ, Cullen WR, Wheater CP (1998) The zonation and ecology of a sand-dune spider community. In: Selden PA (ed) Proceedings of the 17th European colloquium of arachnology, Edinburgh 1997. Source: http://www.european-arachnology.org/proceedings/17th/34BellB.pdf. Accessed 14 Aug 2012

Berg P (2006) The important role of trees in combating coastal erosion, wind and salt spray – a New Zealand case study. NZ Forestry Limited. Source: http://www.fao.org/forestry/11283-1-0.pdf. Accessed 02 Feb 2012

Binggeli P (1994) Historical impact of rabbits on Irish vegetation. In: Proceedings of the Irish Botanists' meeting, University of Ulster, pp 5–6

Black SH, Vaughan DM (2005) Species profile: *Euphilotes enoptes smithi*. In: Shepherd MD, Vaughan DM, Black SH (eds) Red list of pollinator insects of North America. CD-ROM Version 1 (May 2005). Source: http://www.xerces.org/wp-content/uploads/2008/09/euphilotes_enoptes_smithi.pdf. Accessed 02 Feb 2012

Bonte D, Baert L, Maelfait J-P (2002) Spider assemblage structure and stability in a heterogenous coastal dune system (Belgium). J Arachnol 30(2):331–343

Bradshaw ME, Doody JP (1978) Population dynamics and biology. In: Clapham AR (ed) The area and its natural history. Collins, Upper Teesdale, pp 48–63

Carter RWG, Curtis TGF, Sheehy-Skeffington MJ (eds) (1992) Coastal dunes. Geomorphology, ecology and management for conservation. Proceedings of the third European dune congress Galway, Ireland, 17th–21st June 1991. A.A. Balkema, Rotterdam, 533 pp

COSEWIC (2006) COSEWIC assessment and status report on the contorted-pod evening-primrose *Camissonia contorta* in Canada. Committee on the Status of Endangered Wildlife in Canada, Ottawa, p 21. Source: http://dsp-psd.pwgsc.gc.ca/Collection/CW69-14-472-2006E.pdf. Accessed 20 May 2011

Delibes M, Rodrigues A, Ferreras P (2000) Action plan for the conservation of the Iberian lynx (*Lynx pardinus*) in Europe. Convention on the conservation of European wildlife and natural habitats, Oslo, pp 1–40

Drees JM (ed) (1997) Coastal dunes, recreation and planning. Proceedings of European seminar, Castricum, The Netherlands. EUCC, Leiden, 111 pp

Duffey E (1968) An ecological analysis of the spider fauna of sand dunes. J Anim Ecol 37:641–674

European Commission (2007) Interpretation manual of European habitats. Natura 2000. European Commission, DG Environment, Nature and Biodiversity, Brussels. Source: http://ec.europa.eu/environment/nature/legislation/habitatsdirective/docs/2007_07_im.pdf. Accessed 20 May 2011

García Nova F, Crawford RMM, Días Barradas MC (eds) (1997) The ecology and conservation of European dunes, vol 38, Series. Universidad de Sevilla, Sevilla, 375 pp

Gimmingham CH, Ritchie W, Willetts BB, Willis AJ (eds) (1989) Coastal sand dunes, vol 96, Proceedings of the Royal Society of Edinburgh, Section B (Biological sciences). Royal Society of Edinburgh, Edinburgh, 313 pp

Goldschmidt MJ, Jacobs M (1956) Underground water in the Haifa-Acco sand dunes and its replenishment. Assoc int d'hydrologie sci, Symp Darcy 2(41):39–56

Green D (2004) Delivering sustainable coasts: connecting science and policy. In: Littoral 2004, 7th international symposium, Aberdeen, Scotland, United Kingdom, EUCC The Coastal Union and Eurocoast, Two Volumes

Grove AT, Rackham O (2001) The nature of the Mediterranean Europe: an ecological history. Yale University Press, New Haven/London, 383 pp

Haeseler V (1989) The situation of the invertebrate fauna of coastal dunes and sandy coasts in the western Mediterranean (France, Spain). In: van der Meulen F, Jungerius PD, Visser J (eds) Perspectives in coastal dune management. SPB Academic Publishing, The Hague, pp 128–131

Haeseler V (1992) Coastal dunes of the southern North Sea as habitats of digger wasps. In: Carter RWG, Curtis TGF, Sheehy-Skeffington MJ (eds) Coastal dunes: geomorphology, ecology and management, proceedings of the third European Dune Congress, Galway, Ireland, 17–21 June 1992. A.A. Balkema, Rotterdam, pp 381–389

Healy MG, Doody JP (eds) (1995) Directions in European coastal management. EUCC/Samara Publishing, Cardigan, 555 pp

Herrier J-L, Mees J, Salman A, Seys J, Van Nieuwenhuyse H, Dobbelaere I (eds) (2005) Dunes and estuaries 2005, proceedings – international conference on nature restoration practices in European coastal habitats, 19–23 September 2005, vol 19, Special publication. VLIZ, Koksijde

Houston JA, Edmondson SE, Rooney PJ (2001) Coastal dune management: shared experience of European conservation practice. Proceedings of the European symposium, coastal dunes of the Atlantic biogeographical region, Southport, northwest England, September 1998. Liverpool University Press, Liverpool, 453 pp
Howe MA, Knight GT, Clee C (2010) The importance of coastal sand dunes for terrestrial invertebrates in Wales and the UK, with particular reference to aculeate Hymenoptera (bees, wasps & ants). J Coast Conserv 14(2):91–102
Jones PS, Healy MG, Williams AT (eds) (1996) Studies in European coastal management. EUCC/Samara Publishing, Cardigan, 292 pp
Kacoliris F (2007) Natural history and conservation of the sand dune lizard (*Liolaemus multimaculatus*). Iguna International Reptile Conservation Foundation Journal 14(2):66–73
Ketner-Oostra R, Sýkora KV (2004) Decline of lichen-diversity in calcium-poor coastal dune vegetation since the 1970s, related to grass and moss encroachment. Phytocoenologia 34:521–549
Kirby P (1992) Habitat management for invertebrates: a practical handbook. Joint Nature Conservation Committee, Peterborough, 150 pp
Klein AHDF, Maia LP (2008) First Brazilian symposium on dunes systems and international symposium on coastal dunes: a report of meeting. J Coast Res 24(6):1355–1356
Kutiel P (2001) Conservation and management of the mediterranean coastal sand dunes in Israel. J Coast Conserv 7(2):183–192
Lawesson JE, Wind P (2002) Oak dune forests in Denmark and their ecology. For Ecol Manage 164(1–3):1–14
Li TSC, Schroeder WR (1996) Sea buckthorn (*Hippophae rhamnoides* L.): a multipurpose plant. Hortic Technol 6(4):370–380
Martinez ML, Psuty N (eds) (2004) Coastal dunes – ecology and conservation, vol 171, Ecological studies. Springer, Berlin, 386 pp
Matthews S, Brandt K (2006) South America invaded: the growing danger of invasive alien species. Global Invasive Species Programme.
Maun MA (2009) The biology of coastal sand dunes. Oxford University Press, Oxford, 265 pp
Moulton N, Corbett K (1999) The sand lizard conservation handbook. English Nature, p 32. Source: http://publications.naturalengland.org.uk/publication/70048. Accessed 14 Aug 2012
Ovesen CL (ed) (1998) Coastal dunes – management, protection and research. EUCC seminar, 19th–22nd August 1997, Skagen, Denmark. National Forest and nature Agency and the Geological Survey of Denmark and Greenland, Copenhagen, 138 pp
Ovesen CL, Vestergaard P (eds) (1992) Danish dunes – monitoring, management and research. Report from a seminar in Northern Jutland, 25th–27th May 1992. Ministry of the Environment, the National Forest and Nature Agency, Copenhagen, 174 pp
Packham JR, Willis AJ (2000) Ecology of dunes, salt marsh and shingle. Chapman and Hall, London, 335 pp
Plantlife International (2003) Looking after rare mosses and liverworts in coastal dune slacks. Back from the brink management series. Source: http://www.plantlife.org.uk/uploads/documents/management-guide-mosses-and-liverworts-coastal-dune-slacks.pdf. Accessed 14 Aug 2012
Pollet M, Grootaert P (1996) An estimation of the natural value of dune habitats using Empidoidea (Diptera). Biodivers Conserv 5(7):859–880
Powell JA (1976) A remarkable new genus of brachypterous moth from coastal sand dunes in California. Ann Entomol Soc Am 69(2):325–339
Powell JA (1981) Endangered habitats for insects: California coastal sand dunes. Atala 6(1–2):41–55
Ranwell DS (1972) Ecology of salt marshes and sand dunes. Chapman and Hall, London, 258 pp
Rhind PM, Jones PS (1999) The floristics and conservation status of sand-dune communities in Wales. J Coast Conserv 5(1):31–42
Rooney P (2010) Changing perspectives in coastal dune management. Special issue. J Coast Conserv 14(2):71–73
Ross S (1992) The Culbin sands – fact and fiction. Centre for Scottish Studies, University of Aberdeen, Aberdeen, 196 pp

Rotheroe M (1993) The larger fungi of Welsh sand dunes. Countryside Council for Wales, p 44. Source: http://www.aber.ac.uk/waxcap/downloads/rotheroe93-welshsanddunefungi.pdf. Accessed 13 Apr 2012

Salisbury E (1952) Downs and dunes, their plant life and its environment. G. Bell and Sons, London, 328 pp

Salman AHPM, Berends H, Bonazountas M (eds) (1995) Coastal management and habitat conservation. Vol. I: proceedings of the 4th EUCC congress, Marathon, Greece, April, 1993. EUCC, Leiden, 501 pp

Salman AHPM, Langeveld MJ, Bonazountas M (eds) (1996) Coastal management and habitat conservation. Vol. II: proceedings of the 4th EUCC congress, Marathon, Greece, April, 1993. EUCC, Leiden, 476 pp

Small E, Catling PM, Li TSC (2002) Blossoming treasures of biodiversity: 5. Sea Buckthorn *(Hippophae rhamnoides)* – an ancient crop with modern virtues. Biodiversity 3(2):25–27

Sridhar KR, Bhagya B (2007) Coastal sand dune vegetation: a potential source of food, fodder and pharmaceuticals. Livest Res Rural Dev, 19/6. Source: http://www.lrrd.org/lrrd19/6/srid19084.htm. Accessed 06 Feb 2012

van der Meulen F, Wanders EAJ (1985) Dynamics and management of some coastal dune woodlands near The Hague, the Netherlands. Plant Ecol 62(1–3):457–465

van der Meulen F, Jungerius PD, Visser J (eds) (1989) Perspectives in coastal dune management. Proceedings of the European symposium, Leiden, September 7–11, 1987, The Hague. SPB Academic Publishing, Den Haag, 333 pp

Veale EM (1957) The rabbit in England. Agric Hist Rev 5(2):85–90

Veloso Gomes F, Taveira Pinto F, das Neves L (2002) In: Littoral 2002, 6th international symposium. Proceedings, a multi-disciplinary symposium on coastal zone research, management and planning. Eurocoast and EUCC the Coastal Union, Three volumes

Wallage-Drees JM (1988) Rabbits in the coastal sand dunes: weighed and counted. Drukkerij Mostert, Leiden, 153 pp

Ward D (2005) Reversing rabbit decline: one of the biggest challenges for nature conservation in Spain and Portugal. Report. Source: http://www.ualberta.ca/~dhik/lsg/report_lynx_rabit.pdf. Accessed 09 Feb 2012

Weeda EJ (1992) Zandviooltje (*Viola rupestris*) in de duinen van Noord-Kennemerland: hoe een dwerg uit de steppetoendra standhoudt temidden van zand, zeewind en konijnen. *Wetenschappelijke mededelingen KNNV*, 206. KNNV Uitgeverij, Utrecht, NL (English summary)

Yoshichika T (1999) Trend of crops introduced to dune and its background. Agric Hortic 74(9):957–964

Chapter 6
Trends and Trade-offs – Beach/Foredune

Abstract A key issue worldwide is the extent to which the beach/foredune is migrating (moving landward). This situation (Physical States 1 and 2) is the most frequently encountered, seaward progradation (Physical State 3) much less so. This chapter considers the ecosystem services and values of the three states identified in Chap. 4 and the trends and trade-offs associated with different management options. It provides an explanation of the way the values of each state change in response to both spatial and temporal driving forces. It reviews the influence of human activities on these values, providing a basis for assessing the need for intervention. The chapter includes a simple 'State Evaluation Model', designed to help decision-making for nature conservation purposes at the beach/foredune interface. The next Chap. 7 considers the trends and trade-offs associated with inland, more stable vegetated dune forms.

6.1 Physical Trends

Identifying the relative movement of the dune front either seaward or landward is a key indicator in assessing its physical condition. It is relatively easy to identify when confronted with a beach devoid of vegetation, no foredune and a steep dune scarp (State 1) as depicted in (Figs. 4.2 and 11.7). However, one mobile dune can look just like another with areas of bare sand and patches of *Ammophila* spp. or other colonising species. Fixed structures can provide a clue as to net movement landward or seaward in areas where the sediment budget is neutral or mildly negative. On the French coast, relicts from the Second World War provide a reference point, suggesting that mobile sand dune forms (State 2) are migrating inland (Fig. 6.1).

On the coast of Albania, it is possible to gauge the extent of accreting State 3 dunes by reference to the 'Pillboxes' that decorate the coast (Fig. 4.4). Erected by Enver Hoxha in the 1950s and 1960s they were supposedly there to repel invasion, although it is not clear from where. Maps can provide decadal times series

J.P. Doody, *Sand Dune Conservation, Management and Restoration*, Coastal Research Library 4, DOI 10.1007/978-94-007-4731-9_6,

Fig. 6.1 A migrating beach/foredune on the coast near Calais. The bunkers set on *top* of the sand dunes in the 1940s, as part of the German defences provide a reference point. In June 1992, they lay partially buried on the beach, due to landward movement of blowouts and mobile foredunes

(Tentsmuir, Fig. 2.12; Pye and Blott 2008). However, it may be difficult to ascertain whether the long-term trend is one of erosion or accretion. Maps, by depicting the dune front or other feature such as Mean High Water Mark as a static line, do not show variation between dates. Time series measurements using fixed-point photography can be a useful approach (Millington et al. 2009). Recent developments have also shown that using a combination of techniques such as GPS,[1] digital photogrammetry[2] and INSAR[3] it is possible to provide a quick and easy methodology for monitoring coastal erosion (Buckley et al. 2002). Monitoring beach width also provides a potential means of predicting erosion or accretion trends. This is partly linked to the width of the beach with narrower, steeper beaches tending towards erosion, whilst wider shallower ones, accretion (Saye et al. 2005). The time interval between records and time of year will also be important. Annual aerial photographs will be more likely to show ephemeral strandline communities if taken during summer months. Seasonal records will show how beach plant communities wax and wane throughout the year.

[1] Global Positioning System, a U.S. space-based radio-navigation system.

[2] Photogrammetry is the technique of measuring objects (2D or 3D) from photographs or imagery taken by video, CCD (charge-coupled device) cameras or radiation sensors.

[3] Interferometric Synthetic Aperture Radar generates high quality terrain elevation maps.

An adequate sediment supply is the key to the state of the beach/foredune interface. Erosion takes place when there is insufficient sediment to maintain the beach against pressures such as climate change (increased storm frequency and intensity) or sea level rise. Human activities, including recreational use, are also "Mechanisms for Change", which militate against accretion (Sect. 6.3 below). As the balance changes and the forces promoting erosion dominate over those promoting accretion, there may be a perceived need for intervention. An assessment of the change in ecosystem services (values) lost or gained helps identify this need.

6.2 Physical States – Trends and Values

It is important to make a distinction between sand dunes that are eroding and those that are mobile. The values (Sect. 4.3) associated with each state change as the coastal sand dune moves from erosion to accretion. In the former, the trend is for loss of habitat and landward movement. In this, State 1, the sand dune has mostly negative values, particularly when associated with pressures causing a reduced sediment supply (Table 4.1). On the other hand, sand dunes where mobility is part of the 'dynamic' of the beach/foredune (State 2), there will be both positive and negative values. Accreting State 3 dunes tend to have mostly positive values.

Establishing the need for management action is by reference to the current state and the extent to which there is a negative trend in values. The following sections provide help on arriving at 'judgements' on the relative merits of moving the habitat from one state to another. This involves assessing the 'value' of the current state of the habitat, in relation to the 'value' of the restored state. As the states change, it is possible to assess the 'benefit' (gain of interest or value) in taking a particular restoration path. This will then inform policy as to the desirability of taking a particular form of remedial action.

6.2.1 State 1 – Eroding

This state occurs when there is a beach sediment deficit as described in Sect. 4.2.1. In their extreme form, they have little if any nature conservation value, as most beach/foredune features are absent. Although the trends are generally negative, erosion can temporarily at least, re-create the beach profile, as movement of material from the eroding foredune to the beach takes place. This can be augmented by material from 'inland' sand dunes no longer in contact with the beach that are undercut by wave action (Figs. 4.2 and 11.8). However, these rarely survive. The sediment, when moved by alongshore drift can result in the growth of new sand dune elsewhere, for example helping to form spits. Depending on the extent of loss of the 'inland' sand dune dealt with in Chap. 7, this might be positive for both nature

Table 6.1 Positive, neutral or negative trends for the beach/foredune values (Sect. 4.3) associated with an eroding (Physical State 1) dune

Value	Trend	Comment
1. Sediment store	Negative	Loss of sediment reservoir
2. Sea defence	Negative	Narrower buffer zone
3. Sea level rise	Negative/neutral	Response depends on the room for 'roll-over'
4. Nutrient recycling	Negative	Loss of material on foreshore
5. Landfall site	Negative	Exposure of buried pipes
6. Invertebrates	Negative	Loss of species & habitat
7. Birds	Negative/neutral	Less space for nesting/feeding
8. Sea turtles	Negative/neutral	Narrower beach, greater possibility of nests being flooded
9. Recreation	Neutral/negative	Narrower beach
10. Military training	Negative	Reduced area for activity
11. Archaeology	Neutral/positive	Exposure of buried artefacts
12. Geomorphological study	Positive/neutral	Active coastal processes
13. Ecological study	Negative	Absence of early succession stages
	Positive	Opportunity to study vectors of change

conservation and sea defence. On the other hand, sand blown landwards may have a negative impact on assets lying in the path of the drifting sand. Table 6.1 provides a summary of the trends.

6.2.2 *State 2 – Dynamic Equilibrium or 'Semi-stable'*

As the beach sediment budget moves from a significant deficit, to one with a mildly negative, neutral or even slightly positive a state of dynamic equilibrium can exist. This will have a range of interests including plants and animals associated with mobile sand, vegetated strandline and foredunes (Fig. 4.3). So long as there is sufficient sediment to maintain the sequence of communities, their inherent nature conservation values will remain. Over short time scales, there may be both 'erosional' and 'accretional' forms resulting from seasonal variation. These may appear to be in more or less the same place on the beach.

However, this dynamic equilibrium is rarely a true picture of the process. In many locations, the sediment supply is sufficient to maintain the beach profile but under the influence of rising sea levels, or in response to storms it moves (shifts) landward. The maintenance of the positive values associated with the plants and animals of this profile survive at the expense of the vegetated sand dune inland. The impact of this depends to some extent on the speed of migration. This also influences the value for sea defence, as a system rapidly migrating inland may appear to be less effective in its ability to buffer storms or cope with sea level rise. In areas with abundant sediment State 3 accreting dunes develop.

Table 6.2 Positive, neutral or negative trends for the beach/foredune values (Sect. 4.3) associated with accreting (Physical State 3) sand dunes

Value	Trend	Comment
1. Sediment store	Positive	Increase in sediment reservoir
2. Sea defence	Positive	Wider buffer zone
3. Sea level rise	Positive	Depends on the room for 'roll-over'
4. Nutrient recycling	Positive	Increase in material on foreshore
5. Landfall site	Neutral/positive	More sand to bury pipes
6. Invertebrates	Positive	Increase in species & habitat
7. Birds	Negative/neutral	Wider shore, greater opportunities for nesting/feeding, more prey
8. Sea turtles	Positive/neutral	Higher beach, less likelihood of nests being flooded
9. Recreation	Positive	Wider beach
10. Military training	Positive/neutral	Increased area for activity
11. Archaeology	Negative/neutral	Burial of artefacts
12. Geomorphological study	Positive	Active coastal processes
13. Ecological study	Positive	Presence of early succession stages

6.2.3 State 3 – Accreting

Accretional forms depend on the availability of sediment. As this increases and the beach sediment budget moves from neutral to positive, accreting sand dunes usually develop. These can be extensive (Fig. 4.4). The trend in values will be positive for most of the ecosystem services (Table 6.2).

Having established how certain activities influence the nature of the sand dune state and whether the trend is positive, negative or neutral for the values associated with the state, a simple model provides a means of conceptualising the move from one to the other.

6.2.4 The Physical State Evaluation Model

The Physical State Evaluation model (Fig. 6.2) is a representation of the beach/foredune interface. At the extremes, a strongly negative sediment budget usually results in erosion whilst a strongly positive one in accretion. Over time State 1 and State 3 coastal sand dunes move landward or seaward respectively. State 2 sand dunes may have both eroding and accreting dune forms depending on the balance between the forcing factors and sediment availability. In this scenario, the beach/foredune retains its sequence of habitats, although these may shift (migrate) across the beach profile.

In the last 100 years, analysis of global patterns of change indicates that most beaches have been retreating (State 1, erosional or State 2, migrating landward), a few showed no net movement and some (State 3) advanced seaward (Bird 2008). The reasons for this are many and derive from the depletion of sediments, including those deposited at the end of the last glacial period.

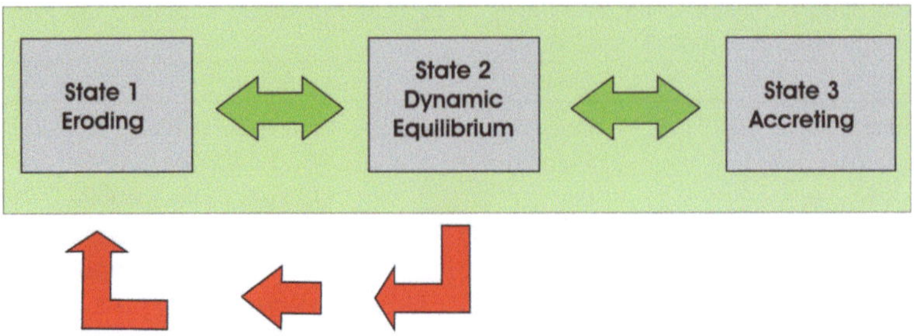

Fig. 6.2 Physical state evaluation model. The *green arrows* imply movement between the states creating a 'dynamic equilibrium'. *Red arrows* indicate the principal route causing loss of nature conservation value. Note the term 'dynamic equilibrium' simply means that the beach profile remains the same even though its position may shift

6.3 Mechanisms for Change

A range of natural processes is important to the development of sand dunes (described in Chap. 1). This section is concerned with those factors (natural and human) that drive the sand-sharing system towards erosion or accretion. Worldwide there has been depletion in the sediment resource such that beach erosion is a major concern. Bird (1996) lists 20 causes. These include natural diminution of material on the sea floor, deposited when the marine transgression of the early Holocene ended. Other 'natural' factors include increase in vegetation in the hinterland or on adjacent inland dunes.

Human intervention adds to these effects, directly by removing sediment from the beach (Sect. 2.6.1). Others are indirect, including river damming (Sect. 2.6.2) and stabilisation of adjacent eroding cliffs. Yet others are highly destructive, replacing the sand dune surface with infrastructure or other land use (Sects. 2.7.1 and 2.8) and by so doing locking up sediment and/or restricting its ability to move. This is particularly important in relation to changes in climate (increase in storm frequency and intensity) and associated changes in sea level. Sediment depletion may cause short-term change whilst sea level trends may affect the balance between erosion and accretion over a much longer timescale.

6.3.1 Offshore Sediment (Aggregate) Extraction

Offshore extraction takes place from a largely finite reservoir of sand deposited during the Pleistocene, or contributed during the late glacial melting of the ice as described in Sect. 1.1. Major uses are for the construction industry and the nourishment of eroding beaches. The question arises as to the impact this has on the stability of adjacent sandy beaches and sand dunes. In particular, does it also deplete the amount of sediment reaching the beach and hence make the foredune more susceptible to erosion?

In the UK, the location of dredging sites is in water depths mostly between 10 and 35 m, some of which can be close to shore. The UK Government view and that of the extraction industry is that there is no evidence to support a causal relationship between offshore extraction and the stability or otherwise of beaches nearby (May 2007).

However, others disagree, and in most of the rest of Europe, strict limits apply and no dredging for sand and gravel takes place within 25 km of the shoreline or in depths less than 20 m. Here commercial exploitation of the resource is restricted because of the damage to the marine environment, shoreline and fishing industry. [For a full exposé of the arguments against offshore extraction, see the UK Marinet Network of Friends of the Earth, web site @ http://www.marinet.org.uk/. This includes a detailed response, dated 11th March 2010, to the House of Commons Treasury Select Committee considering "the Management of the Crown Estate". It provides evidence of the adverse impact of commercial offshore dredging on the "seabed, its ecosystem and the related coastline".]

A study of the German Baltic Sea coast concluded that although there are few examples of direct cause and effect between coastal erosion and marine sand extraction, there is sufficient evidence to raise concerns. Very small changes in bathymetry (1–2 m) can affect the patterns of erosion and accretion on the coast. As there is only a very limited supply of new sediment from cliff erosion, any marine extraction will have a negative effect on the overall sediment budget (Kortekaas et al. 2010). In France, models suggest that extraction from a depth of 1–8 m and 1–6 km offshore may have adverse consequences for the shoreline (Cayocca and du Gardin 2005). It is unclear if this depletes the availability of sediment for onshore movement to restock sandy beaches. However, the void rapidly fills with sand (Desprez 2000; Kenny and Rees 1996) suggesting there is an overall loss of this finite resource.

In the United States of America, there is a link between sand mining in the surf zone and erosion in Southern Monterey Bay, one of the most intensively mined shorelines. Although there is some correlation with the occurrence of El Niño, cessation of most mining in 1990 resulted in reduced erosion rates in parts of the southern bay (Thornton et al. 2006). In New Zealand, the weak recovery of the Pakiri-Mangawhai area following severe erosion may be due to near shore coastal sand mining. Studies concluded that there is a high risk of this adversely affecting coastal processes and landforms (Hesp and Hilton 1996; Hilton and Hesp 1996).

The economic benefits of offshore extraction can be substantial. In the Netherlands, for example the new Port of Rotterdam (Maasvlakte 2) requires enormous quantities of sand derived from offshore deposits. The scheme (due for completion in 2013) provides for a 20 % expansion of the existing port and covers an area of 2,000 ha including 1,000 ha for industrial development. Mitigation of the possible nature conservation effects will include a 25,000 ha seabed protection area and the creation of 35 ha of new sand dune (Barker 2011). Monitoring will assess the long-term environmental implications, although once built it is difficult to see how any problems that occur can be reversed, in such a large project.

A direct link between offshore sand and gravel extraction and foreshore erosion is unproven in many cases. However, there is sufficient evidence to suggest that extraction is adding to the overall depletion of sediment available for accretion onshore.

6.3.2 Foreshore Sand Mining

Onshore mining differs from offshore extraction in that it usually takes place on a smaller and localised scale. However, this form of sand extraction can have an impact on the stability of the beach/foredune because of the cumulative nature of the activity. It is often illegal and can cause significant losses, especially on smaller sites such as those in Northern Ireland (Carter et al. 1992) and Scotland (Ritchie and Mather 1977).

In Jersey, mining of beach sand together with construction of a sea wall and reduced sediment input from offshore deposits has exacerbated erosion of an important tourist beach (Cooper and Pethick 2005). Sediment loss from some smaller sand dunes can reach a point where, even when the activity ceases, rehabilitation does not occur. In the Azores, Portugal near Santa Barbara an undisturbed active beach/foredune system was present in the 1950s. Mining between the 1960s and 1980s removed a considerable volume of sand. Although this ceased in 1995, erosion continues, suggesting that there is a natural 'tipping point' and that there is no longer sufficient sediment to re-establish the protective system. As a result cliff erosion behind the beach continued. The overall profit from the mined sand was much lower than the gain in recreation and amenity value that would have occurred if mining had not taken place (Borges et al. 2002).

In New Zealand, sand mining devastated an area of sand dune with indigenous sand binder *Desmoschoenus spiralis*, in the extensive sand dunes at Kaitorete Spit, Canterbury, New Zealand. The species now only survives in a few untouched areas. Recolonisation is limited due to the presence of *Ammophila arenaria*, which has invaded the older mined dunes, displacing *D. spiralis* or excluding it from re-invading. The remaining mined area has developed sparse sand-plain vegetation, the result of lateral sand movement (Partridge 1992).

In Ghana, the impact of sand mining has been costly. Despite the fact that it helped to boost industrial output from 17.4 % in 1986 to 20.8 % in 1993, it also accelerated coastal environmental degradation in many areas. As a result, the government was compelled to spend millions of dollars to combat erosion by the sea (Mensah 1997). It is also a problem in Tanzania, where it results in coastal erosion affecting social and economic development and the environment (Masalu 2002).

These examples illustrate the way seemingly benign operations have a cumulative and ultimately serious adverse impact on the sand dunes and their nature conservation values, as well as wider effects on coastal stability.

6.3.3 River Damming

Sediment derived from erosion in the hinterland is an important source of sediment for habitat development especially at the mouth of river deltas. Interruption of this sediment supply can have serious implications. River damming (Sect. 2.6.2) depletes the sediment delivery to the coast reducing the buffering capacity of the

Fig. 6.3 Eroding dune front, Ebro Delta, Spain. The walkway over the sand dune has collapsed as the dune has eroded, September 1996

beach/foredune during storms. In the longer term, the loss of sediment increases the vulnerability of the coast to sea level rise. In a study of the 33 areas chosen to represent deltas the world, in the past decade 85 % experienced severe flooding resulting from sediment compaction and sediment trapping in reservoirs upstream (Syvitski et al. 2009). Another study covering 40 deltas came to the same conclusion, and of nine documented examples all bar one cited reduction in sediment load as one of the principal effects (Ericson et al. 2006, Table 1). Reservoir construction may be the most important factor influencing exchange of sediment between rivers and the sea (Walling and Fang 2003; Walling 2009).

The scale of loss is significant for deltas especially in micro-tidal seas such as the Mediterranean. According to Vörösmarty et al. (2003), sediment retention behind river dams is approximately 50 %, compared with the global mean of 25–30 %. In the Nile Delta, the construction of the Delta Barrage in 1868 and other dams, including the Aswan High Dam, have reduced the amount of sediment by 60 % (Stanley and Warne 1983). This resulted in much of the delta coastline eroding at rates of up to 5–8 m per year, in places exceeding 240 m per year (McCully 1996). More recent estimates suggest loss of sediment carried by the Nile river may be as much as 98 %, resulting in erosion of the delta of 125–175 m per year (Chen 2005).

In the Ebro Delta, Spain there was a tenfold reduction in sediment delivery to the delta because of dam construction. This led to subsidence and coastal erosion (Marino 1992; Fig. 6.3). In the Rhone Delta, southern France over a period of

Fig. 6.4 Beach parking along a dune front, Ynylas, Wales in 2007

150 years the sediment load decreased by a factor of four. This was due to the natural decrease of the frequency of major floods (at the end of the Little Ice Age). Since the 1950s, reforestation in the catchment area, dam construction and dredging have caused a further sediment loss (Sabatier et al. 2006), as it has in many of the other Mediterranean deltas.

On the Mississippi Delta, dams and reservoirs built since 1950 have trapped at least 50 % of the river's sediment (Meade 1995). With less material feeding it, the delta plain has been experiencing erosion. River dams have had a similar impact on the coast of California (Slagel and Griggs 2008).

In Iceland, the extensive destabilisation of vast areas of cultivated land elicited the conventional response and major rehabilitation has taken place to reduce sediment loss. However, when combined with control of melt water from glaciers through dam construction, it was predicted that the further reduction in sediment supply to the coast would cause rapid shoreline retreat (Greipsson and El-Mayas 1996).

6.3.4 Recreation and Beach Cleaning

Sand dunes are often associated with recreational beaches. In some areas, parking occurs directly on the beach (Fig. 6.4). People also demand clean beaches and beach cleaning (grooming or raking) to remove strandline debris takes place in popular recreational areas. In combination, these cause compaction of the sand and remove all traces of strandline vegetation. In addition, they prevent the development of strandline and pioneer dunes, exposing the more stable dunes behind. It also alters

Fig. 6.5 Recreational footpath along the crest of a narrow eroding spit, Dawlish Warren, south Devon, England, 2009

the physical characteristics of the beach, strandline and foredune habitats including micro-topography and zonation (Nordstrom 2000). The use of machinery is most damaging, removing all the material on the beach including associated invertebrates. There appears to be no irony in the companies that advertise such beach cleaning machines, with the provision of services such as sand nourishment and sand fencing to stabilise eroding dunes.

Even non-mechanical cleaning resulting in the removal of large logs can have a damaging impact on beach fauna such as the woodlouse *Armadillidium album* or ground beetles *Broscus cephalotes* and *Nebria complanata* (Wheater 1999). Widespread seaweed removal (wrack) also causes immediate reductions in the abundance and diversity of semi-terrestrial crustaceans, insects and their predators (Dugan et al. 2003; Deidun et al. 2009). In southern California, unnaturally wide beaches develop where seasonal cleaning takes place and have virtually no biological interest. This is particularly true for those bordering developed coasts where the demand for clean beaches is high (Nordstrom 2000). On Australia's Gold Coast, there was a clear association between reduced Ghost Crabs *Ocypode* spp. populations and heavy recreational pressure. Beaches with fewer visitors and hence cleaned less frequently, had higher numbers of crabs (Noriega et al. 2012).

High visitor numbers can lead to the destruction of the foredune by trampling. Foredune ridges also offer vantage points for walking and experiencing seascapes. These create paths that further exacerbate loss of habitat and reduce resilience to sand movement (Fig. 6.5). Excavation in the foredune ridges to create 'sun-traps' can also cause erosion problems.

6.3.5 *Off Road Vehicles (ORV)*

Off-road vehicles are especially damaging to sandy beach invertebrate communities. In monitoring the overlap between invertebrate habitat zones and areas of ORV traffic, Schlacher and Thompson (2007) found the majority of invertebrates they sampled (65 %) lived in areas with vehicle traffic. Ghost crabs *Ocypode cordimana* suffered a high mortality rate due to night driving because they construct and live in beach burrows during the day and are mostly active at night. Beaches with fewer ORVs had higher crab populations (Schlacher et al. 2007). The Surf Clam *Donax deltoides* also suffered from the effects of traffic on beaches in Australia. As the number of vehicle passes increased, so did the number of clams killed (Schlacher et al. 2008).

Overall ORVs cause:

- Damage and loss of coastal vegetation, leading to dune erosion;
- Damage to nesting sites of adult marine turtles and death of hatchlings;
- Ruts in the sand, which can impede the passage of adult turtles and hatchlings to the sea;
- Damage and loss of nests and nesting sites for shore birds;
- Disturbance to migratory waders during feeding and roosting.

6.3.6 *Military Activity*

Physical damage to beaches and foredunes was prevalent during the war years, especially the Second World War (Sect. 12.1.3). Amphibious landing practice often took place on beaches in Britain, especially during preparations for the invasion of France in 1944. Sites affected included Gullane, East Lothian, Scotland where training in heavy military vehicle recovery resulted in almost total loss of the dune in a major part of the site (Ranwell and Boar 1986). Studland Bay dunes in Dorset was the site of a practice invasion (Operation Smash) in April 1944, watched over by Churchill and Montgomery, according to a painting held in the nearby Manor House Hotel. Similarly Camber Sands, Kent, UK was used for military exercises from 1939 including practising beach landing manoeuvres in preparation for D-Day. By 1945, most of the vegetation had gone (Pizzey 1975). Extensive military activity also took place on Braunton Burrows, north Devon, England (Sect. 12.3.2).

6.3.7 *Sea Defence*

Planting *Ammophila arenaria* has taken place to combat erosion in many parts of the world because of its ability to trap sand and build new or replacement sand dunes. However, outside its native range it has become invasive in many areas. Due to this,

Fig. 6.6 Hard defences 'protecting' sea front houses built on sand, Jutland, Denmark 1992. It appears as though sand from the beach has been used to augment the dune ridge seaward of the houses

recent approaches to restoring eroding sand dunes in the United States of America, South Africa and New Zealand have moved towards replacing *Ammophila arenaria* (and *A. breviligulata* in northwest America) with native species (Sect. 8.2).

Sand fences and the like often accompany planting sand-stabilising plants. However, if placed on eroding beaches with a sediment deficit they may not be effective in preventing the landward movement of sand dunes in response to storms and rising sea levels. The temptation is therefore to replace them with coastal defence structures, including rock armouring (Fig. 6.6). Generally, this causes a narrowing and steepening of the beach profile, as for example in the United Kingdom, where engineering structures 'protect' long stretches of eroding coastline (Masselink and Russell 2010). Thus rather than "saving the beach", this approach may be "killing them" (Neal et al. 2007). On sandy shores, this can in turn result in loss of strandline debris, invertebrates and habitat for shorebirds (Dugan and Hubbard 2006).

6.3.8 *Grazing*

Grazing occurs on foredune vegetation in some parts of the world where it can have a negative impact. In the United States of America, feral horses graze the Assateague

Island Dunes. Driven to the coast in the summer months by biting insects, they graze predominantly on *Ammophila breviligulata*. Recent increase in the population has caused significant degradation of the sand-trapping vegetation, resulting in the need for fencing to control access to the beach vegetation and prevent erosion (De Stoppelaire et al. 2004). In Australia, grazing by Wallabies negatively affected the frontal dune with adverse effects on the establishment of some perennial plants (Ramsey and Wilson 1997).

6.3.9 Climate and Sea Level Change

According to the Intergovernmental Panel on Climate Change (IPCC) "Fourth Assessment Report: Climate Change 2007", the temperature of the oceans increased by 0.1 °C between 1961 and 2003, from the surface to a depth of 700 m. During the same period, the average rate of sea level rise was 1.8 ± 0.5 mm per year. For the twentieth century, the average rate was 1.7 ± 0.5 mm per year, consistent with the Third Assessment Report estimate of 1–2 mm. This will have a significant impact on beaches and foredune throughout the world, although there will be much regional variation. This is likely to lead to increased risk of erosion in most coastal areas, except where the land is rising faster than sea level or there is an abundant sediment supply. Current IPCC sea level rise predictions suggest that the rate of beach erosion will continue to accelerate.

The same is true with the predicted increased frequency and severity of storms. For example since 1968, Atlantic storm frequency has increased and caused an acceleration of erosion on Magilligan foredunes, Northern Ireland with a maximum loss of 5.4 m between 1978 and 1979. A combination of wave, wind and tides affects the pattern of erosion, which involves foredune scarping and slope failure (Carter and Stone 1988). On the Sefton coast, northwest England changes in the frequency and magnitude of storms, surges and resulting high tides provided the main drivers of change when looked at over a period of 50 years (Pye and Blott 2008).

The key point is that if global warming and sea level rise accelerate, erosional trends are likely to increase. This could move existing State 2 systems towards an eroding State 1 with the loss of vegetated strandline and foredune. Even if the beach/foredune system survives, its migration landwards under the influence of these driving forces could accelerate. In areas where buildings or other development has taken place behind the dune front, sediment is effectively 'locked up' within the system. This prevents remobilisation of the inland dune and hence the possibility of contributing sediment to the beach. As sea level continues to rise, the presence of infrastructure restricts landward migration in response to storms and wave attack. As a result, the beach, foredune and inland dune may be lost altogether, creating the ultimate 'sand dune squeeze' (Sect. 12.2.1).

6.4 Issues and Outcomes of Intervention

Many of the mechanisms for change identified above involve loss of sediment, disruption or removal of the colonising strandline and/or destabilisation of the foredune. These generally result in State 1 sand dunes, which elicit a characteristic response involving attempts to prevent erosion and landward progression of the beach. The precise response will depend on the focus for management i.e. whether for flood defence to protect assets or for nature conservation. Preventing erosion has been the mantra for many years. Historically, unstable dunes appeared as a threat to human activities or assets. Overwhelming agricultural land, farms and other infrastructure by blowing sand and the threat of flooding have resulted in a wide variety of measures to stabilise and 'protect' them. 'Engineering' dunes to prevent sand movement, by creating structures which defend the land from the sea, has been the principal management objective.

Even in dune areas identified for their conservation significance 'protecting' the dune from erosion has often been a significant preoccupation, especially where recreational use is a key destabilising factor. Taking an ecological perspective and learning lessons from the way in which the beach/foredune responds to natural perturbations in the environment, suggests that they may be more resilient than previously thought (Doody 1989, 2001). Thus, it is important to consider the pressures driving sand mobility before embarking on remedial action to prevent it.

6.4.1 Physical State 1 – Eroding

The combination of a naturally depleted sediment supply and human intervention contribute to the reduction of sediment entering the beach/foredune sand – sharing system. Where this results in a negative sediment budget an eroding beach/foredune is likely, which may develop a 'cliff' (Fig. 4.2). All beaches and foredunes are vulnerable, including those on spits and barrier islands when beach levels are falling due to loss of sediment and/or sea level rise. A breach can expose resources behind the dune to wave attack and further erosion and flooding.

In the Netherlands sea defence is the main justification for attempting to control or even reverse erosion trends. Here the dunes provide a defence for large numbers of people, particularly in the central coastal zone of the country. Intervention to prevent the dunes failing is a primary objective, and strengthening the dune front a major preoccupation. Beach nourishment, together with standard stabilisation techniques are the principal methods employed. The Ministry of Transport, Public Works and Water Management established a 'basal coastline' as at the 1st January 1990. This involved a twin-track approach described as "*soft where possible and hard where necessary*", maintaining the shore in a 'dynamic equilibrium'. Within these

Fig. 6.7 'Fixed' steep-faced coastal dune defences in the Netherlands, 1987

constraints, there is room for the development of dune dynamics. Despite this, in practice, opportunities for erosion and with it the re-creation of strandline and mobile dunes are limited (Fig. 6.7).

The barrier island dunes on the eastern coast of the United States of America protect numerous settlements, including those on the New Jersey shore. Here the imperative for those living on the shore where "Our beaches are eroding, sinking, washing out right under our houses, hotels, bridges" (Kaufman and Pilkey 1983) is to prevent this erosion. This can be acute in areas where the natural progression of the islands is landward under the dual effects of restricted sediment supply and sea level rise (Psuty and Ofiara 2002). Reinforcing the shoreline in order to protect real estate is an ongoing commitment. However, erosion, especially under the influence of hurricanes and other storms is difficult to prevent. Erecting hard sea defences can simply make matters worse (Sect. 6.3.3).

It is possible to make a distinction between coastal areas where sand dunes protect buildings or other economically valuable assets, and those where the hinterland is undeveloped. In the former, it may be appropriate and economically viable to seek to stabilise the foredunes through remedial action or to nourish the beach. In the latter, from a nature conservation point of view, this will probably only be appropriate if the losses represent a significant proportion of a valued inland dune habitat. Even then, interference with the structure and function of the system may be prohibitively expensive and ineffective in long term.

6.4.2 Physical State 2 – Dynamic Equilibrium or 'Semi-stable'

Generally, mobile beach/foredune systems are likely to be an acceptable or even 'desired' state from a nature conservation perspective. Values associated with sea defence remain, although prone to reversal when storms occur. Dynamic conditions continue to support specialist plants and invertebrates, even if the dune front moves inland or for limited period seawards. For this state two things are important when considering whether to take action:

1. Does the foreshore have the full sequence of beach, strandline and foredune? If not, make sure that their absence is not related to seasonal variation;
2. Is the sequence moving landward, threatening assets of nature conservation or other values? If so, will protection of the assets inland outweigh the loss of values associated with stabilisation of the dynamic landform?

Determining whether these are short-term cycles or long-term trends is a key to deciding the need for action. In these areas, a question arises as to the value of reversing erosion trends when it may be better to leave the dune to its own devises. Typically, eroding foredunes move landward but maintain their shape, height and width (Ranwell 1958). Thus with room for migration to take place the zones remain, albeit in a different place (Psuty 2004) and there is little or no need to intervene.

In this state the beach/foredune interface supports a range of 'values' (Sect. 4.3), particularly those associated with a flora and fauna specifically adapted to rapidly changing physical conditions. The values associated with economic, socio-economic, biological and nature conservation interests are all likely to remain positive. From a coastal defence perspective, although small-scale erosion may cause problems locally, the dynamic nature of the habitat may be acceptable. Exceptions may occur when in the medium to long-term landward movement is such that it threatens to overwhelm inland dunes. However, even then there may be little point in intervening.

6.4.3 Physical State 3 – Accreting

Seaward accreting foredunes are perhaps the most valued from a nature conservation perspective. Under most circumstances it ensures the development and survival of the early stages in dune succession, together with their plants and animals. Values associated with both nature conservation and sea defence interests are likely to be positive. The misplaced use of alien species on the upper shore may result in displacement of native species and with them the loss of some of the nature conservation and aesthetic values. When this occurs, it may be appropriate to consider controlling these invasive species (Chap. 8).

Generally, little or no intervention is required. However, actively accreting sand dunes may only be part of a much wider system, confined to the distal end of spits

Fig. 6.8 Dawlish Warren, accreting foreshore at the distal (eastern) end of the spit. Note the marker for a wooden groyne erected in the 1970s, and now buried by accreting sand, June 2009. Figure 6.5 shows dune erosion to the west releasing sand for transport to the spit, also June 2009

and bars. It is important to recognise when rapid accretion in one part of a site may be at the expense of erosion at another. At Dawlish Warren, south Devon for example, a railway built by Brunel in the 1850s cut off the sediment supply to the dune system from eroding cliffs further along the coast. Wooden groynes arrested the resulting erosion for a time but erosion resumed and by 2009 the central part of the spit had become very narrow (Fig. 6.5). The displaced sediment accumulated at the distal end of the spit creating a rapidly accreting strandline and foredune (Fig. 6.8).

6.5 Conclusion

The sediment supply to the beaches of the world is diminishing, partly because of natural processes. Human activities, such as river damming, sand mining and offshore extraction are further drivers of erosion. Other activities destabilise the beach, cause compaction or in other ways interfere with its natural dynamic. These may lead to the loss of the early stages of dune development and their nature conservation values. Preventing those activities that impinge directly on the beach, such as sand mining and beach cleaning will help retain or restore the plant and animal communities. However, this is only likely to be effective when there is a positive or mildly negative sediment budget.

Coastal sediment depletion is most acute when the foredune disappears and the dunes inland erode. Under these circumstances remedial action to prevent erosion using traditional sand-trapping methods (Sect. 9.2) are likely to be ineffective in the medium term. The most obvious means of preventing erosion and promoting accretion lies in re-establishing the 'natural' sediment supply to the beach. However, this may not be possible because of the long-term natural diminution of sediment from the sea floor, and losses due to human activities. The economic value of offshore sediments and the need for freshwater storage behind river dams will militate against any policy change that would help restore sediment delivery to the coast.

It seems likely, therefore, that there will continue to be a general trend towards beach and foredune erosion. Unhampered by landward features such as coastal infrastructure, agricultural or rising land, the trend will be migration 'roll over' of the sequence of beach/strandline/foredune communities. Whether or not these communities remain, the loss of the vegetated sand dune inland will continue. The next chapter deals with the trends and trade-offs within the vegetated inland dune, including mobile forms and wet slacks.

References

Barker P (2011) Good progress at Rotterdam's Maasvlakte 2. Marit J, http://www.maritimejournal.com/features101/marine-civils/port,-harbour-and-marine-construction/good-progress-at-rotterdams-maasvlakte-2. Accessed 28 Nov 2011

Bird ECF (1996) Beach management. Wiley, Chichester, 281 pp

Bird E (2008) Coastal geomorphology: an introduction, 2nd edn. Wiley, Chichester, 322 pp

Borges P, Andrade C, Freitas MC (2002) Dune, bluff and beach erosion due to exhaustive sand mining – the case of Santa Barbara beach, São Miguel (Azores, Portugal). J Coast Res 36 (Special Issue):89–95

Buckley SJ, Mills JP, Clarke PJ, Edwards SJ, Pethick JS, Mitchell HL (2002) Synergy of GPS, digital photogrammetry and insar in coastal environments. Presented at the seventh international conference on remote sensing for marine and coastal environments, Miami, Florida, 20–22 May 2002

Carter RWG, Stone GW (1988) Mechanisms associated with the erosion of sand dune cliffs, Magilligan, Northern Ireland. Earth Surf Process Landf 14(1):1–10

Carter RWG, Eastwood DA, Bradshaw P (1992) Small scale sediment removal from beaches in Northern Ireland: environmental impact, community perception and conservation management. Aquat Conserv: Mar Freshw Ecosyst 2(1):95–113

Cayocca F, du Gardin B (2005) Assessing the impact of sand extraction on the shore stability: project for a methodological framework. In: European marine sand and gravel – shaping the future, EMSAGG Conference, Delft University, The Netherlands, 20–21 Feb 2003

Chen J (2005) Dams, effect on coasts. In: Schwartz M (ed) Encyclopaedia of coastal sciences. Springer, Dordrecht, pp 357–359

Cooper NJ, Pethick JS (2005) Sediment budget approach to addressing coastal erosion problems in St. Ouen's Bay, Jersey, Channel Islands. J Coast Res 21(1):112–122

De Stoppelaire GH, Gillespie TW, Brock JC, Tobin GA (2004) Use of remote sensing techniques to determine the effects of grazing on vegetation cover and dune elevation at Assateague Island National Seashore: impact of horses. Environ Manage 34(5):642–649

Deidun A, Saliba S, Schembri PJ (2009) Considerations on the ecological role of wrack accumulation on sandy beaches in the Maltese Islands and recommendations for their conservation management. J Coast Res, Special Issue 56, Proceedings of the 10th international coastal symposium, Lisbon, Portugal, pp 410–414

Desprez M (2000) Physical and biological impact of marine aggregate extraction along the French coast of the Eastern English Channel: short- and long-term post-dredging restoration. J Mar Sci: Journal du Conseil 57(5):1428–1438

Doody JP (1989) Management for nature conservation. In: Gimmingham CH, Ritchie W, Willetts BB, Willis AJ (eds) Coastal sand dunes, vol 96B, Proceedings of the Royal Society of Edinburgh, pp 247–265

Doody JP (2001) Coastal conservation and management: an ecological perspective, vol 13, Conservation biology series. Kluwer Academic Publishers, Boston, 306 pp

Dugan JE, Hubbard DM (2006) The ecological responses of coastal armouring on exposed sandy beaches. Shore and Beach 74(1):10–16

Dugan JE, Hubbard DM, McCrary MD, Pierson MO (2003) The response of macrofauna communities and shorebirds to macrophyte wrack subsidies on exposed sandy beaches of southern California. Estuar, Coast, Shelf Sci 58S:25–40

Ericson JP, Vörösmarty CJ, Dingman SL, Ward LG, Meybeck M (2006) Effective sea-level rise and deltas: causes of change and human dimension implications. Glob Planet Change 50:63–82

Greipsson S, El-Mayas H (1996) The stabilisation of coastal sands in Iceland. In: Jones PS, Healy MG, Williams AT (eds) Studies in European coastal management. EUCC/Samara Publishing, Cardigan, pp 93–100

Hesp PA, Hilton MJ (1996) Nearshore-surfzone system limits and the impacts of sand extraction. J Coast Res 12(3):726–747

Hilton MJ, Hesp P (1996) Determining the limits of beach-nearshore sand systems and the impact of offshore coastal sand mining. J Coast Res 12(2):496–519

Kaufman W, Pilkey OH (1983) The beaches are moving: the drowning of America's shoreline. Duke University Press, Durham, 335 pp

Kenny AJ, Rees HL (1996) The effects of marine gravel extraction on the macrobenthos: results 2 years post-dredging. Mar Pollut Bull 32(8–9):615–622

Kortekaas S, Bagdanaviciute I, Gyssels P, Alonso Huerta JM, Héquette A (2010) Assessment of the effects of marine aggregate extraction on the coastline: an example from the German Baltic Sea coast. J Coast Res 51(Special Issue):205–214

Marino MG (1992) Carbon dioxide increase, sea-level rise and impacts on the western Mediterranean: the Ebro Delta case. In: Tooley MJ et al (eds) Impacts of sea-level rise on European coastal lowlands, vol 27, The Institute of British Geographers, Special publications, series. Blackwell, Oxford/Cambridge, pp 124–135

Masalu DCP (2002) Coastal erosion and its social and environmental aspects in Tanzania: a case study in illegal sand mining. Coast Manag 30(4):347–359

Masselink G, Russell P (2010) Coastal erosion in MCCIP annual report card 2010–11, MCCIP Science Review, p 18. Source: www.mccip.org.uk/arc. Accessed 20 May 2011

May V (2007) Impacts on beach and cliff stability associated with extraction of sediments from beaches and offshore: understanding the communities' responses to perceived risk. In: Proceedings of the Coastal Zone 07, Portland, Oregon. Source: http://www.csc.noaa.gov/cz/CZ07_Proceedings/PDFs/Monday_Abstracts/3107.May.pdf. Accessed 21 Jan 2011

McCully P (1996) Silenced rivers: the ecology and politics of large dams. Zed Books, London, 350 pp

Meade RH (ed) (1995) Contaminants in the Mississippi river, 1987–92. U.S. Geological Survey Circular, 1133, Reston, Virginia. Source: http://pubs.usgs.gov/circ/circ1133/. Accessed 14 Apr 2011

Mensah JV (1997) Causes and effects of coastal sand mining in Ghana. Singap J Trop Geogr 18(1):69–88

Millington JA, Booth CA, Fullen MA, Moore GM, Trueman IC, Worsley AT, Richardson N, Baltranaite E (2009) The role of long-term landscape photography as a tool in dune management. J Environ Eng Landsc Manag 17/4:Ia–Ih

Neal WJ, Pilkey OH, Kelley JT (2007) Atlantic coast beaches. A guide to ripples, dunes, and other natural features of the seashore. Mountain Press Publishing Company, Missoula, 250 pp

Nordstrom KF (2000) Beaches and dunes on developed coasts. Cambridge University Press, Cambridge, 356 pp

Noriega R, Schlacher TA, Smeuninx B (2012) Reductions in ghost crab populations reflect urbanization of beaches and dunes. J Coast Res 28(1):123–131

Partridge TR (1992) Vegetation recovery following sand mining on coastal dunes at Kaitorete Spit, Canterbury, New Zealand. Biol Conserv 61(1):59–71

Pizzey JM (1975) Assessment of dune stabilisation at Camber, Sussex, using air photographs. Biol Conserv 7:275–288

Psuty NP (2004) The coastal foredune: a morphological basis for regional coastal dune development. In: Martínez M, Psuty NP (eds) Coastal dunes: ecology and conservation. Springer, Berlin, pp 11–27

Psuty NP, Ofiara DD (2002) Coastal hazard management – lessons and future directions from New Jersey. Rutgers University Press, New Brunswick, 429 pp

Pye K, Blott SJ (2008) Decadal-scale variation in dune erosion and accretion rates: an investigation of the significance of changing storm tide frequency and magnitude on the Sefton coast, UK. Geomorphology 102(3–4):652–666

Ramsey DL, Wilson JC (1997) The impact of grazing by macropods on coastal foredune vegetation in southeast Queensland. Aust J Ecol 22(3):288–297

Ranwell DS (1958) Movement of vegetated sand dunes at Newborough Warren, Anglesey. J Ecol 46(1):83–100

Ranwell DS, Boar R (1986) Coast dune management guide. Institute of Terrestrial Ecology, HMSO, London, 105 pp

Ritchie W, Mather AS (1977) The beaches of the highlands and islands of Scotland. Commissioned by the Countryside Commission for Scotland 1977. Reprinted 2005 by Scottish Natural Heritage as Commissioned report no 048. Source: http://www.snh.org.uk/pdfs/publications/commissioned_reports/ReportNo048.pdf. Accessed 20 Nov 2010

Sabatiera F, Mailleta G, Provansala M, Fleurya T-J, Suanezd S, Vella C (2006) Sediment budget of the Rhône delta shoreface since the middle of the nineteenth century. Mar Geol 234(1–4):143–157

Saye SE, van der Wal D, Pye K, Blott SJ (2005) Beach-dune morphological relationships and erosion/accretion: an investigation at five sites in England and Wales using LIDAR data. Geomorphology 72(1–4):128–155

Schlacher T, Thompson L (2007) Exposure of fauna to off-road vehicle traffic on sandy beaches. Coast Manag 35:567–583

Schlacher T, Thompson L, Price S (2007) Vehicles versus conservation of invertebrates on sandy beaches: quantifying direct mortalities inflicted by off-road vehicles on ghost crabs. Mar Ecol 28:354–367

Schlacher T, Thompson L, Walker S (2008) Mortalities caused by off-road vehicles (ORVs) to a key member of the sandy beach assemblages, the surf clam *Donax deltoides*. Hydrobiologia 610:345–350

Slagel MJ, Griggs GB (2008) Cumulative losses of sand to the California coast by dam impoundment. J Coast Res 24:571–584

Stanley DJ, Warne AG (1983) Nile Delta: recent geological evolution and human impact. Science 260(5108):628–634

Syvitski JPM, Kettner AJ, Overeem I, Hutton EWH, Hannon MT, Bra-kenridge GR, Day J, Vörösmarty C, Saito Y, Giosan L, Nicholls RJ (2009) Sinking deltas due to human activities. Nat Geosci 2:681–686

Thornton EB, Sallenger A, Conforto Sesto J, Egley L, McGeee T, Parsons R (2006) Sand mining impacts on long-term dune erosion in southern Monterey Bay. Mar Geol 229(1–2):45–58

Vörösmarty CV, Meybeck M, Fekete B, Sharma K, Green P, Syvitski J (2003) Anthropogenic sediment retention: major global impact from registered river impoundments. Glob Planet Change 39:169–190

Walling DE (2009) The impact of global change on erosion and sediment transport by rivers: current progress and future challenges. The United Nations World Water Assessment Programme, Scientific paper, UNESCO, Paris, p 26. Source: http://unesdoc.unesco.org/images/0018/001850/185078e.pdf. Accessed 10 Mar 2012

Walling DE, Fang D (2003) Recent trends in the suspended sediment loads of the world's rivers. Glob Planet Change 39:111–126

Wheater CP (1999) The ecology of urban habitats. Habitat guides. Routledge, London, 188 pp

Chapter 7
Trends and Trade-offs – Inland Vegetated Dune

Abstract In the absence of human intervention and with a reasonable sediment supply, the typical communities will include all the successional vegetation described in Chap. 1, from pioneer dunes to scrub and woodland. However, most coastal dunes have had some form of human interference in the past. The introduction of domesticated stock more than 5,000 years ago helped modify dune vegetation in some parts of the world (Sect. 2.4.2). This chapter is concerned with the 'mechanisms for change' on these vegetated sand dunes. The way the stands of vegetation react in terms of species composition, height and structure particularly to the effects of different grazing regimes, is a key part of the discussion.

A 'State Evaluation Model', mostly based on European examples, provides a summary of the possible pathways to restoring degraded vegetation. This model relates to the surface of dunes inland from the beach/dune interface. In addition to grazing it also discusses the stabilisation of areas historically laid bare by sand drift (Sect. 2.2.) through afforestation (Sect. 2.3.2). It takes into account the extent that native woodland survives and the value of allowing scrub and secondary woodland to develop. The trends associated with recreational activities and water relationships complete the analysis.

7.1 Vegetation Trends

Vegetation cover develops as plants help trap sand and deposition takes place. Wind is a key factor in driving beach sand onshore (Sect. 1.2.3). Low soil moisture in sand dunes limits plant growth, even in areas of relatively high but seasonal rainfall, as it percolates rapidly through the porous sand especially in the first 5 cm (Maun 2009). There is a climatic gradient with vegetation covering and stabilising the mobile sand in temperate regions, as described in Chap. 1. With lower rainfall and higher potential evaporation, vegetation becomes increasingly sparse and eventually gives way to desert/mobile dunes. Although, even here it is only in hyper-arid or overgrazed

J.P. Doody, *Sand Dune Conservation, Management and Restoration*,
Coastal Research Library 4, DOI 10.1007/978-94-007-4731-9_7,

areas that vegetation is completely absent (Pye and Tsoar 1990). Over the last 2,000 years or so climatic variation has also played an important part in determining the presence of sand dunes and the nature of the vegetation (Sect. 1.1.5). Against this background are the modifying effects of human intervention on the vegetation, including the introduction of grazing animals and afforestation. The following sections describe the trends associated with these uses.

7.2 Vegetated States – Trends and Values

Historically grazing played a significant role in helping to determine the nature of vegetation on sand dunes in northwest Europe (State 2 dune grassland or State 3 dune heath, described in Sect. 1.3.3 above). Movement away from the grazing levels that sustain these communities, usually represent negative trends in nature conservation values. These range from situations where the vegetation breaks up and the sand becomes mobile again, to those where dune grassland and heath becomes scrub and woodland. Overgrazed (Vegetated State 1) and undergrazed or abandoned formerly grazed (Vegetated State 4) are representative of these extremes. Planting non-native trees (Afforested State 5) whether on bare sand or vegetated dune completely changes the nature of the dune surface. In the absence of grazing, scrub can develop into woodland but the extent that this is a desirable trend depends on whether it involves native trees (State 6) or not.

7.2.1 State 1 – Heavily Grazed

Although grazing is important to the development and conservation of vegetated sand dunes in some temperate regions of the world, overgrazing can be detrimental. Native grazing animals such as the European Brown Hare *Lepus europaeus* may have a limited impact on the vegetation. On the other hand, high densities of domesticated stock (including horses) and introduced rabbits can cause concern with a loss of structural diversity (Fig. 5.2), species richness and eventual erosion of the vegetated surface (Fig. 5.3). The trends associated with many values on heavily grazed sand dunes are mostly negative (Table 7.1).

7.2.2 State 2 – Moderately Grazed Dune Grassland and State 3 Dune Heath

Moderately grazed inland dunes are to some extent, the 'desired state' supporting typical and rare species associated with dune grassland or heathland. At these levels, the right balance is set between the control of coarse grasses and scrub and the

Table 7.1 Positive, neutral or negative trends for some of the values associated with overgrazed inland dunes (State 1)

Value	Trend	Comment
1. Sediment store	Positive/neutral	Unaffected by grazing
2. Organic material	Negative/neutral	Limited availability for export
3. Water quality	Negative	Potential for raised nutrient levels
4. Sea defence	Negative/neutral	Potential for destabilisation
5. Freshwater	Neutral/negative	Potential for contamination
6. Woodland	Negative	Grazing prevents sapling growth
7. Invertebrates	Negative	Loss of structural diversity
8. Birds	Negative	Loss of nesting/feeding habitat
9. Reptiles	Negative	Loss of cover, basking and foraging areas
10. Rare plants	Negative	Loss of seed production
11. Ecological study	Negative	Loss of 'natural' succession
12. Geomorphology	Neutral/positive	Potential for an increase in the rate of mobilisation by coastal processes
13. Recreation	Negative/neutral	Restrictions on access due to the potential for an increase in instability
14. Archaeology	Neutral/positive	Erosion can expose archaeological features

maintenance of low-growing 'open' vegetation. The vegetation has some structural diversity and because of the absence of dense scrub, the potential to provide conditions for development of high biological biodiversity. Note that domesticated stock and rabbits are not the only grazing animals. There is a remarkable diversity of others, including at different trophic levels (Fig. 1.11) voles, insects, especially caterpillars of butterflies and moths, and snails. Nature conservation managers strive to maintain or restore these vegetated states, as they are likely to support most of the recognised nature conservation values (Sect. 5.4). Included within these two states are dune slacks. These often occur in hollows created by blowouts. Where rabbits are amongst the grazing animals, small-scale disturbance allows plants (including rare annuals) and animals (such as burrowing wasps and bees) especially psammophilic[1] species to colonise. Some internal mobility may be beneficial to the survival of the full range of plants and animals associated with these two states.

7.2.3 State 4 – Abandoned, Formerly Grazed, Scrub Dominated

Undergrazed inland dunes occur when traditional grazing management on dune grassland or heath ceases, or reaches a level that is insufficient to prevent succession to scrub and woodland. This state may occur more quickly when changes in climate, such as increased rainfall favour vegetation growth. In addition, nutrient enrichment

[1] Organisms that live or thrive in sandy soil.

Fig. 7.1 Scrub invasion at Tentsmuir Point, northeast Scotland. The *top picture* (**a**) taken in May 1980, shows the dune soon after the erection of the fence to keep out rabbits from the species rich dune slacks on the right. The *bottom picture* (**b**), taken only 3 years later in August, shows the same area when scrub invasion had become established inside the fence. The rabbit population had crashed in the intervening years

from atmospheric deposition (Sects. 2.4.3 and 7.7.2 below) may also accelerate this change. Soon after the reduction or cessation of grazing, plants begin to flower more prolifically. This can give the impression of enhanced nature conservation values as some of the rarer plants suppressed by grazing become more noticeable (Doody 1989). However, this rapidly gives way to the growth of coarse grasses and scrub development, which happened in the United Kingdom from the early 1960s/1970s (Fig. 7.1) and continues to the present day.

Table 7.2 Positive, neutral or negative trends and values associated with abandoned formerly grazed, scrub dominated inland dunes (State 4)

Value	Trend	Comment
1. Sediment store	Neutral/negative	Sediment 'locked' in stable dunes
2. Organic material	Positive/neutral	More available for export
3. Water quality	Positive/neutral	Reduced nutrient levels, initially
4. Freshwater supply	Negative	Increased evapotranspiration, lowered groundwater table,
5. Woodland	Positive	Regeneration
6. Landfall site	Neutral	Unlikely to affect this activity
7. Military training	Neutral/positive	Restraints disappear
8. Invertebrates	Positive	Increase in structural diversity
	Negative	Loss of bare sand
9. Birds	Positive	Increase in scrub nesting habitat
	Negative	Decrease in species of open habitats
10. Reptiles	Positive	Increase in cover for shelter and foraging
	Negative	Loss of open habitats for basking and foraging
11. Rare dune plants	Negative	Increased competition, loss of dune slacks
12. Ecological study	Positive	'Natural' succession
13. Geomorphology	Negative	Absence/reduction dune mobility
14. Recreation	Neutral/negative	Restrictions on access
15. Archaeology	Negative	Exposure of archaeology reduced
	Positive	Protects features in situ

As the development of the coarse grassland, scrub and woodland proceeds the values associated with typical open sand dunes deteriorate. These include loss of specialist and rare plants as well as their associated animals. Amongst other species are reductions in numbers of breeding Wheatear *Oenanthe oenanthe* and Skylark *Alauda arvensis* as the area of short open vegetation decreases. By contrast, species nesting in scrub showed an increase in breeding pairs (Baeyens and Martinez 2004). The trends associated with some economic values are not as clear (Table 7.2).

7.2.4 State 5 – Afforested Dune

Afforestation continues the dune stabilisation process but with the deliberate introduction of alien tree species. As the canopy closes, with the progressive shading of the vegetation, plant and animal diversity decreases. The dune topography is often still present, although 'fossilised' and in just a few years most of the typical and rare sand dune vegetation and associated animal species are lost. The overall effect is to destroy the open sand dune communities, replacing them with a carpet of inert needles (Sect. 5.2.4; Fig. 5.5).

There is some compensation by way of woodland plants and animals, but this is often limited from a nature conservation perspective when compared with the species of 'open' vegetated sand dunes (Table 7.3).

Table 7.3 Positive, neutral or negative trends and values associated with afforested inland dunes (State 5)

Value	Trend	Comment
1. Sediment store	Neutral/negative	Sediment 'locked' in stable dunes
2. Organic material	Neutral/positive	More available for export
3. Water quality	Neutral/negative	Increased nutrient levels, soil acidification
4. Sea defence	Positive/neutral	Increased stabilisation
5. Freshwater supply	Neutral/negative	Lowering water levels through increased evapotranspiration
6. Timber	Positive	Mature trees, economic value
7. Landfall site	Negative	Forest inhibits excavation
8. Military training	Positive/neutral	Different training options available
9. Invertebrates	Negative	Loss of habitat and species diversity. Possibility for a different suite of plants and animals to invade.
10. Birds	Negative/neutral	
11. Reptiles	Negative/neutral	
12. Rare plants	Negative/neutral	
13. Ecological study	Negative	'Natural' succession disappears
14. Geomorphology study	Negative	Absence of coastal processes
15. Recreation	Neutral/positive	New opportunities arise
16. Archaeology	Negative	Archaeology sites hidden

7.2.5 State 6 – Native Dune Woodland

Woodland composed of native tree species is the final stage in 'natural' dune succession (Sect. 1.3.6). Present day native coastal dune woodland, at least in northern Europe, is rare or poorly represented. This is because of the long history of human interference including afforestation and the relatively high proportion of alien species in developing scrub (Provoost and Van Landuyt 2001).

In some areas, 'native' dune woodland is extensive and apparently 'natural', as in the dunes of the Coto Doñana, Spain. However, here there is evidence of extensive agricultural use. Thick and diverse forest with *Quercus suber* and *Juniperus oxycedrus ssp. macrocarpa* were the dominant species in the thirteenth and fourteenth centuries. After 1636, the introduction of livestock and clearance of the native woodland resulted in degradation of the forest, accelerated by pig farming. Trial plantations using *Pinus pinea* and later *Eucalyptus* spp. sought to stabilise the mobile dunes with some success. Re-forestation of the shifting sands of the seventeenth century took place after 1805. Today, *Pinus pinea* is the characteristic tree, giving the appearance of native woodland but actually replacing the original forest (Garcia Nova 1997).

7.2.6 The Vegetation State Evaluation Model

Having established how certain activities influence the nature of the sand dune state and whether the trend is positive, negative or neutral for the values associated with the state, a simple model provides a means of conceptualising the interaction

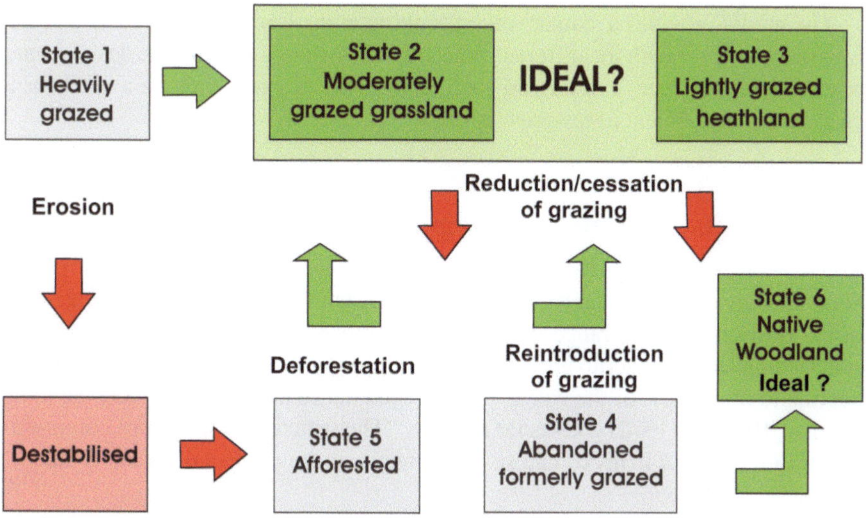

Fig. 7.2 A simple, vegetation 'State Evaluation Model' relating to grazing on inland coastal sand dunes. *Red arrows* indicate the main undesirable trends; *green arrows* the pathways to restoring 'ideal' states from a nature conservation perspective. State 5 largely relates to inland dunes where there has been a history of excessive grazing, which combined with other destabilising agents has caused loss of vegetated dune creating extensive areas of blown sand. These have subsequently been planted with alien trees to prevent blowing sand, depriving the system of the opportunity for 'natural' restoration to dune grassland or heath. State 6 derived from invading scrub could become an ideal state because of the rarity of native woodland in some parts of the world

between the states. A revised 'state evaluation model', developed as part of a coastal habitat restoration guide (Doody 2003) helps tease out the balance between the various factors forcing change and provides a basis for assessing grazing as a tool for managing sand dunes in temperate regions of Europe.

As has been described earlier, the nature of sand dune vegetation varies according to climatic conditions, the underlying nature of the sediment (acidic versus calcareous), exposure, hydrology etc. (Chap. 1). The four vegetation states (States 1–4) that describe different grazing regimes are clearly interlinked. The 'State Evaluation Model' (Fig. 7.2) attempts to show this. Sand dune grassland or heath, as described in Sect. 5.2, represents the 'ideal' situation when grazing pressure is optimum. Where it is not and scrub develops at the expense of dune grassland or heath, intervention may be required.

Inland dunes with high stock numbers, heavy rabbit infestation or direct human influences, can cause destabilisation of the dune surface and erosion. Planting trees (Afforested State 5) to prevent sand movement has resulted in extensive biologically degraded sand dunes. This vegetation state represents a marked departure from the 'ideal' vegetated state. However, restoring these areas to mobile dune, or where appropriate grassland or heath, represents a major nature conservation opportunity. Abandoned formerly grazed grassland or heath may revert, via native scrub, to Native Woodland (State 6). The extent to which this is acceptable depends on the values attached to the original grassland or heath and the species invading.

The model provides a simple way of looking at the forcing mechanisms and the way these interact with the different states. It is possible to identify the desired state based on the attributes (values) of each, and hence assess if there is a need for a change in management to restore lost values.

7.3 Rabbits – Agents of Change

Sand dunes are not the main habitat for the European Rabbit *Oryctolagus cuniculus*. In its 'restricted' native range in Spain and North Africa, it builds warrens in soft soil, finding shelter in scrub growing on rocky ground. However, following its introduction to other parts of the world it found sand dunes particularly conducive to its survival. Ranwell (1972) goes as far as to say "The structure of sand dune communities in Europe, prior to myxomatosis was effectively the product of intensive rabbit-grazing".

7.3.1 European Rabbit Oryctolagus cuniculus

In northern Europe, where its introduction was as a valuable source of food and fur, it was cultivated in 'warrens' particularly on sand dunes and sandy heaths. At some sites, its food and fur value was such that it replaced sheep grazing (Sect. 5.3.4). Sheep and cattle grazing continued on many sand dunes but rabbits invaded almost all but the most inaccessible, only to become a problem as their population increased.

Irish rabbit populations appear to have caused cycles of erosion and accretion. From the early nineteenth to the early twentieth century because of their value for meat and fur, rabbit numbers were under control. However, by 1924 they were at plague proportions and attempts to control them largely failed. During the war years 1940–1945, the populations in Europe fell again due to hunting for food. However, in the period up to the mid 1950s, they were again causing erosion problems in Ireland and the United Kingdom (Binggeli 1994; Thompson and Worden 1956). In other parts of Europe, the value of rabbits as a food source also waned and as their population increased, they again appeared to be agents of destruction (Van Dam 2001).

As far as sand dunes are concerned, their burrowing activities have helped initiate small and large-scale sand dune mobility. Introduced worldwide, rabbits are something of a paradox. For hunters they are fair game but for farmers and horticulturalists, they are a pest. From a nature conservation perspective, they can be both positive and negative. On the negative side, at plague proportions, they can cause excessive erosion. However, at moderate levels and/or where domesticated stock grazing is low or absent (vegetated State 4) they can be a positive agent for management.

Rabbits have small delicate mouths and therefore seek out palatable fodder. They thus tend to graze softer grasses, newly emerged shoots and seedlings of invasive scrub species such as Hawthorn *Crataegus monogyna* and *Hippophaë rhamnoides*.

More mature plants tend not to be palatable although they will eat the bark of some woody species, especially in harsh winters. They have thus helped to control the invasion of scrub in the absence of domestic stock. They appear not to eat some species of scrub, such as elder, which occurs around burrow entrances in Ireland (Binggeli 1994). They also leave Common Ragwort *Senecio jacobaea*, poisonous to cattle, alone.

7.3.2 *Oryctolagus cuniculus* and Myxomatosis

Myxomatosis, a viral disease, was successfully introduced to Australia from Brazil in 1950 to control the infestation of rabbits. Following this, further introductions occurred in other parts of the world where rabbit numbers had also grown to pest proportions. First introduced to France in 1952 and then southeast England in 1953, it took only 1 year to spread to most parts of the British Isles (Fenner and Fantini 1999). The impact was dramatic, and resulted in some areas in a mortality rate of 99.8% (Thompson and Worden 1956).

Rabbits, besides their dramatic impact on farm crops, have also affected natural vegetation where their rapid decline led to equally rapid changes in vegetation. Studies at a number of nature reserves in Great Britain, including one sand dune, revealed that immediately following 1953, there was a rapid increase in cover and growth of scrub (Thomas 1960). Although this change slowed and was even reversed on some sites where numbers recovered, the trend continued into 1962 (Thomas 1963). At Newborough Warren, Anglesey, *Crataegus monogyna* rapidly colonised the dune following the demise of rabbits in 1954. In addition to the loss of species of the open dune vegetation, there was also a considerable increase in soil fertility (Hodgkin 1984).

Other more subtle changes also took place. *Erica cinerea*, unpalatable to rabbits, became dominant over *Calluna vulgaris* in Northern Ireland prior to myxomatosis. Following the crash in numbers *C. vulgaris* gained dominance, a reversal of the previous situation (Binggeli 1994).

The impact of rabbits appears clear; at high densities, they can cause the destruction of vegetation, excessive erosion and deflation of the dune surface. At low levels and in the absence of domesticated stock, the sand dune becomes more and more stable, eventually reverting to secondary scrub and woodland. Rabbit grazing is restricted to the remaining areas of open palatable dune grassland, once scrub has developed. Here they continue to help maintain close-cropped plant communities leaving the adjacent vegetation to develop into impenetrable dense scrub (Fig. 7.3). Moderate population numbers can be beneficial, as their grazing and burrowing activities help maintain a mosaic of open and closed vegetation (Isermann et al. 2010).

Numbers have fluctuated considerably but there appears to have been a general tendency towards higher populations from the 1990s. From a nature conservation perspective in Europe, this is a generally positive trend as rabbits help control or even reverse dune stabilisation (Sect. 9.5.7).

Fig. 7.3 Impact of rabbit grazing on Gullane dunes, southeast Scotland, August 1988. Note the abrupt demarcation between the rabbit grazed grassland, *Hippophaë rhamnoides* scrub and rabbit-disturbed foreground

7.4 Domesticated Stock – Too Much or Too Little?

Traditionally, grazing took place on dunes mainly in winter (cattle and sheep brought from the 'uplands' to the coast for shelter and fodder). As different types of domesticated animals graze in different ways, it is useful to have some insight into how they affect the vegetation:

1. Cattle curl their tongue around the plant pulling it towards their mouth and then tear it off. They thus tend to leave structurally more diverse and taller vegetation;
2. Sheep 'bite' the plants between the lower and upper incisor teeth allowing much closer cropping. This helps to create a low, structurally impoverished sward;
3. Ponies are similar to sheep in that they 'nip' the grass off between their upper and lower incisor teeth. This enables them to graze close to the ground, again helping to create low-growing vegetation.

Other differences include tolerance to some distasteful species. For example, *Senecio jacobaea* is poisonous to cattle but grazed by sheep without any apparent adverse effect. Cattle and horses will not eat vegetation 'tainted' by dung or urine from their own species, although they are not so discriminating when it comes to other species.

Chapter 9 covers this in more detail, specifically how to achieve the right balance between animal numbers and nature conservation requirements. The next sections deal with some of the main effects of changes in levels of grazing by domesticated stock.

7.4.1 Heavily Grazed (Overgrazed)

Sand dunes can have excessive numbers of stock, causing a loss of structural and species diversity as described above (Sect. 5.2.1). Although this is increasingly the exception rather than the rule, where it occurs it may require remedial action. Northam Burrows, Devon, southwest England has common grazing rights extending over an area of approximately 253 ha. These rights allow grazing for up to 1,200 sheep and 100 ponies. This amounts to a stocking level of 1.09 Livestock Unit (LSU)[2] year round. In 2004 grazing levels equated to 0.64 LSU/ha, predominantly in June and July when the highest densities of 0.85 LSU/ha were reached. The lowest levels were 0.32 LSU/ha in February. Grazing intensity at these levels is high, with the stock concentrating on the open more palatable grassland. Less palatable scrub and rush-dominated areas are largely left untouched, although the ponies trample some of this vegetation (Fig. 5.2).

Other than the fenced foredunes, there are no stock fences; animals are free to graze unimpeded across the whole site. This type of grazing, at a high stocking density and year round, causes the sharp discontinuities observed between the different vegetation types shown in Fig. 5.2. Dunging can also exacerbate the effects of grazing by increasing the nutrient levels in the soil. At its most extreme, this can cause extensive loss of diversity, especially when the animals are receiving supplementary feed. This is particularly important on dune heath where the herbage grows relatively slowly and can only support a low stock density. Ross Links, Northumberland on the north east coast of England is a case in point (Fig. 7.4). Here fences, erected in the mid 1960s, were within the main dune and part of an expanding livestock enterprise. Supplementary feeding during the winter resulted in changes to the 'natural' sand dune vegetation, namely:

1. Eutrophication of dune soils and growth of nitrophilous species, most damaging immediately around feeding stations;
2. Overgrazing and loss of species diversity in the vicinity of feeding stations, but tailing off as the distance from them increases;
3. Undergrazing in more remote areas where animals have no need to search for fodder.

[2] The livestock unit, abbreviated as **LSU** (or sometimes as **LU**), is a reference unit and is the grazing equivalent of one adult dairy cow producing 3,000 kg of milk annually, without additional concentrated foodstuffs. European Commission, Eurostat: http://epp.eurostat.ec.europa.eu/statistics_explained/index.php/Glossary:Livestock_unit_%28LSU%29.

Fig. 7.4 Cattle poaching and eutrophication caused by excessive dunging surrounding a feeding station, Ross Links, Northumberland March 2005

Animals exert trampling pressure on dune soils as they graze. Although the heavier cattle might appear to have a greater impact than lighter sheep, this may not be the case. A 440,000 g cow has a surface area in contact with the ground of 300 g/ cm^2, which exerts a pressure of 1,467 g/cm^2. By contrast, an 80,000 g sheep has a much smaller contact area, exerting a pressure of 941 g/cm^2 (Liddle 1997). However, it has been estimated that at 'normal' stocking rates a cow will tread an acre of pasture three or four times a year; a sheep six to ten times. As a result, sheep can potentially exert two to three times more pressure on a pasture than cattle (Frame 2000, Table 16.2). This may be especially significant on the softer substrates associated with sand dunes (Ranwell 1972).

7.4.2 Overgrazing, Deflation and Machair

Overgrazing, often accompanied by other destabilising activities, has caused erosion and deflation of vegetated dune surfaces in the past (Sect. 2.2). It is less frequent today, although it is still a major issue on some exposed northern and western European dunes. Sheet erosion occurs within the body of the dune and can lead to the exposure of large areas of bare sand (Fig. 5.3). An eroding (Physical State 1) beach or a shifting (Physical State 2) beach /foredune may accompany this. Where both forms are present on the same sand dune, they are indicative of severe erosional pressures.

Fig. 7.5 The effects of overgrazing and sheep rubbing on Scottish machair, September 1988. Note the collapsed turf in the foreground, which can help with the colonisation of exposed sand

Machair is a special sand dune habitat, where human use determines the nature of form and function, especially in the cultivated areas of the Outer Hebrides. Grazed by sheep and cattle in Ireland, on mainland Scotland and some uncultivated areas on the islands off the west coast, it is often closely cropped and species rich. Grazing and cycles of erosion provide the driving forces for change. Long-term machair development relies upon ongoing aeolian activity and changes in climate can influence erosion (Chap. 11).

The rabbit introduced to the Outer Hebrides in the late 1700s, through their scraping and burrowing activities, may initiate erosion by exposing the sub-surface to wind (Angus and Elliott 1992). Coupled with high sheep numbers, grazing pressures in the 1980s and 1990s may have also played a significant role in promoting erosion of the machair surface. Once the surface vegetation is lost, wind removes some of the exposed sand to create hollows. These become attractive to sheep, which lie in them and rub against the sides helping to create small cliffs. The effect is to cause the collapse of the vegetated surface, exposing the sand to further erosion (Fig. 7.5).

It is argued that historically the scale of erosion has been much more prevalent than today (Angus and Elliot 1992). The nature of this change is visible in western Ireland, where sheep grazing coupled with high rabbit numbers has lead to severe erosion. In places on exposed coasts, the machair surface vegetation can disappear altogether (Fig. 5.3) exposing ancient middens (Fig. 2.1).

Fig. 7.6 Sandscale Haws, Cumbria, England grazing exclosure, April 1982 after approximately 10 years with no grazing animals. Note the growth of tall grasses and scrub, contrasting with the surrounding grazed dune slack

7.4.3 Reduction or Cessation of Grazing by Domesticated Stock

Lack of grazing in northwest Europe is the most frequently encountered state leading to stabilisation and scrub encroachment. Many factors influence the loss or reduction in grazing pressure. These include changes in the economics of stock utilisation on marginal habitats, including those where there is a move away from traditional management regimes. Change of use, such as military training, may be incompatible with stock management. Loss of native or introduced herbivores, such as the Rabbit because of disease, control or eradication is also important. A reduction in grazing pressure inevitably results in the growth of course grasses, scrub and woodland at the expense of the more open mature grassland and heath as described above. Today many European sand dune sites deemed to be of value for nature conservation, have little or no bare sand and extensive scrub. Establishing workable grazing regimes is a key issue (Chap. 9).

When there is a reduction or cessation of grazing, the first casualties are species sensitive to overcrowding. Sandscale Haws National Nature Reserve in Cumbria grazed since the twelfth century by sheep, was also a rabbit warren. In order to assess the impact of removing stock on nature conservation interests, a small enclosure was set up in the 1970s (Fig. 7.6). This helped to establish the pattern of succession following the removal of grazing animals. Within the first 5 years, the number of plant species appeared to increase. However, this may just have been an initial response to the removal of the constraints imposed by grazing pressure, which allowed species to flower. After 10 years, the species diversity was much lower and scrub had begun to shade out the open dune vegetation.

A study of dune grasslands in two coastal dune areas in the Netherlands showed similar effects. In the absence of grazing, 'grass-encroachment' resulted in a loss of plant species, notably therophytes, bryophytes and lichens, as the amount of daylight reaching the ground floor vegetation diminished. In ungrazed areas, open communities declined from 77 to 17% cover and tall grass cover increased from 3 to 53% in only 6 years (Kooijman and van der Meulen 1996).

Also in the Netherlands, cessation of grazing resulted in the rapid development of scrub and woodland at several dune sites near Oostvoorne with a decline in dune species richness. The spread of coarse grassland, scrub and woodland occurred rapidly from 1910 when there was considerable bare sand due to excessive grazing pressure. The succession was from bare sand via yellow *Ammophila arenaria* dunes, low and tall scrub phases to woodland. In wetter areas behind the yellow dune, development was from marsh to tall wet scrub. Overall, only 4% of the vegetation communities remained unchanged during the period (van der Maarel et al. 1985). This progression was interrupted on two occasions:

1. During the Second World War when reversion to earlier less stable forms took place due to surface disturbance;
2. 1953 when a storm surge broke through the foredunes re-creating dune slacks (Van Dorp et al. 1985).

Hippophaë rhamnoides, is a particularly invasive species (Sect. 8.3), which even in locations where it is native can expand rapidly when grazing pressure decreases. Scrub invasion in the Netherlands from 1934 in Oostvoorne, South Holland, followed a general pattern of stabilisation with *H. rhamnoides* playing a significant role (van der Maarel et al. 1985; van Dorp et al. 1985). Together with *Ligustrum vulgare*, the Sweet Briar *Rosa rubiginosa* and *Sambucus nigra* (all already well developed in 1934) the area of *H. rhamnoides* remained constant until the 1970s. After this, other species such as *Crataegus monogyna* and Buckthorn *Rhamnus catharticus* with low woodland, dominated by *Betula pendula* and *Quercus robur* took over as succession continued to take place (Provoost et al. 2011).

Although native in the United Kingdom, *Crataegus monogyna* scrub can be equally problematic. At Newborough Warren, Wales it became the most abundant invading scrub species after myxomatosis wiped out the rabbit population in 1954 (Hodgkin 1984).

In Israel, sand dune stability in the absence of grazing resulted in fewer annual plants and small mammals, including *Gerbillus andersoni allenbyi* (an endemic sand-living gerbil). Scrub removal was one of the means of taking remedial action (Kutiel et al. 2000).

The absence of domesticated stock does not necessarily mean loss of nature conservation values. Native coastal scrub has values as a natural stage in sand dune succession (Sect. 1.3.5). The garigue and maquis in the Mediterranean (Sect. 1.6.2) or dune scrub in the USA (Sect. 1.6.3) are an important part of the succession.

Table 7.4 Summary of the number of invertebrate species distributed across the beach/foredune and inland dune, derived from studies in Wales and the United Kingdom (Howe et al. 2010)

Habitat	No of species
Strandline & beach flora	33
Marram zone	36
Bare & sparsely vegetated sand	172
Dune slacks	50
Dune grassland	61
Dune heath	1
Dune scrub	24
Bee & wasp nests in sand	20
Dune fungi	4
Dung & carrion	23
Unknown habitat association	38
Total	462

Grazing is not such a feature in either habitat, as it is in the dune grasslands and heathlands of northwest Europe. In the Mediterranean grazing, when it occurs, is with sheep and goats. It is extensive and usually at a low intensity.

Coastal scrub supports a wide variety of species. Even where species are invasive they may have important attributes of value to wildlife, even the often much despised *Hippophaë rhamnoides*. This shrub provides shelter and winter berries for several species of *Tardus* (blackbirds and thrushes) amongst a variety of other values and uses (Sect. 5.3.2). However, despite this the loss of grassland and heath as it invades, represents an overall diminution of nature conservation values. Many other shrub species can provide landfall sites for migratory birds, shelter and winter-feeding for these and a variety of other animals. However, scrub is not restricted to sand dunes and the species that inhabit the vegetation are mostly widespread and/or common. In most cases, justification for retaining it is therefore less persuasive than for its removal.

7.4.4 Grazing and Invertebrates

Sand dune invertebrates are particularly prevalent in the early stages of dune development and on bare and sparsely vegetated inland dunes (Table 7.4). Thus, where the level of grazing (or rabbit burrowing) leads to destabilisation, this can be positive for these species. On the other hand, cessation or reduction of grazing pressure and the reversion to scrub reduces the availability of open areas upon which many of these species depend.

Plant structural diversity is important for invertebrates. Losses due to heavy grazing reduce the number of niches for animal species to occupy. This applies both to the vegetation mosaic and to individual plants. Thus, consumption of plants by larger herbivores results in loss of associated invertebrate fauna. For those species feeding on the above ground parts of individual plants, this may be especially important.

Some species have distinct preferences for different zones at different times of the year. Animals of the drift line move inland in the autumn, not surprising perhaps as this habitat is often lost to autumn and winter storms. Some insects breeding in open areas overwinter in tussocks of vegetation. The presence of a mosaic of semi-mobile habitats, including all stages in succession is best for the diversity of invertebrates (Kirby 1992). Thus, overgrazed or undergrazed (vegetated State 1 and State 4) dunes are the least favourable for many species of invertebrates.

The type and structure of vegetation is also important. Dune hummocks with *Salix repens* provide a suitable habitat for a specialist ant-catching spider *Theridion saxatile*, not found in other habitats. For this species, the steep sides of the hummocks appear to help in web construction (Duffey 1968). It is important, therefore, to have all stages of inland dune development including bare sand. With greater stability and loss of bare sand, the number of sand dune specialist invertebrates declines (Howe et al. 2010).

7.4.5 Dune Slacks

Dune slacks (Sect. 1.3.4) rely on the water table being at or near the dune surface. They become naturally stabilised as litter builds up in the absence of grazing. Dune slack succession has at least four stages:

1. A pioneer phase with species that grow on bare soil, which may be covered with a thin layer of green algae and microbial mats;
2. Colonisation by plants adapted to very low nutrient availability;
3. Development of a moss layer with typical dune slack species;
4. Rapid accumulation of organic matter, increase in tall grasses and shrubs (Grootjans et al. 2002).

Rarer plants (e.g. *Liparis loeselli*) and animals Natterjack Toad require early stages in dune slack development. The most species-rich communities, sustained by grazing, belong to stages 2 and 3. A reduction or cessation of grazing lifts the constraints on plant growth, allowing the resumption of the process of succession to scrub. Stage 4 thus represents the degenerating phase, which leads to the decline of both the rare and more typical dune slack species and growth of scrub, notably *Salix repens*.

Grazing may slow down the succession but atmospheric nitrogen deposition, by increasing plant growth, has the reverse effect. Dune slack acidification occurs as precipitation and water draining from adjacent slopes enters the system. This not only changes the nature of the vegetation, but may also make the pools unsuitable for breeding amphibians. Evidence from inland heathland shows the decline of Natterjack Toad in two large ponds in Great Britain. Over a period of 50 years, the decline was attributable to acidification through atmospheric pollution (Beebee et al. 1990). This process also accelerated the decline already caused by cessation of grazing and increase in forestry activities. This in turn improved the competitive advantage of the Common Toad *Bufo bufo*. These losses also took place in dune slacks, although here they were less pronounced (Beebee 1977).

An increase in scrub and woodland took place on the West Frisian Islands, because of natural succession to less calcareous habitats and drinking water abstraction. This caused the bryological value of calcareous wet dune slacks to deteriorate, with the loss of 7 out of 24 species Red Data List species (van Tooren and Bruin 2004).

7.5 Afforestation – Friend or Foe?

Afforestation has had a major impact on sand dunes in many regions of the world (Sect. 2.3.2). Unlike native woodland, the planted alien trees do not have a large number of species associated with them. This represents a final, if artificial, stage in the development of stable sand dunes. These forests have commercial value for timber (Sect. 5.3.1) and can be important recreational areas. They may also have some nature conservation values. For the dune conservation manager these provide only limited compensation for the loss of open vegetated sand dune habitat. However, it is important to understand these values and the trade-offs associated with sand dune restoration, including removal of the forest.

7.5.1 Conservation Values

Dunes planted with alien trees can support a few typical and rare dune plant species, some even thriving in the forest. These include Dune Helleborine *Epipactis leptochila* var. *dunensis*, Newborough Warren, Anglesey in Wales and on the Sefton Coast, northwest England; *Pyrola rotundifolia*, Culbin Forest, Grampian Region and Coralroot Orchid *Corallorhiza trifida* in Tentsmuir Forest, southeast Scotland. Culbin Forest also supports a significant lichen flora, which includes over 30 species of the genus *Cladonia*. These include species with a northerly distribution better represented here than anywhere else in Britain (Gilbert 2004).

Although not restricted to coastal woodlands, the native Red Squirrel *Sciurus vulgaris* occurs in pine plantations on sand dunes in the UK. Reintroduced to the forest at Newborough Warren, Anglesey in 2003 it is one of only a few Welsh populations. The site also has woodland mosses, ferns and orchids; unplanted slacks and some road verges retain species of the original vegetated dune (Hill and Wallace 1989). Plantations on the Sefton coast are also refuges for Red Squirrel in an otherwise depleted English population. As we will see (Sect. 9.7.2), charismatic species such as this can create local opposition to the removal of the forest in order to restore dune habitat. In addition to the rare plants, Culbin Forest in Scotland also includes several less common animals such as Pine Martin *Martes martes* and three species of roosting bat.

A few older established forests planted with alien trees can develop biological diversity. In Denmark, forest originally planted in the mid nineteenth century, was mainly with conifers (Mugo Pine *Pinus mugo*, White spruce *Picea glauca* and European Silver Fir *Abies alba*) and some broad-leaved species such as oak further inland. After 150 years, when combined with the planting of a wider variety of trees including indigenous species, the plantations have developed nature conservation and recreational values (Wilkie 2002). Older plantings of well-spaced trees can also provide habitat for some larger animals such as wild boar in Continental Europe.

7.5.2 Peripheral Influences

In addition to the loss of open dune habitat, as the forest develops, two other factors come into play:

1. Changes in hydrology as the evapotranspiration from the canopy causes a lowering of the water table both within the forest and on nearby dunes;
2. Invasion of self-sown pine trees into the sand dunes adjacent to the forest.

Hydrological changes include ground water levels that are as much as 1 m lower beneath the pine plantations than on adjacent dunes in the Ainsdale National Nature Reserve on the Merseyside coast, England (Clarke 1980). At Newborough Warren, the impact of the pine forest also appears to have lowered ground water levels. However, it is difficult to establish direct cause and effect because of the complex geology affecting dune water relations at this site (Stratford et al. 2007). In the Netherlands, evapotranspiration from coniferous woodland was much greater than under bare dunes (Bakker 1990) again leading to changes in water relations nearby.

Self-sown pines may also result in an unplanned forest adjacent to the original planting. In the absence of grazing, or at low stock densities, the invasion of open dune increases with the presence of a readily available seed source. Non-native pines together with native trees such as *Betula* spp. may be especially successful. In addition, a lower water table can adversely affect species-rich dune slacks, by facilitating the invasion by birch and pine because of reduced winter flooding in the dune hollows (Doody 1989).

7.5.3 Amenity Verses Nature Conservation Values

Many large artificial dune forests are commercial operations. Recreational use is often also important. This may include car parks, footpaths and picnic areas. However, these activities are not specific to coastal marginal habitats such as sand dunes. It is clear from a nature conservation perspective that overall, despite the amenity values associated with the planted forest and the 'replacement' conservation values, these interests do not compensate for the change in the dune landscape and loss of diversity, especially of specialist dune species. Today most nature conservationists look to opportunities for restoring open sand dune vegetation by removing forestry plantations. This results in conflict with recreational interests and potential loss of charismatic animals such as the Red Squirrel. As a result, in some areas forest plantings with non-native species continues to the detriment of open dune vegetation. Chapter 9 considers the trade-offs in more detail as well as options for re-creating mature 'native' forests.

7.6 Recreation

Sand dunes and associated sandy beaches are a significant recreational resource (Sect. 5.5.1). Whilst for many people the attraction is sand and sea, the landscape and wildlife values are also important as part of the recreational experience. Activities impacting on the dunes come in a variety of forms and include trampling, lighting fires, excavating sun-traps, pony riding, vehicular traffic and in higher dune ridges 'sand sliding'. The impact of these activities varies with their scale and the intensity of use. They may have severe adverse consequences for both landscape and wildlife values.

7.6.1 Trampling

Trampling can create problems on dunes. In some areas where access to the beach is via specific well-worn tracks, erosion can result. However, before physical destruction of the dune vegetation takes place there may be subtle changes in its species composition caused by trampling (Liddle and Greig-Smith 1975; Liddle 1975). Trampling affects species diversity, the balance between monocotyledons and dicotyledons and soil bulk density. Generally, towards the centre of a path the degree of compaction leads to a decrease in the number of species, favouring monocotyledons, which also have lower biomass. Towards the margins of the path, the reverse tends to be true (Liddle and Grieg-Smith 1975).

Studies on the Island of Skalingen, Denmark showed how depth and width increased with the number of passes along a path. There was also a reduction of 50% in heathland vegetation cover when there were more than 200 passes representing a relatively low carrying capacity (Hylgaard and Liddle 1981). In the United States of America trampling pressure reduced the overall species diversity in coastal dune vegetation as well as causing a shift in the horizontal distribution of the vegetation. It seems to have favoured the more robust *Ammophila breviligulata*, resulting in an expansion of the foredune at the expense of vegetation at the back of the dune (McDonnell 1981).

Inevitably, as trampling and other uses intensify, destruction of the surface vegetation follows, exposing the underlying sand to the action of wind and rain. This in turn may make the dunes mobile again. These effects can appear to be serious, for example trampling damage to beaches and dunes in Scotland (Ritchie and Mather 1984) affected 52% of all complexes. Of these 7% were 'severely' damaged. The situation in Lithuania between 1982 and 1994 appeared to be even worse with total vegetation cover within the dunes declining by 50% and on beaches by 100%, due to recreational use (Olšauskas 1995).

Of course, this is only part of the story and whilst these effects may appear to be highly damaging they are often restricted in their overall impact on the sand dune system. Where grazing has ceased on some dunes light trampling may open up the vegetation, contributing to the restoration of biological diversity. In other areas, it might also act as a restraining force on the growth of coarse grasses, for example along the edges of footpaths facilitating the survival of typical dune grassland or heath.

7.6.2 Other Recreational Activities

In addition to trampling, driving cars and other vehicles onto sand dunes can have a significant impact. When paths coalesce or combine with other destabilising activities, significant losses can occur. On the sand dunes of Tywyn Aberffraw, Anglesey in north Wales from 1940 to 1993 the rate of parabolic dune migration ranged from 0 to 3.6 m per year (Bailey and Bristow 2004). Increased use of the motor car from the 1960s onwards and their unrestricted access to the site, in combination with grazing by cattle and rabbits caused an extensive network of paths to develop. In 1960, there were 3.2 km of tracks and 2.2 km of footpaths, but by 1970, this had increased to 11.7 km of tracks and 16.5 km of footpaths (Liddle and Grieg-Smith 1975; Ranwell and Boar 1986.).

Access routes can create the most visible signs of erosion. This occurs when access to the beach from car parks, caravan sites or other recreational facilities is via a sand dune. Continuous pressure from pedestrians can cause severe erosion, although this is usually restricted to a defined path, which provides the easiest route to the beach (Fig. 7.7). Trail bikes and off-road vehicles such as jeeps on the other hand, seek out steep slopes and are not so restricted. They also exert pressures in excess of 2,000 g/cm^2; much greater for example than the 1,467 g/cm^2 for cattle (Liddle 1997). The long term effects of these activities depend on the extent of restriction on access, zonation or the provision of walkways and the like (Chap. 9).

Note that these and other uses can also introduce a level of disturbance, which helps compensate for the lack of grazing pressure at some sites. Sand sliding on steep bare slopes may help create, maintain, or expand erosional features. Depending on the view taken of mobile sand, these activities can either destroy or help to sustain important sand dune features (Sect. 7.8). Lighting fires or digging hollows to create sheltered areas can also be damaging, especially in dune slacks when they are dry. The danger of fire spreading to the surrounding vegetation and the amount of rubbish left behind is an additional management problem.

Fig. 7.7 Erosion resulting from access between the caravans and the beach, Ynyslas West Wales, July 1986

7.7 Water Relationships

Dunes with natural drainage into the system and/or where they develop over a hard substrate can have an extremely complex hydrology. On the other hand, large dune systems with their own catchments may be more or less isolated and less complex (Packham and Willis 1997). In coastal locations, water comes from three basic sources:

1. Saline water, derived from the sea, generally lying below any freshwater lens and fluctuating with the tide;
2. Freshwater aquifer derived from seepage from the substratum and/or surface runoff from the land;
3. Precipitation.

Dune water table height fluctuates according to the input from the above sources, especially between summer and winter. Sites with their own catchments may develop a domed water table, which is 6–7 m higher in the centre of the system than on the shore or inland, as is the situation at Braunton Burrows, southwest England (Packham and Willis 1997). In isolated systems the domed water table depends primarily on precipitation with little or no input from the land. Capillary action helps create a 'fringe' nearer the sand surface where water is accessible to plants with short root systems.

Dune slacks depend on a combination of water flow fluctuations and succession, which result in changing vegetation patterns (Sect. 1.3.4). Modified by human activities notably water abstraction, the fall in water levels accelerates the natural tendency of the slack to become dominated by species such as *Salix repens* (Sect. 2.7.2). There are two detailed reports on the hydrology of sand dunes in Europe (Davy et al. 2006, 2010) both available to download. See also the European Union report on humid dune slacks (Houston 2008).

7.7.1 Water Abstraction

Water abstraction can have serious consequences for sand dune vegetation, especially dune slacks. In the Netherlands sand dunes became a significant source of drinking water in the first half of the twentieth century due to the growing demand from the large Dutch cities (Amsterdam, Rotterdam and The Hague). Degradation and/or loss of dune slack vegetation became one of the more obvious manifestations of this abstraction, depleting the populations of dune slack species including many rarities such as *Parnassia palustris*. Infiltrated water from the rivers Rhine and Meuse was, from 1957, used to recharge the dune aquifer in an attempt to reverse the loss of habitat. However, although there was an increase in open water and other wetland communities the dune slacks did not recover because of the high nutrient status of the infiltrated water (van Dijk and Grootjans 1993). From the 1980s, the inflow of nutrients has abated, as the water supply companies purified the river water before pumping it into the dunes. As a result, there has been a recovery of dune slack vegetation although not without additional management (Sect. 9.4.1).

In the Coto Doñana National Park in southern Spain, a tourist development (Matalascañas) began in the 1960s expanded considerably between 1980 and 1990. This required a drinking water supply. Together with increasing agricultural irrigation, which also took water from the reserve, this caused a reduction in dune water levels (Muñoz-Reinoso 2001) and adversely affected small 'dune ponds' (Zunzunegui et al. 1998). To a more limited extent, abstraction for other purposes is also damaging to the water relations within any dune. This includes use for irrigating 'links' golf courses (Chap. 10).

7.7.2 Nitrogen Enrichment

Sand dunes tend to be deficient in a variety of nutrients. Of these, the mineral nutrient status is probably most significant. For example, at Braunton Burrows the sparse growth of vegetation rested on the deficiency of N and P and a lesser to K in the dune soils (Willis and Yem 1961; Willis 1963). Factors such as climate or soil moisture were of a lesser importance. Additional N inputs stimulate plant growth, and atmospheric nitrogen deposition can have a significant impact on sand dune vegetation.

By enriching dune soils nitrogen also has an acidifying effect and causes eutrophication. In turn this results in a decline in the species richness of dune grassland as the above ground biomass increases (Jones et al. 2004). It is implicated in increasing rates of vegetation succession and soil development particularly in the early stages of succession. Although nitrogen derived from biological fixation is roughly four times greater than from atmospheric deposition the latter still represents a significant source, especially in dry dunes (Jones et al. 2008).

In northern Europe, this has become increasingly problematic for the rare lichen-rich dune grassland and heath and has led to the loss of species diversity and specialist interest on many sites, especially on coastal sand dunes. In the Netherlands, communities dominated by *Corynephorus canescens* and *Carex arenaria* have a dense layer of lichens including *Cladonia* spp. and *Cladina* spp. Since the 1970s, there has been a relatively rapid change with an impoverishment of the vegetation and loss of rarer lichens in particular (Veer 1997; Veer and Kooijman 1997). Experimental studies confirm this effect. It seems that nitrogen deposition caused by industrial pollution, car exhausts and other sources increase growth rate and accumulation of organic matter in the soil. This favours plants that respond to higher nitrogen levels such as *Carex arenaria* (Ketner-Oostra and Sýkora 2004; Remke 2009).

The Nitrogen critical load (damage threshold) for sand dunes in the United Kingdom is 10–20 kg N per ha per year. (Jones et al. 2004). This was the level reached on the sand dunes in the West Frisian Islands of the Netherlands in 1990. Although it has reduced since then, near to intensive agriculture it still attains this level in some areas (Ketner-Oostra 2006). In the Baltic even low to medium nitrogen deposition affects dune vegetation, promoting the dominance of taller graminoids such as *Carex arenaria* over the lichen-rich short grass vegetation (Remke et al. 2009). This in turn accelerates the loss of habitat suitable for sand-mining bees and other specialist invertebrates and vertebrates of bare sand, with knock-on effects higher up the food chain. Some of these species are prey items for the Red-backed Shrike *Lanius collurio* and their loss is implicated in the demise of this species from sand dunes in the Netherlands and Denmark (Domburg et al. 2005). This effect is in addition to scrub development resulting from reduction or cessation of grazing pressure.

7.8 Conclusion

The bulk of this chapter has been concerned with describing changes in vegetation in relation to grazing pressures. The vegetation model (Sect. 7.2.6) identifies dune grassland or dune heath as the 'desired state', with too much or too little grazing leading to loss of nature conservation values. Although there are examples of present day grazing pressures causing erosion these are relatively limited. Most coastal vegetated inland dunes formerly grazed by domesticated stock, lack this management and have developing scrub or woodland replacing open grassland or heath. Other factors such as afforestation, water relations and eutrophication contribute to this process. Dune slacks also loose rare plants and animals from this water dependent habitat. As the vegetation closes in, opportunities for species of bare sandy habitats are also lost.

Maintaining or restoring appropriate grazing management is one of the means of achieving the desired state of dune grassland or heath. However, managing the inland dune relying on grazing alone will not always fully address problems associated with regressive scrub development (Sect. 7.4.3). Neither will it restore those dunes afforested with alien trees. Although scrub removal or deforestation will be important, follow up treatment will always be required. Chapter 8 describes some significant alien invaders of sand dunes, including shrub species and specific methods of control. Chapter 9 covers the detailed management options for both the beach/foredune and inland dune.

References

Angus S, Elliott MM (1992) Erosion in Scottish machair with particular reference to the Outer Hebrides. In: Carter RWG, Curtis TGF, Sheehy-Skeffington MJ (eds) Coastal dunes: geomorphology, ecology and management for conservation. A.A. Balkema, Rotterdam, pp 93–112

Baeyens G, Martinez ML (2004) Dune animals: from exploitation to protection. In: Martinez ML, Psuty NP (eds) Coastal dunes, ecology and conservation, vol 171, Ecological studies. Springer, Berlin/Heidelberg, pp 279–296

Bailey SD, Bristow CS (2004) Migration of parabolic dunes at Aberffraw, Anglesey, north Wales. Geomorphology 59(1–4):165–174

Bakker TWM (1990) The geohydrology of coastal dunes. In: Bakker TW, Jungerius PD, Klijn JA (eds) Dunes of the European coasts, vol 18, Catena supplement. Catena, Cremlingen, pp 109–119

Beebee TJC (1977) Environmental change as a cause of natterjack toad (*Bufo calamita*) declines in Britain. Biol Conserv 11(2):87–102

Beebee TJC, Flower RJ, Stevenson AC, Patrick ST, Appleby PG, Fletcher C, Marsh C, Natkanski J, Rippey B, Battarbe RW (1990) Decline of the natterjack toad *Bufo calamita* in Britain: palaeoecological, documentary and experimental evidence for breeding site acidification. Biol Conserv 53(1):1–20

Binggeli P (1994) Historical impact of rabbits on Irish vegetation. In: Proceedings of the Irish Botanists', Meeting. University of Ulster, Coleraine, pp 5–6

Clarke D (1980) The groundwater balance of a coastal dune system: a study of the water table conditions in Ainsdale Sand Dunes National Nature Reserve. Unpublished Ph. D. thesis, University of Liverpool

Davy AJ, Grootjans AP, Hiscock K, Peterson J (2006) Development of eco-hydrological guidelines for dune habitats – phase 1. English Nature Research reports no 696, Peterborough. Source: http://publications.naturalengland.org.uk/publication/61018. Accessed 15 Aug 2012

Davy AJ, Hiscock KM, Jones MLM, Low R, Robins NS, Stratford C (2010) Protecting the plant communities and rare species of dune wetland systems: ecohydrological guidelines for wet dune habitats. Phase 2. Bristol, UK, Environment agency, p 113. Source: http://nora.nerc.ac.uk/9926/. Accessed 30 Dec 2011

Domburg P, Erisman JW, de Haan BJ, de Vries W, Kros J, Sanders K (2005) The Dutch nitrogen cascade in the European perspective. Energy research Centre of the Netherlands (ECN) Biomass, Coal and Environmental Research, p 109. Source: http://www.ecn.nl/docs/library/report/2005/c05007.pdf. Accessed 30 Dec 2011

Doody JP (1989) Management for nature conservation. In: Gimmingham CH, Ritchie W, Willetts BB, Willis AJ (eds) Coastal sand dunes, vol 96B, Proceedings of the Royal Society of Edinburgh. Royal Society of Edinburgh, Edinburgh, pp 247–265

Doody JP (2003) Coastal habitat restoration, towards good practice, Introductory report. Part of the LIFE funded 'Living with the Sea Project' Contract no. FST20-48-006 01/0384, Natural England. Source: http://webarchive.nationalarchives.gov.uk/20091011154212/http://www.eclife.naturalengland.org.uk/project_details/good_practice_guide/habitatcrr/ENRestore/AccompanyingReport.pdf. Accessed 31 Dec 2011. The good practice guide is available in the United Kingdom National Archive http://webarchive.nationalarchives.gov.uk/20091011154212/http://www.eclife.naturalengland.org.uk/project_details/default.asp

Duffey E (1968) An ecological analysis of the spider fauna of sand dunes. J Anim Ecol 37:641–674

Fenner F, Fantini B (1999) Biological control of vertebrate pests: the history of myxomatosis, an experiment in evolution. CABI Publishing/CAB International, Wallingford, 352 pp

Frame J (2000) Improved grassland management. Farming Press/Miller Freeman, Tonbridge, 338 pp

Garcia Nova F (1997) The ecosystems of Doñana National Park. In: Garcia Nova F, Crawford RMM, Diáz Barradas MC (eds) The ecology and conservation of European dunes. Universidat de Savilla, Sevilla, pp 87–116

Gilbert O (2004) Lichens, Naturally Scottish. Scottish Natural Heritage, Battleby, Perth

Grootjans AP, Geelen HWT, Jansen AJM, Lammerts EJ (2002) Restoration of coastal dune slacks in the Netherlands. Hydrobiologia 478(1–3):181–203

Hill MO, Wallace HL (1989) Vegetation and environment in afforested sand dunes at Newborough, Anglesey. Forestry 62(3):249–267

Hodgkin SE (1984) Scrub encroachment and its effects on soil fertility on Newborough Warren, Anglesey, Wales. Biol Conserv 29(2):99–119

Houston JA (2008) Management of Natura 2000 habitats. 2190 Humid dune slacks. European Commission. Source: http://ec.europa.eu/environment/nature/natura2000/management/habitats/pdf/2190_Humid_dune_slacks.pdf. Accessed 30 Dec 2011

Howe MA, Knight GT, Clee C (2010) The importance of coastal sand dunes for terrestrial invertebrates in Wales and the UK, with particular reference to aculeate Hymenoptera (bees, wasps & ants). J Coast Conserv 14(2):91–102

Hylgaard T, Liddle MJ (1981) The effect of human trampling on a sand dune ecosystem dominated by *Empetrum nigrum*. J Appl Ecol 18(2):559–569

Isermann M, Koehler H, Mühl M (2010) Interactive effects of rabbit grazing and environmental factors on plant species-richness on dunes of Norderney. J Coast Conserv 14(2):103–114

Jones MLM, Wallace HL, Norris D, Brittain SA, Haria S, Jones RE, Rhind PM, Reynolds BR, Emmett BA (2004) Changes in vegetation and soil characteristics in coastal sand dunes along a gradient of atmospheric nitrogen deposition. Plant Biol 6(5):598–605

Jones MLM, Sowerby A, Williams DL, Jones RE (2008) Factors controlling soil development in sand dunes: evidence from a coastal dune soil chronosequence. Plant Soil 307(1–2):219–234

Ketner-Oostra R (2006) Lichen-rich coastal and inland sand dunes (Corynephorion) in the Netherlands: vegetation dynamic and nature management. Ph. D. thesis, Wageningen University and Research Centre

Ketner-Oostra R, Sýkora KV (2004) Decline of lichen-diversity in calcium-poor coastal dune vegetation since the 1970s, related to grass and moss encroachment. Phytocoenologia 34:521–549

Kirby P (1992) Habitat management for invertebrates: a practical handbook. Joint Nature Conservation Committee, Peterborough, 150 pp

Kooijman AM, van der Meulen F (1996) Grazing as a control against 'grass-encroachment' in dry dune grasslands in the Netherlands. Landsc Urban Plan 34(3–4):323–333

Kutiel P, Peled Y, Geffen E (2000) The effect of removing shrub cover on annual plants and small mammals in a coastal sand dune ecosystem. Biol Conserv 94(2):235–242

Liddle MJ (1975) A selective review of the ecological effects of human trampling on natural ecosystems. Biol Conserv 7:17–36

Liddle MJ (1997) Recreation ecology: the ecological impact of outdoor recreation and ecotourism. Chapman and Hall, London, 641 pp

Liddle MJ, Grieg-Smith P (1975) A survey of tracks and paths in a sand dune ecosystem. II. Vegetation. J Appl Ecol 12(3):909–930

Maun MA (2009) The biology of coastal sand dunes. Oxford University Press, Oxford, 265 pp
McDonnell MJ (1981) Trampling effects on coastal dune vegetation in the Parker River National Wildlife Refuge, Massachusetts, USA. Biol Conserv 21(4):289–301
Muñoz-Reinoso JC (2001) Vegetation changes and groundwater abstraction in SW Doñana, Spain. J Hydrol 242(3–4):197–209
Olšauskas AM (1995) Influence of recreation on flora stability on the Lithuanian coastal dunes. In: Healy MG, Doody JP (eds) Direction in European coastal management. Samara Publishing, Cardigan, pp 103–105
Packham JR, Willis AJ (1997) Ecology of dunes, salt marsh and shingle. Chapman and Hall, London, p 335
Provoost S, Van Landuyt W (2001) The flora of Flemish coastal dunes (Belgium) in a changing landscape. In: Houston JA, Edmondson SE, Rooney PJ (eds) Coastal dune management, shared experience of European conservation practice. Liverpool University Press, Liverpool, pp 381–401
Provoost S, Jones LML, Edmondson SE (2011) Changes in landscape and vegetation of coastal dunes in northwest Europe: a review. J Coast Conserv 15(1):207–222
Pye K, Tsoar H (1990) Aeolian sand and sand dunes. Unwin Hyman, London, 396 pp
Ranwell DS (1972) Ecology of salt marshes and sand dunes. Chapman and Hall, London, 258 pp
Ranwell DS, Boar R (1986) Coast dune management guide. Institute of Terrestrial Ecology, HMSO, London, 105 pp
Remke E (2009) Impact of atmospheric nitrogen deposition on lichen-rich, coastal dune grasslands. Ph. D. thesis, Radboud University Nijmegen. Source: http://www.waddenacademie.knaw.nl/fileadmin/inhoud/pdf/06-wadweten/Proefschriften/Thesis_Eva_Remke_2009_EBOOK_version_01.pdf#page=99. Accessed 17 Nov 2011
Remke E, Brouwer E, Kooijman A, Blindow I, Esselink H, Roelofs JGM (2009) Even low to medium nitrogen deposition impacts vegetation of dry, coastal dunes around the Baltic Sea. Environ Pollut 157(3):792–800
Ritchie W, Mather AS (1984) The beaches of Scotland. Commissioned by the Countryside Commission for Scotland 1984. Reprinted 2005 by Scottish Natural Heritage as Commissioned report no 109
Stratford CJ, Ratcliffe J, Hughes A, Roberts J, Robins N (2007) Complex interaction between shallow groundwater and changing woodland, surface water, grazing and other influences in partly wooded duneland in Anglesey, Wales. In: Proceedings of the CDXXXV congress international association of hydrogeologists: groundwater and ecosystems. International Association of Hydrogeologists, pp 1–10. Source: http://www.thedotenvironment.com/DOCS/Complex%20Interactions%20at%20Newborough%20Warren%20Stratford%20et%20al.pdf. Accessed 25 Nov 10
Thomas AS (1960) Changes in vegetation since the advent of myxomatosis. J Ecol 48(2):287–306
Thomas AS (1963) Further changes in vegetation since the advent of myxomatosis. J Ecol 51(1):151–183
Thompson HV, Worden AN (1956) The rabbit. Collins New Naturalist, London, 240 pp
Van Dam PJEM (2001) Status loss due to ecological success. Landscape change and the spread of the rabbit. Innov:Eur J Soc Sci 14(2):157–170
van der Maarel E, Boot R, van Dorp R, Rijntjes J (1985) Vegetation succession on the dunes near Oostvoorne, The Netherlands; a comparison of the vegetation in 1959 and 1980. Plant Ecol 58(3):137–187
van Dijk HWJ, Grootjans AP (1993) Wet dune slacks: decline and new opportunities. Hydrobiologia 265:281–304
Van Dorp D, Boot R, van der Maarel E (1985) Vegetation succession on the dunes near Oostvoorne, The Netherlands, since 1934, interpreted from air photographs and vegetation maps. Vegetatio 58(3):123–136
van Tooren B, Bruin K (2004) Bryophytes of the sand dunes of the West Frisian Islands. Lindbergia 29(1):51–63
Veer MAC (1997) Nitrogen availability in relation to vegetation changes resulting from grass encroachment in Dutch dry dunes. J Coast Conserv 3(1):41–48

Veer MAC, Kooijman AM (1997) Effects of grass-encroachment on vegetation and soil in Dutch dry dune grasslands. Plant Soil 192:119–128

Wilkie ML (2002) From dune to forest: biological diversity in plantations established to control drifting sand. Unasylva, Special issue 209, Forest Biological Diversity, 53:64–69

Willis AJ (1963) Braunton Burrows: the effects on the vegetation of the addition of mineral nutrients to the dune soils. J Ecol 51:353–374

Willis AJ, Yem EW (1961) Braunton Burrows: mineral nutrient status of the dune soils. J Ecol 49:377–390

Zunzunegui M, Diaz Barradas MC, Garcia Novo F (1998) Vegetation fluctuation in Mediterranean dune ponds in relation to rainfall variation and water extraction. Appl Veg Sci 1(1):151–160

Chapter 8
Alien Plant Invasion

Abstract The introduction of non-native plants and animals by humans has occurred over the centuries both by accident and as a deliberate policy. Due to of the absence of natural control mechanisms many of these species become invasive. This can be at such a rate that the often biologically diverse indigenous habitats and species become overwhelmed. Coastal sand dunes are naturally dynamic and hence offer a range of opportunities for colonisation. This chapter provides a review of the main species invading sand dunes and their impact on the functioning of the habitat and its native fauna and flora. The chapter includes some positive aspects of species invasion (including reasons for introduction), as well as a general review of the effects and efficacy of control. It deals with individual species interactions rather than the more general question of scrub invasion dealt with in Chap. 7.

8.1 Introduction

Alien species,[1] which spread at the expense of native species or cause economic or other damage, can become invasive.[2] Elton in his classic book on the "Ecology of Invasions by Plants and Animals" (Elton 2000) describes seven 'invaders' as examples

[1] **Alien** – "A species, sub-species or lower taxon occurring outside of the historically known range it occupies naturally and outside its dispersal potential as a result of direct or indirect introduction or care by humans. It includes any part, gametes or propagule that might survive and subsequently reproduce. Synonyms are non-native, non-indigenous, foreign and exotic" (Scalera and Zaghi 2004, p 4).

[2] **Invasive** – "A species that is able to establish stable populations, colonizing irreversibly and spreading rapidly in entirely natural or semi-natural ecosystems. Biological invasions may also be a natural phenomenon, determining natural range expansions or contractions, without direct interventions by humans, although sometimes they may be fostered by possible human related environmental changes" (Scalera and Zaghi 2004, p 4).

J.P. Doody, *Sand Dune Conservation, Management and Restoration*,
Coastal Research Library 4, DOI 10.1007/978-94-007-4731-9_8,

of the way species brought from one country have 'exploded' in another. One of them is coastal, namely *Spartina* spp. (Doody 2008, Chap. 9). In some ways, this is illustrative of the importance that coastal areas play in providing a gateway for alien invasion. Their proximity to the sea provides a transportation route and the means for species to 'hitch a ride' on sea-going vessels. The dynamic nature of many coastal habitats, especially sand dunes, makes them an ideal repository for species that have newly arrived by providing opportunities for establishment. These include pioneer species such as *Ammophila arenaria* (outside Europe) shrubs including *Hippophaë rhamnoides* and trees (e.g. *Pinus* spp.) introduced to stabilise blowing sand.

It is possible to make a distinction between those species arriving because of human activity and those arriving 'naturally'. In Great Britain, for example, probably more than half of the vascular plants occurring on sand dunes are from introductions 'directly or indirectly' resulting from human activity or by birds. Those 'introduced' by humans include species associated with forestry operations or invasions from surrounding arable land (Ranwell 1972a, b). In the Gulf of Mexico only about 10 % of the dune flora consists of coastal species, the rest are either plants naturally colonising disturbed ground or species from other communities including adjacent agricultural land (Castillo and Moreno-Casasola 1996).

The number of aliens varies from continent to continent and from site to site. In Europe, a study "Delivering Alien Invasive Species Inventories for Europe (DAISIE)" provides a source of information on biological invasions. This has identified 11,000 alien mammals, fish, birds, plants, insects and other species (Daisie 2009). Of these there are 5,789 plant species classified as alien. This represents a threefold increase in species arriving in Europe over the last 25 years (Lambdon et al. 2008).

In the United States of America, of the 50,000 non-native alien plant and animal species, many are invasive. They cause a considerable amount of environmental and economic damage amounting to many billions of dollars a year. In one sand dune complex within the Humboldt Bay National Wildlife Refuge alone, there are no fewer than 23 invasive species considered a high priority for control. These occur across the range of habitats from pioneer dynamic communities, to grasslands, swales (dune slacks) and woodland. (Humboldt Bay National Wildlife Refuge Complex 2011 http://www.fws.gov/humboldtbay/invasives.html).

A few of these are widespread and have had significant adverse impact on sand dunes in other parts of the world. They form two broad types:

1. Those introduced intentionally by humans, which can be divided in coastal areas in two further main groups:

   a. Those introduced to help stabilise eroding sand dunes such as *Ammophila arenaria*, *Hippophaë rhamnoides* and *Acacia* spp.;
   b. Species introduced for amenity purposes such as landscaping and gardening including *Carpobrotus* spp. and *Rosa rugosa*.

2. Species accidentally introduced with transport, travel, goods etc. such as Narrow-leaved Ragwort *Senecio inaequidens*.

For a general review of plant invasion in relation to underlying processes and patterns of invasion from transport, colonization and establishment to landscape spread see Theoharides and Dukes (2007).

A key characteristic of many invasive species is that they are highly efficient at nitrogen fixation. They can enrich the soil to such an extent that they facilitate invasion by other alien species (Maron and Connors 1996).The discussion that follows includes some species that have had up until now, a significant adverse impact on sand dunes worldwide.

8.2 *Ammophila* spp.

Ammophila arenaria is one of two principal species of pioneer beach grasses, the other being *A. breviligulata* from North America. Both are exceptionally well adapted to burial by sand and grow rapidly to create the extensive foredunes described in previous chapters. Introduced as a sand binding agent because of this ability, the former species occurs in many parts of the world (Fig. 8.1).

Initially heralded as having a significant and positive effect in reversing dune erosion, it has become a major threat to native foreshore plants and animals in parts

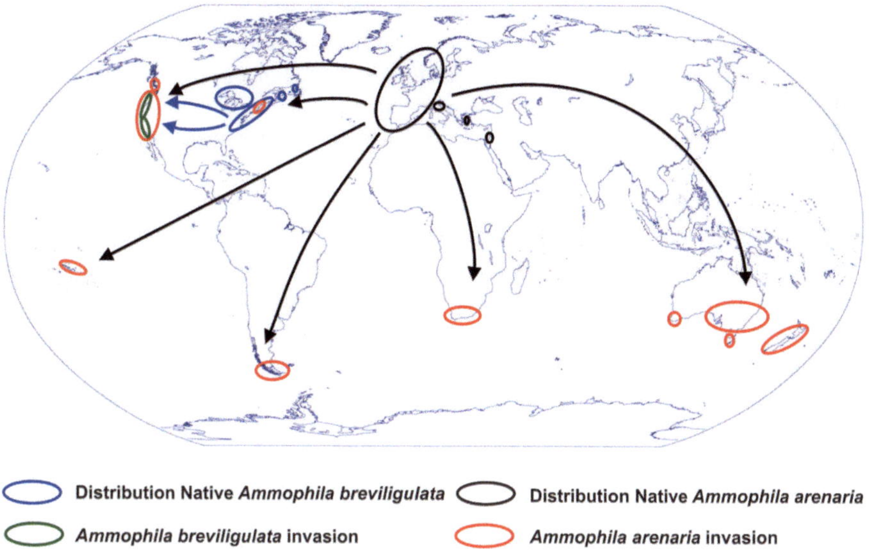

Fig. 8.1 *Ammophila arenaria* and *A. breviligulata* invasion

of the world outside its native range. To anyone familiar with sand dunes in Europe, where *Ammophila arenaria* is a natural and significant component of any developing sand dune system, it is strange to see it viewed as an alien, invasive species.

8.2.1 Invasion of North America

Introduced to stabilise eroding sand dunes in the 1800s, *Ammophila arenaria* spread along much of the west coast of North America, where it replaced the previously dominant American Dunegrass *Leymus mollis*. In California, for example, on the North Spit, Humboldt Bay between 1939 and 1989 its area cover increased by 574 %. This seems to be typical of its rapid rate of spread in this area, where it has replaced many of the native foreshore plant species (Buell et al. 1995). Amongst these, it also suppressed both the diversity (using the Shannon-Weaver information index) and density of sand-burrowing arthropods such as the Dune Beetle *Coelus ciliatus* on sand dunes in central California, even at relatively low densities. Sand stabilisation and reduced food availability at or near the surface as *A. arenaria* becomes established may be partly responsible for this effect (Slobodchikoff and Doyen 1977).

Following its introduction and colonisation in the late 1800s, *A. arenaria* expanded farther north into Oregon. High foredunes rapidly developed not only creating dunes where none had existed before, but also having an adverse impact on the native vegetation by outcompeting colonising strandline plants such as Pink Sand-verbena *Abronia umbellata* (Wiedemann and Pickart 1996, 2004). The introduction of *Ammophila breviligulata*, native on the eastern seaboard of the United States of America and around the Great Lakes, although on a smaller scale and much later (1935) than *A. arenaria* also helped create new foredunes. Unnoticed at first, this species has overtaken *A. arenaria* and replaced it in many areas as the primary coloniser.

This change in species appears to have had little impact on the already depleted species diversity but it has created sand dunes that are 50 % lower than *A. arenaria* (Seabloom and Wiedemann 1994). This process continues, and at the turn of the twenty first century, *A. breviligulata* had expanded slowly into Washington where it is the main dune building species across the whole area (Hacker et al. 2012). Its discovery (2001) in British Columbia, Canada on the west coast of Vancouver Island extends its influence further along the coast (Page 2001). Even in its native range on the east coast of United States of America, continual and deliberate planting can reduce the abundance and diversity of native annual herbs (Cheplick 2005).

In addition to its adverse impact on native plants and animals of the foreshore, the presence of a high foredune with *A. arenaria* encouraged building along the Oregon coast. It helped protect homes, roads and other infrastructure, as well as serving as a barrier against flooding. The much lower dunes created by *A. breviligulata* are less effective in this respect and may lead to an eventual decline in their coastal protection function (Hacker et al. 2012).

8.2.2 Invasion of Australia and New Zealand

In Australia and New Zealand, although not so widely reported, a similar situation occurs. Originally, the introduction of *Ammophila arenaria* to New Zealand was part of a stabilisation process designed to stop the threat from the 'evil' drifting sands (Sect. 2.3.1). The rate at which it traps sand and builds dunes appears to exceed the threshold of tolerance of local native species, in particular Pingao *Desmoschoenus spiralis* the main native foredune plant. In effect it becomes buried by the dune-forming processes associated with *A. arenaria* in a similar way to those on the west coast of the United States of America (Hilton et al. 2005). In Australia *A. arenaria* as well as the upper beach coloniser Russian Wheatgrass *Thinopyrum junceiforme*, a native of North America, had the same effect, creating large evenly vegetated sand dunes. In both New Zealand and Australia, the effect was the displacement of indigenous foredune vegetation (Hilton et al. 2006).

8.2.3 *Ammophila arenaria* in South Africa

Introduced into South Africa in the late 1800s, *Ammophila arenaria* has helped stabilise mobile sand dunes along the Cape coast. It has proved to be more effective at overcoming sand drift than native species, again because of its ability to trap sand. An appraisal of its status and potential threat to native foreshore species was undertaken (Lubke et al. 1995). This and subsequent work concluded that its ecological tolerance, vigorous rhizomatous reproduction and potential to set viable seed (it was thought not to propagate from seedlings in South Africa) might give cause for concern in the future, especially considering its invasiveness on sand dunes in North America, Australia and New Zealand. A further concern was its adverse impact on the aesthetics of the natural coastal dune environment as the much larger and higher dunes of *A. arenaria* replace the small hummock dunes formed by indigenous species (Hertling and Lubke 2000). However, despite this, further studies suggest that although it has a slightly negative impact it does not dominate over other, mainly native species, as it does on the North American Pacific coast. In addition, *A. arenaria* appears to provide temporary stability, allowing its replacement by indigenous plants (Lubke 2004).

8.2.4 Nematodes and Pathogens

Soil borne fungi and nematode assemblages associated with *Ammophila arenaria* in the Netherlands included 47 species of fungi and ten genera of plant-parasitic nematodes. Stands more than 10 years old had significantly different communities of soil organisms from recently established stands of 3 years old (de Rooij-van der Goes et al. 1995). Outside its natural range, *A. arenaria* becomes more abundant and survives longer in the dune. The alien species appeared to have a competitive advantage because of the absence of feeding by specialist nematodes (van der Putten et al. 2005). However, the suppression of plant growth may be due to a much more complex series of relationships between endoparasitic nematodes and other components in the soil community (Brinkman et al. 2005).

Thus, whilst some studies concluded that release from native soil pathogens might explain invasion of exotic plant species, at least one other does not. In California, a greenhouse experiment on the soil community found negative effects on seed germination, seedling survival and plant growth (Beckstead and Parker 2003). Despite this *A. arenaria* is highly aggressive indicating the pathogens are less effective. A mathematical model suggests that rather than being free from soil pathogens *A. arenaria* accumulates them resulting in exclusion of native plant species (Eppinga et al. 2006).

Dune nematodes *Pratylenchus dunensis* and *P. brzeskii* needed very high densities to produce damage in *A. arenaria*, whilst the more wide-ranging *P. penetrans* needed very few specimens. Thus, although all three nematodes have a negative effect on plant growth, they do so at very different densities (Peña et al. 2008). Fungi are also capable of reducing plant susceptibility to plant-parasitic nematodes and may be important in influencing the vigour of natural populations of *A. breviligulata* (Little and Maun 1996). Studies on the interaction between root-feeding nematodes, arbuscular mycorrhizal fungi and *Ammophila arenaria* suggest that the fungi are crucial for the control of root-feeding nematodes in natural systems (Peña et al. 2006).

Pathogenic nematodes may cause the extensive dieback of *Ammophila breviligulata* observed along the north and mid Atlantic coast of the United States from 1980 to 1995. Complete loss of sand dune vegetation can occur, as *A. breviligulata* disappears from the sward leaving bare sand for up to 5 years. The application of fertilisers and limestone help recolonisation (Seliskar and Huettel 1993; Seliskar 1995).

In South Africa where *A. arenaria* is not so invasive, the role of pathogens is difficult to unravel. Studies indicate that native species can compete effectively with *A. arenaria* and that this is partly because they host potential soil pathogens that may negatively affect the invader (Knevel et al. 2004). These complex and conflicting results and other studies suggest there is much more research needed in this area.

8.3 *Hippophaë rhamnoides* L. Sea Buckthorn

Hippophaë rhamnoides L. is a spiny shrub native to Europe and Asia. In north and west Europe, one of the sub-species is largely a coastal plant, with a native distribution on the coasts of southeastern England, France, the continental shores of the North Sea, Norway and around most of the Baltic Sea. In the rest of its range in the east including China and Russia, other sub-species occur in uplands above 1,200 m. There are nine geographically defined sub-species, some of which are very restricted in their distribution (Small et al. 2002). Although the coastal subspecies occurs

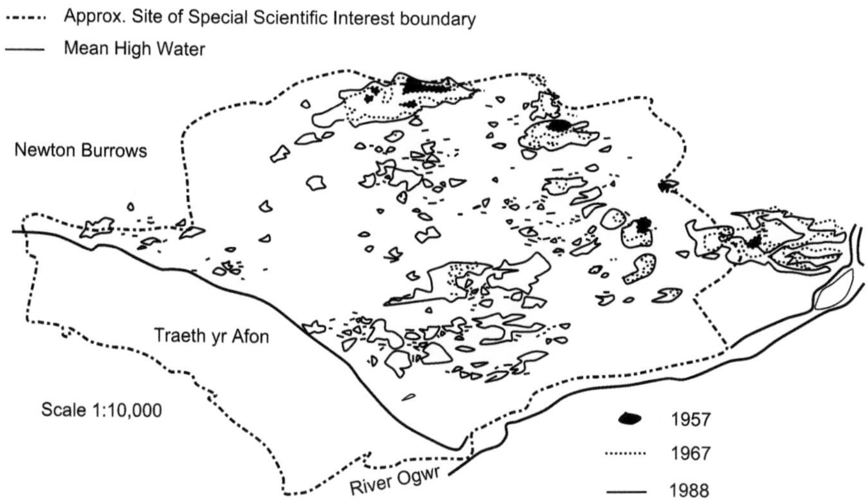

Fig. 8.2 Invasion by *Hippophaë rhamnoides* of Merthyr Mawr Warren 1957–1988, (Adapted from Dargie (1992))

elsewhere in the World as an introduced species, the following discussion relates mainly to its impact on sand dunes in United Kingdom and Ireland.

8.3.1 Invasion

Often introduced for stabilisation, amenity or as a barrier to prevent access *Hippophaë rhamnoides* has spread rapidly, but with varying impacts on sand dunes. The progress of invasion follows a similar pattern throughout the United Kingdom and Ireland. At Merthyr Mawr Warren, south Wales *H. rhamnoides* increased in stages from 2.4 ha in 1957 to 53.8 ha in 1991 (Dargie 1992) and 60.9 ha in 1996 (Richards and Burningham 2011). The pattern of invasion involved a few initial colonies, which supplied seed for further establishment (Fig. 8.2). As the new colonies increased in number (there were 276 in 1988) they coalesced to form mono-specific stands of vegetation. By 1991, the number of separate colonies had fallen to 82 because of this process (Dargie 1992).

On the Sefton coast northwest England, the Lords of the Manor introduced it from eastern England around 1900 to help stabilise blowing sand. The loss of rabbits in the 1950s left the species unchecked and it spread rapidly on many parts of the sand dunes. At Eskmeal Dunes, Cumbria, England *Hippophaë rhamnoides* planted in the 1950s, also spread rapidly when grazing ceased in the late 1970s. Ironically, planting took place partly because of disturbance to vegetation, which resulted in blowouts. The Ministry of Defence took over the site in 1980, when further dune stabilisation took place. Since then *H. rhamnoides* has continued to increase to

Fig. 8.3 Eskmeals Dunes *Hippophaë rhamnoides* in1985. The pattern of invasion was similar to Merthyr Mawr

Table 8.1 Sea buckthorn cover (% of total area) from 1949 to 1989 at Portstewart dunes Northern Ireland (After Binggeli et al. 1992)

Portstewart		
Year	Bare sand	*Hippophaë rhamnoides*
1949	31.2	1.6
1963	30.7	5.5
1975	11.0	9.8
1989	7.5	13.1

cover large areas of the site with a dense scrub canopy (Fig. 8.3). In 2000, the Cumbria Wildlife Trust began a gradual process of scrub removal and re-instating a degree of mobility in order to rehabilitate the dunes.

Introduction also took place at two sites in Northern Ireland for sand stabilisation, in the late nineteenth century on the Murlough Dunes and around 1900 at Portstewart. It spread slowly at first by means of vegetative propagation. However, following the outbreak of myxomatosis in the 1950s it again spread rapidly through seed germination and establishment, as well as by vegetative growth (Whatmough 1995). The invasion showed an increase in area with a corresponding decrease in bare sand (Table 8.1).

Table 8.2 Changes in percentage (estimated) vegetation cover and bare sand between 1942 and 1975, at Murlough spit, an area which has accreted since 1834 (After Binggeli 1992)

Murlough spit				
Year	Bare sand	Typical dune vegetation	*Hippophaë rhamnoides*	*Acer pseudoplatanus*
1942	0	74.5	24.0	1.5
1963	1.4	10.8	82.3	5.5
1975	7.4	0	67.1	25.5

Hippophaë rhamnoides also acts as a precursor to the development of other, native shrubs. For example, on the Portstewart sand dunes, *Sambucus nigra* regenerated below *H. rhamnoides*, reaching canopy height about 15 years after initial establishment. Subsequent incremental growth is small due to intolerance of growing shoots to exposure from wind and salt spray (Binggeli et al. 1992). In only 33 years, *H. rhamnoides* replaced typical *Ammophila arenaria* vegetation acting as a precursor for Sycamore *Acer pseudoplatanus* woodland (Table 8.2).

In the United Kingdom, this rapid spread results in a number of concerns for nature conservation by:

- Increasing the nutrient status of the soils, replacing botanically rich plant communities including uncommon or rare plant species, typically requiring nutrient poor conditions;
- Shading and ultimately replacing, sand dune vegetation (Isermann et al. 2007);
- Aiding the expansion of native woody plants such as *Clematis vitalba*, *Sambucus nigra* and *Acer pseudoplatanus*.

In mitigation it also:

- Helps stabilise mobile sand;
- Can help control visitors;
- Is a food source for wintering birds (especially thrush family);
- Increases abundance of soil organisms.

8.3.2 Natural Succession

Hippophaë rhamnoides appears as a natural component of the sand dunes in the Netherlands, Germany and Denmark, and in most cases it does not create extensive mono-specific stands characteristic of invaded sites elsewhere. This is probably because the lime content of the soil is lower than in regions where invasion takes place. Thus in acid sand dunes with a pH value around 5.5, the shrub appears to die much earlier due to the nematode predation on the root hairs. This makes it more difficult for the plant to take up phosphorous (Zoon et al. 1993). Large numbers of the nematode *Longidorus dunensis*, a species known to cause root deformations, is one of the species involved (Oremus and Otten 1981). Their numbers were higher in the degenerative stage of natural development of soil and vegetation (Zoon et al. 1993.).

8.4 Wattle *Acacia* spp.

Some species such as *Acacia* produce large quantities of seeds, which can also survive fire. They tolerate a wide range of soil types and have extensive root systems. Introduced as stabilising agents in the dune fields of South Africa in 1845, they include Orange Wattle or Golden Wreath Wattle *Acacia saligna* and Red-eyed Wattle or Western Coastal Wattle *A. cyclops* native to southwest Australia. In a relatively short space of time dense stands of impenetrable tall shrubs and short trees formed. Even though by the late 1940s it became apparent that this was at the expense of the indigenous vegetation, planting continued until 1974 (Avis 1989).

In Portugal *Acacia* spp., e.g. Golden Wattle *Acacia longifolia* was introduced to stabilise eroding sand dunes (Marchante et al. 2003). Since its introduction, it has spread in many areas, adversely affecting ecosystem processes and displacing native vegetation. Older established stands show a more pronounced impact with an increase in organic matter in the accumulated litter (Marchante et al. 2008).

Also planted in Turkey it was part of a dune stabilisation programme begun in 1955. By 1991, *Acacia* spp. had invaded sand dunes outside the main plantations completely replacing native dune vegetation (Fig. 8.4).

Fig. 8.4 Invasive *Acacia* spp. scrub colonising sand dunes along the southern coast of Turkey near Adana in 1991

In Israel, between the years 1965 and 1999, *Acacia saligna* increased in area by 166 % at an annual growth rate of 2.92 %. The invasion took place in three stages:

1. Colonisation, which lasted for 20 years;
2. Establishment during the following 8 years;
3. Expansion from 1990 onwards.

Sand quarries and depressions between dunes, characterised by relatively high soil moisture, were colonised by *Acacia saligna*, but not mobile dunes. In the invaded habitats, dense mono-specific stands crowd out all the native plants (Bar et al. 2004).

8.5 Other Invasive Species

This chapter has dealt with the most widespread examples of non-native invasive species. There are many others, which can have significant if local effects on native sand dune vegetation. Two of these, *Rosa rugosa* and *Carpobrotus* spp. originate from amenity planting and are amongst the top 100 most invasive species in Europe (DAISIE 2009) where they are widely distributed. Others have a more limited distribution, although where they do occur they can cause significant problems for coastal sand dunes.

8.5.1 *Japanese Rose Rosa rugosa*

Rosa rugosa native to China is a spiny low-growing plant, which mainly colonises old stabilised coastal sand dunes. Here it occurs with other woody species creating a scrub habitat, eventually developing into dune forest. Introduced from eastern Asia around 1796 the species is widely distributed on the coasts of the North and Baltic Seas, as well as the northeast Atlantic around the whole of the UK and in Ireland (Bruun 2005). Outside its native range, *R. rugosa* can create dense thickets of impenetrable monospecific stands of vegetation up to 2 m high. It is at its most prolific on sand dunes around the coasts of Europe, but also occurs in a wide range of other habitats (Weidema 2006; Isermann 2008a). In Denmark, for example, it has been a common plant used for landscaping e.g. along highways and in cities, which together with coastal sand dunes are favoured areas for colonisation and spread (Jørgensen and Kollmann 2008). It is tolerant to salt and thus not only invades coastal areas including sand dunes (Fig. 8.5) but also roadsides where salt helps prevent icing of road surfaces in winter.

In common with many other alien species, it also displaces the natural flora of beach and dune vegetation. It has broader environmental tolerances and invades both dune grassland and heath (Isermann 2008a). Stands of *Rosa rugosa* are species-poor, irrespective of the dune type in which the shrub is established, due to its dense canopy and shading effect (Isermann 2008b). It reproduces readily from seed especially on disturbed ground.

Fig. 8.5 Invasive *Rosa rugosa* on foredunes in Denmark, June 1999

In the USA, for example around Cape Cod, it does not have the same problem status as in Europe although it is a 'noxious weed' in some states. It may also be invasive in some regions (Dickerson and Miller 2002)

8.5.2 Hottentot Fig Carpobrotus spp.

The Hottentot Fig is the common name for a number of succulent species also called Iceplant. Two species native to South Africa which appear in the literature are, *Carpobrotus edulis* and *C. acinaciformis*. They frequently occur on sand dunes, but also invade maritime cliffs. Plants grow on a wide range of soil types and are resistant to drought, but sensitive to frost. Hence, in Europe plants occur largely in the southwest and around the Mediterranean.

In the Mediterranean, it is highly competitive and rapidly replaces native sand dune vegetation. It modifies soil properties by increasing N and organic C and reducing soil pH. In dune habitats, it also prevents sand movement, hindering the natural successional processes. Hybrids are very vigorous and may lead to intensified invasion.

A review of alien plants (including *Carpobrotus* spp.) on Mediterranean islands found one third of the sites sampled were on sand dunes (the rest were on rocky coast). It had become a threat to indigenous, endemic or rare coastal plants on all the

Fig. 8.6 *Carpobrotus edulis* carpeting sand dune 'lows' in southern Portugal, May 1992

islands surveyed (Suehs et al. 2001). It forms impenetrable mats and in some areas can carpet sand dunes, overwhelming the diverse native plant communities (Fig. 8.6). In the central western coast of Italy, invasion of early stages in dune succession was almost exclusively by *Carpobrotus aff. acinaciformis*. The species has a high cover value and displaces native species, especially in transitional communities (Carboni et al. 2010).

It is also invasive in the United States of America. Introduced in the 1900s, viable seed production and successful competition for water resources helped *Carpobrotus edulis* displace native shrubs in California (D'Antonio 1990; D'Antonio and Mahall 1991). This effect was, as in other parts of the world, enhanced by the build up of litter and lowering of soil pH. This also had an inhibiting effect on the germination, survival, growth and reproduction of rare native species such as the annual Manyleaf Gilia *Gilia millefoliata* (Conser and Connor 2009).

8.5.3 Yellow Bush Lupine Lupinus arboreus

Lupinus arboreus is native to southern and central California, USA. Outside its relatively restricted range in the United States of America, it becomes invasive, as it has in Northern California. Its repeated introduction on many dune systems as a sand stabiliser took place during the early to mid-1900s. The seeds are long-lived, germinate freely and have high survival rates. It matures in 1–2 years and lives for up to 7 years (Davidson and Barbour 1977). It was already present on the dune systems in Humboldt Bay in 1908, and by 1939 it covered 98 ha. By 1998, the area had increased to over 400 ha when it dominated 28 % of the total vegetation cover. At an ecosystem-level, its nitrogen fixing ability facilitates colonisation to form dense, mat-like vegetation. After a few years, further stabilisation and elevated nitrogen levels resulted in invasion by non-native weedy grasses and shrubs. This caused the loss of native plants and associated animals (Pickart et al. 1998; Pickart 2004).

In its native range, competition and herbivory are both important sources of mortality for *Lupinus arboreus* seedlings (Maron 1997). The locally abundant caterpillar of the Western Tussock Moth *Orgyia vetusta* feeds almost exclusively on *Lupinus arboreus* (in the north) and *L. chamissonis* (in the south). Whilst it can cause complete defoliation, the shrubs survive and recover to produce seeds in the following growing season (Harrison et al. 2005). Ghost Moth caterpillars *Hepialus californicus* feed directly on the roots and can kill individual plants even at quite low densities (Maron 1998). Entomopathogenic nematodes (commonly called roundworms) attack the caterpillars reducing their impact (Harrison et al. 2005). Also influenced by climatic factors, these interactions help to explain the rapid fluctuations in the populations of lupine bushes.

8.5.4 Rhododendron *Rhododendron ponticum* L.

Two sub-species of *Rhododendron ponticum* have disjunct distributions in Europe, along the shores of southwest Spain and southern Portugal (*R. ponticum* L. ssp. *ponticum*) and the Black Sea coast, Turkey, Lebanon, Bulgaria and the Caucasus (*R. ponticum* ssp. *baeticum*). Introduced to northwest Europe it has coastal populations in the British Isles, The Netherlands, Belgium and France. It is only reported as an invasive weed on one sand dune site in England, Winterton Dunes, Suffolk where it came to dominate some areas at the expense of the native dune heath (Fuller and Boorman 1977). It is much more invasive on inland heaths and in woodland where it is a major and continuing problem in the British Isles.

8.5.5 Locally Invasive Species

The invasive Heath Star Moss *Campylopus introflexus* is a plant of the southern hemisphere, first seen in England and western France in the 1940s. It is an invasive moss in Europe and North America, although of particular concern on dune heath in the Netherlands and southern Baltic Sea (Klinck 2010). It appears to be part of a process involving nitrogen enrichment, which facilitates invasion by the moss and the replacement of specialist *Cladonia* lichens and *Corynephorus canescens* (Ketner-Oostra and Sýkora 2004). The dense carpets that develop change the habitat

Fig. 8.7 *Clematis vitalba* invading and smothering *Hippophaë rhamnoides*, Saltfleetby Nature Reserve, Lincolnshire, England, September 2009

and can have a deleterious impact on some endangered beetles and spiders on acidic coastal dunes (Schirmel et al. 2011). Cape Ivy *Delairea odorata*, an invasive plant of coastal dune scrub in United States of America originally from South Africa, lowers the species richness of native plants as its cover increases (Alvarez and Cushman 2002).

Pirri-pirri Bur *Acaena novae-zelandiae* is found on the dunes at Lindisfarne, northeast England. It occupies a sizeable area and threatens to displace some of the native flora. It also poses a threat to dogs, as burs can become caught in their fur. Old Man's Beard *Clematis vitalba* appears to be invasive on some dunes. This is a native plant in Britain, but smothers scrub at sites such as Saltfleetby National Nature Reserve on the east Lincolnshire coast (Fig. 8.7) and Braunton Burrows, North Devon.

Prunus serotina began to invade pine woodlands adjacent to the Amsterdam Water Supply dunes in the 1980s. In the 1990s it spread into the dunes, and from 2004 to 2006 its area increased from 323 to 483 ha, a 25 % increase. By 2009, it had become mainly a problem in *Hippophaë rhamnoides* stands, where it could constitute 80 % of the habitat (reported in Houston and Edmondson 2009). It is also problematic in wooded areas and *Calluna vulgaris* heathland including sand dunes in west Jutland, Denmark (Starfinger 2010).

Garden escapes adjacent to sand dunes that become a repository for garden waste, are another source of exotic species (Edmondson 2009). These plants rarely reach proportions that threaten the survival of native species, although they can be visually dominant. The top five most commonly occurring plants along the housing boundary on the dunes on the Sefton coast northwest England are Snow-in-Summer *Cerastium tomentosum*, Russian Vine *Fallopia baldschuanica*, Montbretia Crocosmia Pottsii Crocosmia Aurea, Red Hot Poker *Kniphofia uvaria* and Spanish *Hyacinthoides hispanica* and Hybrid Bluebells *Hyacinthoides x massartian* (Edmondson 2010).

8.6 Control

The next chapter deals with approaches to control in more detail. This section covers some of the effective methods of control for specific invasive alien species.

8.6.1 Hippophaë rhamnoides Control and Management

Controlling *Hippophaë rhamnoides* has become a key management issue on many of the sites mentioned above, especially in the United Kingdom and Ireland. In the 1970s, the '*Hippophaë* Study Group' convened by The Nature Conservancy suggested three management options (Ranwell 1972a, b):

1. "Maintain zero *Hippophaë rhamnoides* population on any sites currently free from plants by uprooting seedlings as they are found;
2. Where establishment is at an early stage, attempt to eliminate *Hippophaë rhamnoides*;
3. Where *Hippophaë rhamnoides* is well established, control so as to maintain habitat diversity (but allowing a proportion of the stands to develop naturally to maturity)."

As it can easily regenerate from root fragments, its eradication is very difficult. Early attempts involved cutting mature stands or uprooting smaller plants by hand. These attempts proved ineffectual as the species regenerated quickly, especially in the absence of follow up treatment. Control was more effective when using general methods such as:

- Cutting of the scrub by hand (time consuming and costly);
- Using a variety of types of machinery;
- Treatment of the stumps with herbicides;
- Burning;
- Removal of litter.

Control has been relatively successful at Merthyr Mawr, South Wales with reduction from a maximum extent of 60.9 ha in 1996 to approximately 23.5 ha in 2006 (Richards and Burningham 2011). The removal of more mature stands was by an excavator fitted with a weed rake, followed up with herbicide foliar spraying on smaller plants (Rhind and Jones 2009). The precise approach adopted depended on the colony size. Priority action focused on small colonies from 0 to 0.39 ha with complete removal and regular follow-up treatment. Colonies from 0.4 to 0.69 ha (intermediate) remain at their existing size, whilst preventing coalescence. Prevented from spreading, stands larger than 0.7 ha were left to mature (Richards and Burningham 2011). Note these approaches mirror the management options suggested by the '*Hippophaë* Study Group' in the 1970s.

At other sites in the United Kingdom, such as Saltfleetby-Theddlethorpe Dunes and Gibraltar Point National Nature Reserves, Lincolnshire, England control is by a combination of mechanical and hand removal, and grazing. This involved the use of a 16-ton excavator on long and wide tracks fitted with a modified 6 ft scrub rake, which pulled the bushes out of the sand (Houston 2011). Extensive grazing with hardy sheep and cattle helps to secure long-term control (Rooney et al. 2011). Removal also formed part of a recovery programme for Natterjack Toad in England and Wales, which included the sand dunes of the Sefton coast northwest England (Denton et al. 1997).

8.6.2 Physical Control

A programme of eradication of *Carpobrotus edulis* by physical control on the Mediterranean island of Menorca was mostly successful, by hand-pulling individual plants and removing buried stems. Removal of plants from coastal grassland, back-dune and coastal scrub is relatively easy. There were difficulties due to the terrain and physical disposal of the plant remains. In addition, it was important to remove all plant remains as these can harbour seeds and become a focus of regeneration. In 2005, the plant only occurred in two areas on the island where local landowners have opposed its removal (Fraga et al. 2006). In Spain control was also largely successful, although in the long term repeated hand-pulling of regenerating plants will be needed (Andreu et al. 2010).

Rosa rugosa control should aim to reduce seed production and dispersal as well as reducing the incidence of human disturbance (Kollmann et al. 2007). Experiments with physical control by burial showed some limited success. Key requirements include:

- Uprooting should include the shrub and a 1 m zone around it;
- Plant fragments should be buried to a depth of at least 0.5 m;
- In eroding sites even deeper burial may be needed;
- Follow-up treatment with hand pulling and harrowing may also be required to prevent regeneration (Kollmann et al. 2011).

Physical removal of invasive plants can leave a void. For example, one consequence of removing *Delairea odorata* from invaded areas in California appears to be the proliferation of other non-native species. Thus, control may not always have the desired effect (Alvarez and Cushman 2002).

Fig. 8.8 *Lupinus arboreus* invasion Dawlish Warren, June 1981 showing the extent of invasion prior to the introduction of an aphid from North America

8.6.3 Biological Control

There have been some reported successes when using biological agents for controlling invasive species. The gall wasp, *Trichilogaster acaciaelongifoliae* and the seed-feeding weevil, *Melanterius ventralis* in South Africa (Dennill and Donnelly 1991), decreased *Acacia longifolia* reproductive potential by more than 90 % (Dennill et al. 1993). A rust fungus also seems to be highly effective against *A. saligna* in South Africa (Wood and Morris 2007). Introducing control agents from the native range of *Rosa rugosa* may also be effective, as most of the specialised insect and fungal enemies are absent from its introduced range (Bruun 2005).

Lupinus arboreus invaded Dawlish Warren, a sand dune nature reserve in south Devon, England (Fig. 8.8). Clearance began in 1978 at a rate of about 5 % per annum. In 1983, there was a major reduction in the population following the introduction of Lupin Aphid *Macrosiphum albifrons* from North America. The plant has since become re-established and remains a threat, along with a number of other invasive shrubs, which require management control to conserve the nature conservation value of the site (Redden 2004). In California, Western Tussock Moth *Orgyia vetusta* caterpillars defoliate seedlings of *Lupinus arboreus* leading to significant mortality (Maron 1997).

Biological control is a contentious issue. Introducing one species to control another can have unforeseen consequences. These examples suggest to some that it has potential to suppress alien species. Van Driesche et al. (2010) argue, "Classical biological control is a powerful tool for suppression of invasive plants and insects in natural ecosystems. It will play an increasingly important part in ecological restoration because it provides a means of permanently suppressing invaders over large landscapes without long-term resource commitments and hence is sustainable. As such, it merits use against many invasive plants and insects that are environmental pests in sensitive landscapes." This may not be such a responsible approach on sand dunes as their dynamic nature provides ideal opportunities for colonisation and invasion. Proper monitoring over a timescale suitable to identify responses of non-target native species will be essential, if past mistakes are not to be repeated.

8.7 Conclusion

The deliberate introduction of species from one place to another is commonplace. The spectre of eroding sand dunes has been a primary cause for concern. *Ammophila arenaria*, *Hippophaë rhamnoides* and *Acacia* spp. are often the main species used to combat erosion because of their ability to grow rapidly, even in soil with low levels of nutrients. Of the other alien invaders, those resulting from introductions via amenity planting *Rosa rugosa* and *Carpobrotus* spp. are the most significant. These species reach proportions where they threaten the survival of native species, especially in the beach/foredune.

A commonly held view is that invasiveness has had a significant impact on native species and this is borne out above. This also seems to be true on a more general worldwide basis, at least for Mediterranean-type ecosystems (Gaertner et al. 2009). In many instances, the decline in native species richness is attributable to alien invasions. Once established the species can be difficult and costly to remove or even control.

In a few instances due to their robustness, alien invaders may be beneficial. They provide some native rare species with shelter or food resources. *Ammophila* spp. is more likely to survive in a future with a rapidly changing climate because of the ability to withstand burial by sand. However, as this chapter has shown, deliberate introduction of alien species can have serious consequences for native plants and animals of sand dune ecosystems. In the light of the adverse consequences on the native fauna and flora of past introductions, any new ones must be carefully appraised.

References

Alvarez ME, Cushman JH (2002) Community level consequences of a plant invasion: effects on three habitats in coastal California. Ecol Appl 12:1434–1444

Andreu J, Manzano-Piedras E, Bartomeus I, Dana ED, Vilà M (2010) Vegetation response after removal of the invasive *Carpobrotus* hybrid complex in Andalucía, Spain. Ecol Restor 28(4):440–448

Avis AM (1989) A review of coastal dune stabilization in the Cape Province of South Africa. Landsc Urban Plan 18(1):55–68

Bar (Kutiel) P, Cohen O, Shoshany M (2004) Invasion rate of the alien species *Acacia saligna* within coastal sand dune habitats in Israel. Isr J Plant Sci 52(2):115–124

Beckstead J, Parker IM (2003) Invasiveness of *Ammophila arenaria*: release from soil-borne pathogens? Ecology 84(11):2824–2831

Binggeli P (1992) Patterns of invasion of sycamore (*Acer pseudoplatanus L.*) in relation to species and ecosystem attributes. D.Phil. thesis, The University of Ulster

Binggeli P, Eakin M, Macfadyen A, Power J, McConnell J (1992) Impact of the alien sea buckthorn (*Hippophaë rhamniodes* L.) on sand dunes ecosystmes in Ireland. In: Carter RWG, Curtis TGF, Sheehy-Skeffington MJ (eds) Coastal dunes: geomorphology, ecology and management, proceedings of the third European Dune Congress, Galway, Ireland, 17–21 June 1992. A.A. Balkema, Rotterdam, pp 325–337

Brinkman EP, Troelstra Sep R, van der Putten WH (2005) Soil feedback effects to the foredune grass *Ammophila arenaria* by endoparasitic root-feeding nematodes and whole soil communities. Soil Biol Biochem 37(11):2077–2087

Bruun HH (2005) *Rosa rugosa* Thunb. ex Murray. J Ecol 93(2):441–470

Buell AC, Pickart AJ, Stuart JD (1995) Introduction history and invasion patterns of *Ammophila arenaria* on the North Coast of California. Conserv Biol 9(6):1587–1593

Carboni M, Santoro R, Acosta ATR (2010) Are some communities of the coastal dune zonation more susceptible to alien plant invasion? J Plant Ecol 3(2):139–147

Castillo SA, Moreno-Casasola P (1996) Coastal sand dune vegetation: an extreme case of species invasion. J Coast Conserv 2(1):13–22

Cheplick GP (2005) Patterns in the distribution of American Beachgrass (*Ammophila breviligulata*) and the density and reproduction of annual plants on a coastal beach. Plant Ecol 180(1):57–67

Conser C, Connor EF (2009) Assessing the residual effects of *Carpobrotus edulis* invasion, implications for restoration. Biol Invasions 11(2):349–358

D'Antonio CM (1990) Seed production and dispersal in the non-native, invasive succulent *Carpobrotus edulis* (Aizoaceae) in coastal strand communities of Central California. J Appl Ecol 27(2):693–702

D'Antonio CM, Mahall BE (1991) Root profiles and competition between the invasive, exotic perennial, *Carpobrotus edulis*, and two native shrub species in California coastal scrub. Am J Bot 78(7):885–894

DAISIE (2009) Handbook of alien species in Europe. Invading nature, vol 3, Springer series in invasion ecology. Springer, Dordrecht, 400 pp

Dargie T (1992) Historical extent, potential threat and control of sea buckthorn *Hippophaë rhamnoides L.* at Merthyr Mawr Warren, South Wales. A report to the Countryside Council for Wales, p 29

Davidson ED, Barbour MG (1977) Germination, establishment and demography of coastal bush lupine (*Lupinus arboreus*) at Bodega Head, California. Ecology 58:592–600

de la Peña E, Echeverría SR, van der Putten WH, Freitas H, Moens M (2006) Mechanism of control of root-feeding nematodes by mycorrhizal fungi in the dune grass *Ammophila arenaria.* New Phytol 169(4):829–840

de la Peña E, Vandegehuchte M, Bonte D, Moens M (2008) Analysis of the specificity of three root-feeders towards grasses in coastal dunes. Plant Soil 310(1–2):113–120

de Rooij-van der Goes PCEM, van der Putten WH, van Dijk C (1995) Analysis of nematodes and soil-borne fungi from *Ammophila arenaria* (Marram grass) in Dutch coastal foredunes by multivariate techniques. Eur J Plant Pathol 101(2):149–162

Dennill GB, Donnelly D (1991) Biological control of *Acacia longifolia* and related weed species (Fabaceae) in South Africa. Agric Ecosyst Environ 37(1–3):115–135

Dennill GB, Donnelly D, Chown SL (1993) Expansion of host-plant range of a biocontrol agent *Trichilogaster acaciaelongifoliae* (Pteromalidae) released against the weed *Acacia longifolia* in South Africa. Agric Ecosyst Environ 43(1):1–10

Denton JS, Hitchings SP, Beebee TJC, Gent A (1997) A recovery program for the Natterjack Toad (*Bufo calamita*) in Britain. Conserv Biol 11(6):1329–1338

Dickerson J, Miller C (2002) Rugosa Rose *Rosa rugosa Thunb.* Plant Fact Sheet, United States Department of Agriculture Natural Resources Conservation Service, p 2. Source: http://plants.usda.gov/factsheet/pdf/fs_roru.pdf. Accessed 25 Mar 2012

Doody JP (2008) Saltmarsh conservation, management and restoration. Coastal Systems and continental margins, vol 12, Springer, Dordrecht, 217 pp

Edmondson S (2009) Focus on alien species. In: Houston J (ed) UK sand dune and shingle network, Seventh Newsletter, Liverpool Hope University. Source: http://www.hope.ac.uk/coast/newsletters/. Accessed 12 Dec 2011

Edmondson S (2010) Alien invaders. Coastlines, Newsletter of the Sefton Coast Partnership, Summer edition, p 16. Source: http://www.seftoncoast.org.uk/pdf/coastlinessummer10.pdf. Accessed 25 Nov 2010

Elton CS (2000) The ecology of invasions by animals and plant. With a New Foreward by Daniel Simberloff, University of Chicago Press, Chicago, p 196

Eppinga MB, Rietkerk M, Dekker SC, De Ruiter PC, Van der Putten WH (2006) Accumulation of local pathogens: a new hypothesis to explain exotic plant invasions. Oikos 114(1):168–176

Fraga P, Estaún I, Olives J, Da Cunha G, Alarcón A, Cots R, Juaneda J, Riudavets X (2006) Eradication of *Carpobrotus* (L.) N.E. Br. in Minorca. In: Brunel S (ed) Invasive plants in the Mediterranean type regions of the World, vol 59, Environmental encounters series. Council of Europe, Strasbourg, pp 289–298

Fuller RM, Boorman LA (1977) The spread and development of *Rhododendron ponticum* L. on dunes at Winterton, Norfolk, in comparison with the invasion by *Hippophaë Rhamnoides* L. at Saltfleetby, Lincolnshire. Biol Conserv 12:83–94

Gaertner M, Breeyen AD, Hui C, Richardson DM (2009) Impacts of alien plant invasions on species richness in Mediterranean-type ecosystems: a meta-analysis. Prog Phys Geogr 33(3):319–338

Hacker SD, Zarnetske P, Seabloom E, Ruggiero P, Mull J, Gerrity S, Jones C (2012) Subtle differences in two non-native congeneric beach grasses significantly affect their colonization, spread, and impact. Oikos 121(1):138–148

Harrison S, Hastings A, Strong DR (2005) Spatial and temporal dynamics of insect outbreaks in a complex multitrophic system: tussock moths, ghost moths and their natural enemies on bush lipines. Ann Zool Fenn 42:409–419

Hertling UM, Lubke RA (2000) Assessing the potential for biological invasion – the case of *Ammophila arenaria* in South Africa. S Afr J Sci 96(9–10):520–528

Hilton M, Duncan M, Jul A (2005) Processes of *Ammophila arenaria* (marram grass) invasion and indigenous species displacement, Stewart Island, New Zealand. J Coast Res 21(1):175–185

Hilton M, Harvey N, Hart A, James K, Arbuckle C (2006) The impact of exotic dune grass species on foredune development in Australia and New Zealand: a case study of *Ammophila arenaria* and *Thinopyrum junceiforme*. Aust Geogr 37(3):313–334

Houston J (2011) Management case studies. In: Rooney PJ, Houston JA, Weaver G (2011) The conservation and management of sea buckthorn (*Hippophaë rhamnoides*) in the UK: report of the workshop at Saltfleetby-Theddlethorpe dunes and Gibraltar point SAC, 17–18 September 2009. Sand dune and shingle network: Occasional Paper no. 3, Liverpool Hope University Press

Houston J, Edmondson S (2009) Conservation and management of coastal dunes in North Holland. Report of the study tour of The Netherlands 9–13 February 2009. Sand dune and shingle network, Occasional paper no. 2, Published by Liverpool Hope University Press. Source: http://www.hope.ac.uk/coast/newsletters/. Accessed 31 Dec 2011

Isermann M (2008a) Effects of *Rosa rugosa* invasion in different coastal dune vegetation types. In: Tokarska-Guzik B, Brock JH, Brundu G, Child L, Daehler CC, Pyšek P (eds) Plant invasions: human perception, ecological impacts and management. Backhuys Publishers, Leiden, pp 289–306

Isermann M (2008b) Classification and habitat characteristics of plant communities invaded by the non-native *Rosa rugosa* Thunb. in NW Europe. Phytocoenologia 38:133–150

Isermann I, Diekmann M, Heemann S (2007) Effects of the expansion by *Hippophaë rhamnoides* on plant species richness in coastal dunes. Appl Veg Sci 10:33–42

Jørgensen RH, Kollmann J (2008) Invasion of coastal dunes by the alien shrub *Rosa rugosa* is associated with roads, tracks and houses. Flora 204:289–297

Ketner-Oostra R, Sýkora KV (2004) Decline of lichen diversity in calcium-poor coastal dune vegetation since the 1970s, related to grass and moss encroachment. Phytocoenologia 34:521–549

Klinck J (2010) NOBANIS – invasive alien species fact sheet – *Campylopus introflexus*. Online Database of the European Network on Invasive Alien Species – NOBANIS. Source: www.nobanis.org. Accessed 14 Feb 2012

Knevel IC, Lans T, Menting FBJ, Hertling UM, van der Putten WH (2004) Release from native root herbivores and biotic resistance by soil pathogens in a new habitat both affect the alien *Ammophila arenaria* in South Africa. J Oecol 141(3):502–510

Kollmann J, Frederiksen L, Vestergaard P, Bruun HH (2007) Limiting factors for seedling emergence and establishment of the invasive non-native *Rosa rugosa* in a coastal dune system. Biol Invasions 9(1):31–42

Kollmann J, Brink-Jensen K, Frandsen SI, Hansen MK (2011) Uprooting and burial of invasive alien plants: a new tool in coastal restoration? Restor Ecol 19(3):371–378

Lambdon PW, Pyšek P, Basnou C et al (2008) Alien flora of Europe: species diversity, temporal trends, geographical patterns and research needs. Preslia 80:101–149

Little LR, Maun MA (1996) The '*Ammophila* problem' revisited: a role for mycorrhizal fungi. J Ecol 84(1):1–7

Lubke RA (2004) Vegetation dynamics and succession on sand dunes of the eastern coasts of Africa. In: Martínez M, Psuty NP (eds) Coastal dunes: ecology and conservation, vol 171, Ecological studies. Springer, Berlin, pp 67–84

Lubke RA, Heartling UM, Avis AM (1995) Is *Ammophila arenaria* (Marram grass) a threat to South African dune fields? J Coast Conserv 1(1):103–108

Marchante H, Marchante E, Freitas H (2003) Invasion of the Portuguese dune ecosystems by the exotic species *Acacia longifolia* (Andrews) Wild: effects at the community level. In: Child L, Brock JH, Brundu G, Prach K, Pyse k K, Wade PM, Williamson M(eds) Plant invasions: ecological threats and management solutions, 6th international conference on the ecology and management of alien plant invasions (EMAPi), 12–15 Sept 2001, Loughborough, UK, pp 75–85

Marchante E, Kjøller A, Struwe S, Freitas H (2008) Short- and long-term impacts of *Acacia longifolia* invasion on the belowground processes of a Mediterranean coastal dune ecosystem. Appl Soil Ecol 40(2):210–217

Maron JL (1997) Interspecific competition and insect herbivory reduce bush lupine (Lupinus arboreus) seedling survival. Oecologia 110(2):284–290

Maron JL (1998) Insect herbivory above and below ground: individual and joint effects on plant fitness. Ecology 79(4):1281–1293

Maron JL, Connors PG (1996) A native nitrogen-fixing shrub facilitates weed invasion. Oecologia 105(3):302–312

Oremus PAI, Otten H (1981) Factors affecting growth and nodulation *of Hippophaë rhamnoides* L. ssp. *Rhamnoides* in soils from two successional stages of dune formation. J Plant Soil 63(3):317–331

Page N (2001) *Ammophila breviligulata (Poaceae)* new to British Columbia. Botanical Electronic News, p 276, Source: http://www.ou.edu/cas/botany-micro/ben/ben276.html. Accessed 06 June 2011

Pickart AJ (2004) Introduced yellow bush lupin (*Lupinus arboreus*) in coastal dunes of Northern California. In: van Santen E, Hill GD (eds) Wild and cultivated lupins from the tropics to the poles, proceedings of the 10th international lupin conference, Laugarvatn, Iceland, 19–24 June 2002. Published International Lupin Association, Canterbury, pp 216–222

Pickart AJ, Miller LM, Duebendorfer TE (1998) Yellow bush lupine invasion in northern California coastal dunes: I. ecology and manual restoration techniques. Restor Ecol 6:59–68

Ranwell DS (1972a) Ecology of salt marshes and sand dunes. Chapman and Hall, London, 258 pp

Ranwell DS (ed) (1972b) The management of Sea Buckthorn *Hippophaë rhamnoides* on selected sites in Great Britain. Report of the *Hippophaë* study group, The Nature Conservancy, Norwich. Source: http://nora.nerc.ac.uk/7928/1/Sea_Buckthorn.pdf. Accessed 21 Dec 2011

Redden D (2004) Dawlish Warren Website. Source: http://www.dawlishwarren.co.uk. Accessed 30 Dec 2011

Rhind P, Jones R (2009) A framework for the management of sand dune systems in Wales. J Coast Conserv 13:15–23

Richards EG, Burningham H (2011) *Hippophaë rhamnoides* on a coastal dune system: a thorny issue? J Coast Conserv 15(1):73–85

Rooney PJ, Houston JA, Weaver G (2011) The conservation and management of Sea Buckthorn *Hippophaë rhamnoides* in the UK: report of the workshop at Saltfleetby-Theddlethorpe dunes and Gibraltar Point SAC, 17–18 September 2009. Sand dune and shingle network: Occasional Paper no. 3, Liverpool Hope University Press. Source: http://www.hope.ac.uk/coast/newsletters/. Accessed 15 Aug 2012

Schirmel J, Timler L, Buchholz S (2011) Impact of the invasive moss *Campylopus introflexus* on carabid beetles (Coleoptera: Carabidae) and spiders (Araneae) in acidic coastal dunes at the southern Baltic Sea. Biol Invasions 13(3):605–620

Seabloom EW, Wiedemann AM (1994) Distribution and effects of *Ammophila breviligulata* Fern. (American beachgrass) on the foredunes of the Washington coast. J Coast Res 10(1):178–188

Seliskar DM (1995) Coastal dune restoration: a strategy for alleviating dieout of *Ammophila breviligulata*. Restor Ecol 3(1):54–60

Seliskar DM, Huettel RN (1993) Nematode involvement in the dieout of *Ammophila breviligulata* (Poaceae) on the Mid-Atlantic coastal dunes of the United States. J Coast Res 9(1):97–103

Slobodchikoff CN, Doyen JT (1977) Effects of *Ammophila arenaria* on sand dune arthropod communities. Ecology 58(5):1171–1175

Small E, Catling PM, Li TSC (2002) Blossoming treasures of biodiversity: 5. Sea Buckthorn (*Hippophaë rhamnoides*) – an ancient crop with modern virtues. Biodiversity 3(2):25–27

Starfinger U (2010). NOBANIS – invasive alien species fact sheet – *Prunus serotina*. – From: Online Database of the North European and Baltic Network on Invasive Alien Species – NOBANIS. Source: www.nobanis.org. Accessed 31 Dec 2011

Suehs CM, Médail F, Affre L (2001) Ecological and genetic features of the invasion by the alien *Carpobrotus* plants in Mediterranean island habitats. In: Brundu G, Brock J, Camarda I, Child L, Wade M (eds) Plant invasions: species ecology and ecosystem management. Backhuys Publishers, Leiden, pp 145–158

Theoharides KA, Dukes JS (2007) Plant invasion across space and time: factors affecting nonindigenous species success during four stages of invasion. New Phytol 176:256–273

van der Putten WH, Yeates GW, Duyts H, Reis CS, Karssen G (2005) Invasive plants and their escape from root herbivory: a worldwide comparison of the root-feeding nematode communities of the dune grass *Ammophila arenaria* in natural and introduced ranges. J Biol Invasions 7(4):733–746

Van Driesche RG, Carruthers RI, Center T, Hoddle MS, Hough-Goldstein J, Morin L, Smith L, Wagner DL et al (2010) Classical biological control for the protection of natural ecosystems. Biol Control Suppl 1:S2–S33

Weidema I (2006) NOBANIS – invasive alien species fact sheet – *Rosa rugosa*. – From: Online Database of the North European and Baltic Network on Invasive Alien Species – NOBANIS. Source: www.nobanis.org. Accessed 31 Dec 2011

Whatmough JA (1995) Case study grazing on sand dunes: the re-introduction of the rabbit *Oryctolagus cuniculus* L. to Murlough NNR, Co. Down. In: Bullock DJ, Harvey HJ (eds) The National Trust and nature conservation 100 years on. Academic, London, pp 39–43

Wiedemann AM, Pickart A (1996) The *Ammophila* problem on the northwest coast of North America. Landsc Urban Plan 34:287–299

Wiedemann AM, Pickart AJ (2004) Temperate zone coastal dunes. In: Martinez M, Psuty N (eds) Coastal dunes: ecology and conservation, vol 171. Springer, Berlin, pp 53–65

Wood AR, Morris MJ (2007) Impact of the gall-forming rust fungus *Uromycladium tepperianum* on the invasive tree *Acacia saligna* in South Africa: 15 years of monitoring. Biol Control 41:68–77

Zoon FC, Troelstra SR, Maas PWT (1993) Ecology of the plant-feeding nematode fauna associated with sea buckthom (*Hippophaë rhamnoides* L. ssp. *rhamnoides*) in different stages of dune succession. Fundam Appl Nematol 16(3):247–258

Chapter 9
Management and Restoration – Applying Best Practice

Abstract This chapter brings together information on the states and values, and the analysis of trends and trade-offs for the beach/foredune (Chaps. 4 and 6) and inland vegetated sand dunes (Chaps. 5 and 7). Two simple State Evaluation Models (Sects. 6.2.4 and 7.2.6) provide a means of visualising the relationship between the states. Based on the current 'value' of each state it is possible to identify if the habitat at each location is in the 'desired state' or not. If intervention is required, this chapter gives details of some of the methods used in habitat management and restoration. These include taking action to control erosion, manipulate grazing regimes or restore and re-create sand dune systems.

9.1 Introduction

Understanding the successional relationships, and the way in which they change in response to human activities and natural forces, is the key to assessing the effect on the integrity and condition (value) of an individual site. The native vegetation and the descriptions in Chap. 2 provide a starting point for this assessment. Sand dunes exist in different forms at different times, depending on the dynamics of the system as a whole. It is important, therefore, to consider both the transitional and temporal changes when deciding whether to intervene or not.

The need is established, based on the descriptions of the states (Chaps. 4 and 5), the trends and trade-offs and mechanisms causing change (Chaps. 6 and 7) and the invasion of alien species (Chap. 8). The most appropriate method of management or restoration i.e. the established ways of moving from the 'existing state' to the 'desired state' include:

1. Restoring the beach/foredune to semi-stable (State 2; Sect. 4.2.2) including preventing erosion by arresting sand movement, promoting accretion by sand nourishment or curtailing those activities that reduce the sediment supply to the beach (Sect. 9.2);

J.P. Doody, *Sand Dune Conservation, Management and Restoration*,
Coastal Research Library 4, DOI 10.1007/978-94-007-4731-9_9,

2. Methods of managing established inland dunes (States 2 or 3), including restoring vegetation (Sect. 9.3), prevent/reverse vegetation deterioration through scrub control (Sect. 9.4) and grazing management (Sect. 9.5);
3. Restoring landscape characteristics e.g. by accommodating recreational use (Sect. 9.6) and deforestation (Sect. 9.7);
4. Creating new sand dunes (Sect. 9.8).

The decision as to whether to intervene or not, will be based on an assessment of the 'cost' (loss of interest or value) or 'benefit' (gain of interest or value) when taking a particular course of action. In the context of the ecosystem services values the use of the terms 'value' and 'cost' do not imply an monetary valuation. Attempts to put an economic value on the natural environment include the recent UK National Ecosystem Assessment (UK NEA). This provides an analysis of the benefits of the natural environment to society and continuing economic prosperity. Chapter 11 covers coastal margins, including sand dunes (Jones 2011).

9.2 Preventing Erosion, Promoting Accretion

A common approach to protecting an eroding foredune (State 1) or one that is shifting inland (State 2) in the Physical State Evaluation Model, is to prevent or reduce sand movement. Encouraging the deposition of sand grains and with it, the establishment of vegetation is the key. Whilst this will occur naturally (Sect. 1.3), overcoming sediment depletion and excessive human activity may require intervention. In some cases, such as beach cleaning, all that is required is the cessation of the activity. Many of the techniques that involve active intervention have been in use for many years, (Sect. 2.3.1, Fig. 2.3). These include erection of fencing to help trap sand, together with planting dune-forming species such as *Ammophila arenaria*. The techniques described next are those normally associated with the beach/foredune. Some are also suitable to help combat erosion of inland dunes (Sect. 9.3).

9.2.1 Marram (Ammophila spp.) Planting

It is not surprising that early attempts to control sand movement not only included laws to prevent the removal of *Ammophila arenaria*, but also to encourage planting. The ability of this species to withstand desiccation and burial by sand is a key to its effectiveness. The techniques for planting are well known. They usually involve uprooting from dune areas nearby where *A. arenaria* is growing vigorously, especially on the lee slopes of back-dunes. The plants should have at least 150 mm of healthy root with two or three nodes on the rhizome, such that it breaks away from the roots buried deep in the sand. *A. arenaria* planting should not take place within 2–3 m of the high tide line, as it is intolerant of exposure to salinity levels above 1%. There should also be an adequate supply of sand. Generally, although a distance of 450 mm between plants is usual, planting the rhizomes closer together can be more effective in areas with a high rate of erosion (Brooks and Agate 1986). *Leymus arenarius* is an important agent, similarly used in stabilising the coastal volcanic sand of Iceland (Greipsson and El-Mayas 1996).

Ammophila breviligulata 'die-back', possibly due to attack by a pathogenic nematode impeded the use of the native on the north and mid-Atlantic coast in restoring dune vegetation in North America. Rehabilitation of moribund stands included application of macronutrient fertilisers (N-P-K) and treatment with dolomitic limestone, which increased the survival of newly introduced plants (Seliskar 1995).

9.2.2 *Ammophila spp. – Too Much of a Good Thing?*

Whilst *Ammophila arenaria* planting is a well tried and tested method of dune restoration, it is important to recognise that it can have severe drawbacks. Introduced on the west coast of America, in New Zealand, Australia and South Africa (Sect. 8.2) it has become invasive with adverse consequences, especially for the native flora of the beach/foredune.

In Oregon and Washington State on the west coast of the United States of America, *Ammophila arenaria* and *A. breviligulata* (both alien there) have been successful in stabilising the eroding foredunes. However, this has been at the expense of the naturally occurring and globally endangered native foredune grassland (see also Sect. 8.2.1). The densely vegetated ridges formed by the *Ammophila* spp. also reduced the available nesting sites of the threatened Western Snowy Plover *Charadrius alexandrinus nivosus*. Twenty years of management to remove the invading species of *Ammophila* had the desired effect of decreasing its area and increasing the population of the plover. However, at the same time it had the undesired effect of reducing the richness and abundance of native foreshore plants (Zarnetske et al. 2010). The repetitive use of machinery to remove the vegetation prevented the establishment of the natural beach profile. Lower intensity regimes would facilitate the establishment of a more natural dynamic morphology as well as providing habitat suitable for the natural and endemic vegetation (Zarnetske et al. 2010).

In California, United States of America restoring foredunes with native species following *Ammophila arenaria* removal proved to be relatively easy. A review of approaches suggests that increasingly, to be cost-effective, larger scale projects should be undertaken (Pickart and Barbour 2007). The use of targeted herbicide treatment is also effective, as is hand removal, which although costly, continues in some places and obviates the need for chemicals (Pickart 2008).

Also in the United States of America, although not affecting coastal populations, *A. breviligulata* transplanted from the Great Lake State of Michigan to Minnesota appears to be a distinctive genetic form of the species. It grows more vigorously, and although flowering much later than the native genotype, it may out-compete plants of the threatened Minnesota population through greater vegetative and sexual reproduction (Holmstrom et al. 2009).

In New Zealand, removal of *Ammophila arenaria* by herbicide treatment (Gallant, grass-specific) has helped create a wider beach and to re-establish open native foredune vegetation. In turn, this has provided nesting sites for the Chatham Island Oystercatcher *Haematopus chathamensis*, which is able to nest higher up the shore making them less susceptible to flooding by high tides and storms (Moore and Davis 2004). However, here *A. arenaria* dunes also provide shelter for a wide variety of dune animals. There is an abundance and diversity of beetle, spider and ant families. The estimated population size of Common Skink *Oligosoma nigriplantare polychroma* and mouse population densities are higher in *A. arenaria* dunes than native vegetation. There is a positive correlation between vegetation cover and species diversity. Consequently, removal of *A. arenaria*, or its mass replacement by native species in the short term, may have an adverse impact on some native fauna (Jamieson 2010).

It is clear that *Ammophila* spp. can have a significant detrimental impact on native strandline and foredune plants and animals (Sect. 8.2). In many instances, control or elimination is appropriate. However, this brief review suggests there are some benefits associated with planting these species. Their control must therefore, take into account the interaction with native species, as well as any unforeseen consequences of removing alien species.

9.2.3 Native Species

The value of native species for erosion control followed the recognition of the adverse impact of *Ammophila arenaria* and to a lesser extent *A. breviligulata* on dune structure, function and species complement. In Florida, there are a number of native plants with suitable characteristics for erosion control, making them favourable for restoring foredune vegetation (Williams 2007). None of them shows the dominating tendencies of non-native species.

In New Zealand, the wholesale destruction of native dune-building species, coupled with a misunderstanding of their value, led to the introduction of a range of alien species to combat erosion (Chap. 8). However, there is increasing recognition of the value of native species in terms of functionally and aesthetic superiority in dune stabilisation. A better understanding of propagation techniques, (Bergin and Kimberley 1999) provides a means of restoring ‘natural’ foredune communities. There have been similar conclusions reached in South Africa, where a review of 17 dune restoration sites, including extensive transverse dune fields, concluded that indigenous species were capable of facilitating dune restoration (Avis 1995).

The Dunes Restoration Trust of New Zealand provides a forum for promoting sand dune rehabilitation using native species. By 2050, it hopes to have the majority of New Zealand dunes restored and sustainably managed using indigenous species.

Fig. 9.1 Grid chestnut paling fencing on a coast apparently with a positive sediment budget, Algarve near Faro 2005

9.2.4 Fencing

Fencing often accompanies planting with non-native and native species when attempting to prevent or reverse trends in sand dune erosion at the beach/dune interface. This includes erecting some form of barrier to impede airflow and increase sand deposition and/or discourage human access. The effectiveness of any physical barrier depends on the location, orientation (in relation to wind direction and strength), and measures to control people. Each form of fencing may have different effects on dune development, for example:

1. Synthetic textile mesh, helps create a relatively wide shallow dune;
2. Brushwood, creates a more irregular shape and steeper profile due to the variation in density of the material;
3. Chestnut paling fencing provides conditions for a steeper dune (Brooks and Agate 1986).

Whatever the methods, the geomorphological context for dune development is important when deciding the appropriate orientation, frequency and location of fences (Ranwell and Boar 1986). Fencing configuration varies depending on the location and direction of the prevailing wind. This can range from straight shore parallel fencing to zigzags or various combinations of boxed layouts as shown for example in Fig. 9.1. Note the use of fencing with metal wire or plastic mesh can

Fig. 9.2 Information, signage and 'barriers' designed to protect eroding sand dunes along an eroding beach near Les Salinas on the south side of the island of Ibiza, Balearic Islands, Spain, May 2008

create problems in the medium to long term. As the fences degrade, they leave behind material that if not permanently covered by sand will be potentially hazardous to introduced domesticated stock or people.

Other approaches are less invasive including symbolic fences to allow aeolian transport while preventing trampling. These allow a more natural dynamic dune form to develop (Nordstrom et al. 2000; Fig. 9.2). This approach may also be beneficial for nesting sea turtles. The area of the beach just above high water provides the location for females to lay their eggs. Fencing along the upper beach may obstruct access to this part of the beach or even obscure it altogether. However, sand trapping fences, constructed to allow spaces between, can successfully restore the dune without impeding access for turtles.

In the medium to long term, one of the least effective and environmentally damaging methods of erosion control include concrete revetments. These may exacerbate or even initiate erosional trends. Scouring around the base of the structures can cause them to sink, as happened with the wartime defences on the French coast (Fig. 6.1). On beaches with limited sediment supply, structures may be undermined and collapse, whilst beach and dune erosion continues (Fig. 9.3).

Where the eroding beach threatens housing or other real estate, this may include additional concrete structures (Fig. 6.5) and gabions. Even from an engineering standpoint, in these situations sand fences may be more appropriate (Phillips and Willetts 1978). However, all forms of protection that successfully encourage

Fig. 9.3 'Protecting' a links golf course and clubhouse on the North Norfolk coast, England, August 1990

accretion near these structures can result in erosion elsewhere, because none of them increase the overall sediment supply (Nordstrom 2004).

9.2.5 Mulching, Thatching and the Use of Fertilisers

Sand dune soils are naturally deficient in nutrients and dry. To aid the process of sand accretion and vegetation establishment, mulching and thatching may take place to help retain moisture and prevent sand blow. This may or may not have accompanying fencing or hessian matting, although planting with *Ammophila* spp. and other sand-binding plants usually takes place. There are a variety of mulches including chopped straw, peat, sewage sludge and seaweed. The use of some of these materials is inappropriate, as they can introduce alien species and additional nutrients into the restored system.

Erosion control using forestry brashings to thatch exposed sand, with conifers preferred because of their flat fan shapes, can also be successful. Material should cover 20–30% of bare sand and is most effective if regularly maintained, excludes human disturbance and includes planting between the brashings (Scottish Natural Heritage 2000). On the east coast of United States of America, a water adsorbing polymer gel placed in the hole prior to planting aids the establishment of Seaoats *Uniola paniculata*, one of the main species colonising dunes on the Florida coast (Williams 2007).

In Iceland, stabilising the volcanic sands with *Leymus arenarius* required the addition of nitrogen fertiliser to be effective (Greipsson and Davy 1997). However, for sand dunes with a diverse plant community at Holkham National Nature Reserve, Norfolk, England fertilisers had an adverse impact on species richness (Boorman and Fuller 1982).

Hydraulic seeding using mulches and fertilisers can be effective in helping to establish seedlings of dune grasses in areas with intensive recreational use. Typically, in the UK a clover and grass seed mixture contains species such as *Trifolium repens*, *Festuca rubra*, *F. ovina* and *Agrostis tenuis*. Rapid colonisation is possible with the application of water-based slurry with added fertiliser and seed. Mulches also help bury the seed and chemical stabilisers reduce seed loss by erosion. At Camber Sands, England the method employed also involved disc-harrowing chopped straw into the surface sand, spraying seed and fertiliser in a water slurry and topping off with a second layer of straw covered with bitumen (Pizzey 1975). These methods have varying degrees of success and are probably most applicable to areas with little nature conservation interest.

9.2.6 Beach Cleaning

Beach cleaning (Sect. 6.3.4) occurs in those locations with high levels of recreational use, especially near urbanisations. In addition to its adverse impact on the foreshore flora and fauna, it can exacerbate erosion. The only requirement in most cases is to cease the activity and allow dune vegetation to re-establish. Left 'ungroomed', and in the absence of other pressures such as car parking and recreational activity, a cleaned beach can gradually recover coastal strand and foredune vegetation. For example, a section of beach in San Buenaventura State Beach, California developed coastal strand vegetation and hummocks extending 20–40 m seaward of the vegetated foredune, less than 4 years after all cleaning had stopped (Dugan and Hubbard 2010). They suggest alongshore stretches or "islands" of beach that are left ungroomed year round could be beneficial for foredune stability, helping to preserve some of the native flora and fauna.

9.2.7 Beach Nourishment

Beach nourishment is a generic term used to describe the placement of sediment on depleted beaches as a means of counteracting erosion. It is usually associated with those beaches that fulfil an important sea defence function, although creating amenity beaches may also be the aim (Fig. 9.4). Whilst all the above approaches can be successful, at least in the short term, to be sustainable sand dunes need sediment derived from sources outside the system. Beach nourishment (beach feeding) is the practical means of providing this.

Engineering models often help determine the most appropriate form of nourishment. However, experience from the USA suggests that practical approaches to nourishment may be equally, if not more successful. In this context, there are three principal types of scheme:

Fig. 9.4 Beach nourishment using volcanic sand to create an amenity beach in Tenerife, 1995. Material is pumped onshore via a pipeline from an offshore vessel

1. Imitate nature – observe beach behaviour over time and assume post-replenishment behaviour will be similar;
2. Kamikaze beach – replenish the beach and see what happens. Careful monitoring will help in the design of any succeeding replenishments;
3. Learn from the past – research what has happened in previous replenishment projects (of similar, nearby schemes) and assume similar behaviour of the new beach (Pilkey et al. 1994).

Beach nourishment is one of the key methods of sea defence in the Netherlands. It forms part of a dynamic approach to coastal defence policy adopted by the Dutch, which seeks to "protect the entire coastline in its 1990 position". The application of 'dynamic preservation' emphasises the "wish for the preservation of the natural dynamics and character of the dune coast". In the Port of Amsterdam, the 'Sand Engine' project (Sect. 2.6.2) has a number of environmental safeguards and includes the creation of a 35 ha sand dune (Sect. 6.3.1).

In the UK, a detailed review of past beach management emulates design approach three (learn from the past), and provides some necessary background information (CIRIA 1996). Updated in 2010, it includes research and summaries of a large number of beach management/recharge schemes carried out since 1996 (CIRIA 2010). In New South Wales, Australia, the NSW Coastline Management Manual suggests the placement of material similar in size or preferably slightly coarser than the natural beach. It also recommends taking care when dredging sand from offshore, in order to ensure that altering the existing wave refraction patterns does not adversely influence the adjacent coastline (New South Wales Government 1990).

Although beach nourishment may appear an environmentally friendly way of combating erosion, this is not always the case. There is evidence to suggest that the approach has significant adverse consequences for the beach and foredunes. Amongst the issues of concern are burial of nearshore shallow reefs and degradation of other beach habitats, depressing nesting in sea turtles and reducing the densities of prey for shorebirds, fishes and crabs (Peterson and Bishop 2005). Other values potentially affected by beach nourishment include nesting sites for rare and vulnerable shorebirds such as *Charadrius melodus* in North America. Their displacement and that of the specialist invertebrates is, however, only temporary and they will return as the system readjusts to the additional sediment placed on the beach. This of course depends on the material having the same grain size and texture as the original beach.

Nourishment sometimes takes place on the inland dune, as well as on the beach. At Talacre north Wales, sea defence issues and rehabilitation of sand dunes for nature conservation combined to provide justification for beach nourishment, as well as feeding sand into the inland dune profile. This is one of several sites forming a pilot study of beach nourishment in Wales (McCue et al. 2010).

9.3 Restoring Vegetation on Inland Eroding Dunes

This section deals with those situations where the requirement is to restore vegetation on inland eroding sand dunes. In the context of the vegetated State Evaluation Model (Fig. 7.2), this means reversing erosion trends resulting from excessive grazing, or other destabilising activities such as rabbit burrowing, trampling or vehicle use. Moving from this situation involves restoring dune grassland (Vegetated States 3) or dune heath (Vegetated State 4) respectively. This section considers those areas of bare sand, sufficiently large to justify remedial action to re-establish vegetation. Some of the sand stabilisation techniques described above will be applicable. However, removal of the agent causing destabilisation may be the only requirement, allowing the dune to heal naturally. The next sections describe examples of additional management action to speed up the process.

9.3.1 Dune Grassland and Heath

Restoring blowouts and other areas of bare sand to dune grassland takes time. The techniques of sand stabilisation described above are equally applicable within the body of the dune. In the past *Ammophila* planting (Sect. 9.2.1), fencing (Sect. 9.2.3), mulching etc. (Sect. 9.2.4) have all been used. In Brittany, western France restoration of three small dunes involved fencing and *Ammophila arenaria* planting similar to that described above. A review of the effectiveness of these actions showed that after 10 years, although the geomophological structure of stable dunes had formed, restoration of species-rich dune grassland had not yet taken place (Rozé and Lemauviel 2004).

On inland dunes, techniques including planting seeds or disk-harrowing rhizome fragments can be effective in speeding up colonisation (van der Laan et al. 1997). Indeed, where living rhizome fragments of grasses including *A. arenaria* are already present and in the absence of grazing (by rabbits) simply fencing an area can result in the re-establishment of healthy vegetation on bare sand (van der Putten and Peters 1995). In this context, it is important to understand the timescales for restoration. What is a long time? Ten years is probably not long enough. However, as we will see in Chap. 12 (Fig. 12.4) scrub can come to dominate bare sand in only 40 years.

9.3.2 Dune Scrub and Native Woodland

Dune scrub is a natural stage in dune succession. However, it often has a negative influence on open dune vegetation and its specialist and rare species in northern Europe, in other parts of the world it is an integral part of the dune habitat (Sect. 1.3.5). Since it regenerates with little help, there is only limited need for active intervention by way of management to promote its colonisation. Removing grazing animals, if present, may be all that is required. Figs. 7.1 and 7.6 show how rapidly scrub can develop in their absence. The natural progression to woodland will follow.

A question arises for both scrub and woodland as to the species composition and the extent to which this form of restoration results in natural and semi-natural habitat. In Europe, the Manual of European Habitats (European Commission 2007) lists two dune woodland types:

1. "Wooded dunes of the Atlantic, Continental and Boreal region". These woodlands include those with long established natural or semi natural deciduous (North Sea and Baltic) or mixed woodland with *Pinus sylvestris* (Baltic) on coastal dunes. Pine forests without a near-natural understory (e.g. Fig. 5.5) are excluded. The Natura 2000 Viewer lists 20 such sites with a minimum percentage cover of 50%: Sweden (6), Estonia (5), Latvia (5), Poland (1), Germany (2) and France (1);
2. "Wooded dunes with *Pinus pinea* and/or *Pinus pinaster*". Dunes with a variety of Maritime pines and *Juniper* spp. including long established pine plantations, with an undergrowth similar to that, which would occur under natural conditions. The Natura 2000 Viewer lists 14 such sites with a minimum percentage cover of 50%: Italy (10), Spain (1) and France (3) (European Environment Agency http://natura2000.eea.europa.eu/#).

The extent of active restoration of native scrub and woodland on bare sand is unknown. There are no obvious examples from the United Kingdom or on the European continent. By contrast, sand dunes planted with non-native species are commonplace. The usual route to restoration is by modifying trends in succession

(Sect. 9.4) and/or the removal of conifer or plantations of other alien species (Sect. 9.7) below. Given the rarity of the native coastal dune woodland and the importance attached to them in Europe, there is a strong argument for restoration in some areas. Whether the substratum is bare sand or vegetated, allowing the natural progression from mobile dune to stable grassland, heathland, scrub and ultimately native woodland may be desirable.

9.4 Modifying Trends in Succession

Changes in a grazing regime (Chap. 7) can cause adverse effects including loss of species diversity. Degradation also occurs because of factors such as changes in hydrology (Sect. 7.7.1) and increased nitrogen deposition (Sect. 7.7.2). These changes can ultimately lead to scrub invasion and woodland development at the expense of biologically richer forms of vegetation.

Where sand stabilisation and scrub development have overwhelmed open species-rich dune turf remedial action is often required. The approach to rehabilitation depends largely on whether the development has been process oriented i.e. through succession or by afforestation. For the former many of the values attributable to moderately grazed dune grassland and heath (State 2 and State 3 respectively; Sect. 7.2.2) survive at least until a full scrub canopy develops (State 4, Sect. 7.2.3). This may take several years allowing opportunities to intervene to reverse the process by scrub clearance together with any accumulated plant material. By contrast, planting non-native trees can have a major impact in a relatively short space of time, with an almost complete loss of dune plants and animals as the forest matures (State 5, Sect. 7.2.4). This section deals with the modification of successional trends including restoring dune slack vegetation, dune grassland and heathland and the removal or control of scrub.

9.4.1 Dune Slacks

Dune slacks naturally accumulate organic matter, lifting the vegetation above the level of periodic flooding that characterises the habitat (Sect. 7.4.5). Restoring this vegetation and its associated animals is achievable by scrub control, introducing or reintroducing grazing (Sect. 9.5). However, these approaches may not be enough when the ground water level is below the optimum for many of the rarer species associated with this habitat.

Sod cutting from dunes, including dune slacks took place on a large scale to cover the fortifications of the German Atlantic Wall during the 1940s. This had the effect of lowering the ground level, setting-back the vegetation to an earlier stage in succession. It appears this may be why many dune slack species still survive along the dune coast of Northwest Europe (Grootjans et al. 2004).

Taking a cue from this, studies in the Netherlands show that restoring dune slacks by top soil removal is achievable. Here projects involved the excavation of up to 30 ha to a depth of 30 cm. The design of the operation was at a scale thought to be large enough to be sustainable and long lasting (Terlouw and Slings 2005). However, large-scale excavations do not necessarily lead to successful restoration. Many projects have been carried out in situations that had depleted seed banks or where hydrological conditions and seed dispersal were less than optimal (Grootjans et al. 2002). Although the soil seed bank contributes to the re-establishment of species in some places, much more important is the presence of local populations of dune slack species. Natural succession and dispersal from the extant populations provide the necessary agents for restoration (Bakker et al. 2005). Elsewhere excavating scrapes had a beneficial effect on the flora of the Sefton Coast dunes, Merseyside, England (Smith 2006).

In Belgium, in a small Flemish Nature Reserve D' Heye (20 ha) humid dune slacks developed following the excavation of an area of fertilised sand dune. Within the excavated areas a number of typical dune slack species, such as Three-nerved Sedge *Carex trinervis* and Glaucous Sedge *Carex flacca* soon appeared (Herrier et al. 2005). The restoration of dune slacks in north France also included sod cutting (Lemoine 2005). Test plots showed good recovery, and as a result a programme of mechanical excavation followed, removing at least 10 cm of the top, A-soil horizon. Creating large pools may be particularly significant for the rare Natterjack Toad (Corbett and Beebee 1975).

9.4.2 Dune Grassland and Heath

The development of coarse grassland usually occurs before shrubs overwhelm dune grassland or heath. However, attempts to intervene at this relatively early stage are limited. Approaches include mowing, with removal of excess nutrients in the cuttings and turf stripping. Trials of a new restoration technique called topsoil inversion or deep ploughing took place in Wales. Here a double-bladed plough buried existing overgrown vegetation to a depth of 80 cm, which was then covered with 40–50 cm of mineral sand. At this depth, the pH and organic matter of the surface layer became comparable to those of mobile dunes. Subsequent development depended on the degree of exposure (to wind) and human disturbance. Although only partially successful, it could provide another management tool for restoring dune vegetation (Jones et al. 2010).

Calcium poor sand dunes support lichen-rich dune heath, a prized and rare habitat in Europe (Sect. 1.3.3). The vegetation is particularly prone to enrichment by nitrogen deposition (Sect. 7.7.2). Studies in the Netherlands concluded, "Burning or removal of the topsoil is not sufficient to regain biodiversity as long as the current N emission still greatly exceeds the critical deposition values for these nutrient-poor ecosystems. The crucial factor to assure the restoration of the former lichen-rich dune communities seems to be in-blowing sand." (Ketner-Oostra and Sýkora 2007, p. 2). Attempts to promote the development of coastal dune heath are limited. Note this is different to the restoration of overgrown heath resulting from scrub encroachment (Sect. 9.7.3). There have been experiments using herbicides on inland non-coastal lowland heath to eradicate Bracken *Pteridium aquilinum*. Methods for restoring degenerate *Calluna vulgaris* on these inland sites also include:

- Eliminating coarse grassland species (mowing alone, and in combination with burning and rotovating);
- Exposing buried seed by rotovating and turf-stripping;
- Reducing bracken *Pteridium aquilinum* density (Marrs 1987a, b; Lowday and Marrs 1992; Marrs and Lowday 1992).

9.4.3 Dune Scrub

Scrub is one of the most frequently encountered problems for nature conservation management on coastal sand dunes in northwest Europe. It has caused the loss of open species-rich calcareous sand dune grassland and dune heath on many sites where grazing by domesticated stock has become uneconomical or is in conflict with other uses (Sect. 7.4.3). Many of the species causing concern are alien plants introduced to combat erosion that have now become invasive (Chap. 8). Whatever their origins, removal of both the scrub and any accumulated litter is a prerequisite for nature conservation management.

Physical scrub removal is the first stage in control. A number of techniques are applicable, ranging from large-scale mechanical uprooting to cutting by hand. Whatever method is applied it is essential to remove as much of the rootstock as possible, since many of the species sucker freely from underground rhizomes, and to treat cut stumps and any regrowth with herbicides. Some specific recommendations for individual species derived from the literature include:

1. Sea Buckthorn *Hippophaë rhamnoides* suckers prolifically, re-generating from coppice stools forming dense, impenetrable stands. Clear large stands of mature scrub using excavators or bulldozers. These can also remove the accumulated nutrient-rich litter layer. Follow up with herbicide treatment to cut stumps and any regrowth (see also Sect. 8.6.1);
2. Blackthorn *Prunus vulgaris* and its close relatives sucker prolifically to form dense impenetrable stands especially after coppicing. Cut stumps to 10–20 cm from the ground, drill holes in the stump surface and fill holes with ammonium sulphamate (brand name Amcide or Root-Out) or glyphosate. This should be carried out in late autumn/winter;
3. European Gorse *Ulex europaeus* germinates readily from seed on bare or disturbed soils, especially after fire. It grows vigorously when cut at an early age and sprouts from surface roots. Cut and paint the stumps with glyphosate;
4. Rhododendron *Rhododendron ponticum* spreads by windblown seed, layering, root suckers, and grows vigorously from cut stumps. Cut to leave 300 mm stumps (cutting

Fig. 9.5 Removing dune scrub and woodland from a former tidal beach plain of the medieval Yser estuary between Nieuwpoort and Oostduinkerke to restore dune slacks in Belgium, picture taken in 2005 (Herrier et al. 2005)

back to ground level is not recommended because of vigorous re-growth). Use a 12–13 mm deep hole for every square inch of the stump's surface. Fill the holes with a solution of ammonium sulphamate (brand name Amcide or Root-Out). Alternatively, cut to leave a 300 mm stump during the winter months and then spray the re-growth with glyphosate (plus a sticking agent) in May or June.

Increasingly there are specialist firms with machinery able to remove scrub, including root systems (Fig. 9.5). Volunteers are also a primary source of labour, suitable for work such as physical removal by hand of seedlings and young plants. It is labour intensive and can be slow, but it prevents unwanted scrub from establishing and is more cost effective and less disruptive in the long term. It is important to be careful and not to leave root fragments in the soil.

Although largely written from a United Kingdom perspective, the Scrub Management Handbook (Bacon 2003) provides a comprehensive description of the value of all types of scrub, methods of management and control.

9.5 Grazing Management

In northwest Europe, grazing by domestic animals is often the preferred management strategy on sites of high nature conservation value, either as a continuation/modification of existing practice, or re-introduction where it has ceased. It is often a key to the successful conservation of dune grassland, heath and dune slacks. This section considers the more detailed requirements associated with grazing management as a tool for maintaining and restoring valuable habitat.

Setting appropriate grazing regimes is the key issue in maintaining beneficial conditions for nature conservation. At too high a density the vegetation becomes overgrazed (State 1) or in their absence or at too low a density (State 4) it becomes rank and the sand dune dominated by scrub (Sect. 7.2). Most of the sites requiring management or restoration suffer from a lack or cessation of grazing (Sect. 7.2.3), and/or because of the introduction of alien shrubs as stabilising agents (Chap. 8).

In areas where scrub has already invaded and created a dense extensive canopy, its removal is required (Sect. 9.4.3). In other areas where grazing intensity is too low to combat scrub invasion, but has not reached the stage requiring physical removal, an increase in grazing levels may be beneficial for wildlife. Species richness (i.e. total number of species) increases and species typically associated with sand dunes become more abundant. This change occurs as grazing removes biomass, opens the sward and inhibits the growth of aggressive species. At the same time, it allows more light to reach the surface layer. Smaller plant species with low competitive ability, such as annuals, biennials and bryophytes benefit. Animals aid seed dispersal, help cause soil disturbance and generally increase diversity in drier sand dune vegetation (Plassmann et al. 2009).

9.5.1 Reducing Grazing Pressure

Overgrazing, usually through intensive use by domesticated stock, can cause loss of species and structural diversity. In addition to stock levels, grazing intensity depends on the size of the site and the mosaic of habitats. Small sites can show a relatively uniform effect, as for example at Northam Burrows, Devon (Fig. 5.2). Animals may graze larger sites differentially, with heavier use in the more palatable grassland areas compared to those with coarse grass, heath or scrub (Fig. 7.3). The precise impact also depends on the type of stock and the location of watering and supplementary feeding areas (Fig. 7.4).

In theory, redressing this situation is relatively easy. All that is required is to reduce the number of stock or periods of grazing. In practice, it is much more difficult. Societal (some grazing rights may extend back over centuries) and economic concerns may militate against bringing about changes for nature conservation reasons, even on sites designated under national or international laws or conventions. Dealing with these issues is not the purpose of this book. However, identifying appropriate grazing regimes for sites and habitats is. This involves determining what to graze (type and breed of animal), when to graze (season) and for how long (period). The following sections provide guidance on these issues.

Fig. 9.6 Highland cattle graze sand dune heath on the island of Texel, the Netherlands, September 1995

9.5.2 Type of Domesticated Stock

In north and west Europe, choosing the right type of domesticated stock for conservation grazing management is an important consideration. Nature conservation managers will go to great lengths to secure the most appropriate breed for their particular requirements. The nature of the terrain, exposure, type of vegetation (e.g. grassland versus scrub) and human activity (e.g. walking with dogs) all have a bearing on the animal best suited to the situation. The following provides information on some of the characteristics that make certain breeds suitable for a particular type of site:

Cattle – Conservation grazing on sand dunes requires breeds that are hardy and small to medium in size and weight. These are well suited for conservation grazing and are less likely to cause damage to sites susceptible to disturbance, such as sand dunes. Those that fulfil these criteria in the United Kingdom include upland beef cattle such as Highland, Galloway, Welsh Black, Beef Shorthorn and Vaynol (Welsh). Species known to have been successful in the United Kingdom and Netherlands on sand dunes are:

1. The **Dexter**, an ancient native Irish cattle breed, is especially suitable, it is hardy and adaptable, produces good quality meat and milk;
2. The **Highland** is also extremely hardy with an extensive ranging behaviour and lends itself to grazing and browsing coarse vegetation on large sites (Fig. 9.6);

3. The **Welsh Black**, is an extremely hardy breed, which can browse scrub so is particularly useful in situations where sand dunes have become overgrown.

Lowland beef cattle including Hereford and Aberdeen Angus, (continental examples include Limousin and Charolais), being only moderately hardy and larger are less suitable for grazing on sensitive swards or wet ground. These modern breeds are reluctant to graze even some of the coarser grass species such as Wood Small-reed *Calamagrostis epigejos*, a species which can invade dune slacks.

Ponies – Ponies and horses are generally selective grazers. There is a variety of breeds native to the British Isles including Highland, Exmoor, Dartmoor, Dales, Fell, Shetland, New Forest, Welsh Mountain and primitive European hardy breeds such as Przewalski, Konik, Fjords, Icelandic and Camargue. All are relatively hardy and resistant to insect bites, unlike domesticated breeds such as Arabs and Thoroughbreds, which have a thin skin.

Donkeys – The Donkey can eat a wide variety of foods but is less hardy than other animals. In coastal dunes of Flanders, Belgium, they did eat a variety of grass species, although browsing woody species was relatively limited (Hoffmann et al. 2001). It is not suitable in many exposed situations as it is relatively intolerant of wet conditions and requires shelter. Donkeys are not therefore, particularly suited to sand dunes.

Sheep – Sheep are especially suited to sand dunes if present at a relatively low density. As with cattle, some of the older breeds are the most useful. Amongst these, the following provide adequate grazing for nature conservation purposes:

1. **Herdwick** are very hardy, placid i.e. not prone to be frightened by dogs, and medium sized. A special feature is their waterproof coat. They can eat coarse vegetation, including some scrub. They also have a strong hefting[1] instinct and are renowned for longevity;
2. **Hebridean** are a minority breed. Hardy sheep, they were once widespread. They have become very popular for use in nature conservation grazing schemes, because of their ability to thrive on poor vegetation;
3. **Soay** are very hardy, but prone to dog worrying. Another ancient breed, they have retained primitive characteristics enabling them to cope in harsh conditions and utilising poor quality forage. They are ideally suited to semi-feral situations on large, ring-fenced sites, where moving stock only involves opening and closing gates between compartments.
4. **Swaledales** are less able to cope with scrub vegetation preferring shorter grasses that are more palatable.

The information is summarised from the Grazing Advice Partnership (GAP/FACT 2009 http://www.grazinganimalsproject.org.uk/), which includes a Grazing Case Studies and UK Grazing Map.

[1] Hefting occurs when a flock of hill sheep recognise and stay within 'their' part of the hillside.

Fig. 9.7 Grazing experiment with Soay sheep at Newborough Warren, National Nature Reserve, Anglesey, Wales, 1980

9.5.3 Grazing and Mowing Experiments

In the United Kingdom, trials began at Newborough Warren in the 1980s to establish the effectiveness of sheep in combating scrub invasion (Fig. 9.7). Accompanied by mowing trials, both were effective but grazing was most beneficial to vascular plants (Hewett 1985). Mowing experiments in Jersey showed that it could reduce competition for rare and sensitive species, such as annuals, and help create swards suitable for invasion by rabbits (Anderson and Romeril 1992).

Management at Braunton Burrows, North Devon also included experiments to assess the influence of grazing. Grazing by Soay sheep and mowing help to prevent loss of plant diversity in vegetation, which had become rank. Grazing also tends to create niches for regeneration, restoring species-rich short turf (Packham and Willis 2001). However, from a practical point of view, mowing as a means of simulating grazing proved to be a very labour intensive process (personal communication, John Breeds, Site Manager). This left grazing as the only practical means of maintaining the conservation value over the whole of this large (approximately 1,350 ha) sand dune site.

Between 1997 and 2003, a further grazing trial on 27 ha of sand dune, with a mix of cattle and sheep took place at Braunton Burrows. Stocking rates varied between years from 0.12 to 0.49 LU/ha/day. Detailed vegetation monitoring followed in July each year. Grazing had a significant beneficial effect on the overall diversity of the sand dune plant community, as well as helping to control shrubs such as *Ligusticum*

Table 9.1 Type of stock and stock numbers used, provide an indication of the wide variety of grazing regimes available to the dune manager (van Dijk 1992)

Site	Animal	Area grazed (ha)	Stock numbers
Terschelling	Moorland sheep	22	22
Zwanenwater	Cattle (& 2 horses)	160	28
Texel	Sheep	800	500–600
Voorne	Icelandic ponies	130	12
	Limousin cattle	30	No information
Middelduinen	Sheep	116	300–500

vulgare. The grazing animals had limited success in controlling dense scrub and brambles. Therefore, a combination of extensive mixed grazing and mechanical scrub control was the most appropriate management, helping to maintain and restore species-rich vegetation (FitzGibbon et al. 2005).

Hebridean sheep, introduced on an experimental basis for nature conservation grazing to Lindisfarne National Nature Reserve, Northumberland in 1993, grazed a small area of dune from September to December. They were moderately successful at controlling rank vegetation, grasses and scrub. However, problems of dog worrying led to their replacement by cattle in 2005. Cattle graze the area in autumn and early spring and include 38 in calf cows of mixed breeds, Hereford and Limousin crosses, in calf to a Limousin bull (GAP/FACT 2009, http://www.grazinganimalsproject.org.uk/).

Domestic stock grazing in the Netherlands had all but died out in the early 1900s. As a result, on many dunes coarse grasses, scrub and woodland came to dominate the vegetation. This led to a number of trials for their reintroduction, in order to combat the loss of dune species (Table 9.1). In Belgium Konik horses and donkeys grazed coarse grasses but failed to provide enough scrub control to be completely effective for nature conservation purposes (Cosyns et al. 2001).

9.5.4 *Grazing Intensity*

A review of the literature suggests that grazing levels to meet nature conservation objectives on calcareous dune grasslands lie within the range 1.5–3.0 sheep (0.25–0.45 LSU[2]) per ha year round. There are no easily accessible recommendations for cattle grazing but 0.2–0.6 (0.2–0.6 LSU) per ha year round represent a 'best guess' (Doody 2001). Much lower levels are appropriate on less productive dune heath. Ranwell and Boar (1986) suggest dune turf can support similar annual grazing intensities of 0.5 cattle and 4.0 sheep per ha. However, they go on to make the point that sheep have a 25–40% greater impact than cattle because of their small feet relative to the weight of the animal and their rate of 'treading' is also much higher (Sect. 7.4.1).

[2] Livestock Unit. Defined in Sect. 7.4.1.

Fig. 9.8 Gibraltar Point, cattle grazing on the Lincolnshire Wildlife Trust, National Nature Reserve, September 2009. In the United Kingdom, this is an increasingly common sight as conservation organisations seek sustainable management on their nature reserves

9.5.5 Conservation Grazing in Practice

There are a significant number of documented examples of grazing for nature conservation purposes on sand dunes in Great Britain. Active information exchange of practical approaches to management occurs under the auspices of the Sand Dune and Shingle Network run from Liverpool Hope University. Newsletters provide summary information for several sites. For example, at Kenfig Dunes National Nature Reserve, Wales 50 cows grazed the site in 2010. The removal of standard beef cows and calves left Highland cattle to graze on the reserve throughout the winter. These hardier animals only required supplementary food for a few particularly cold days. Site management at Tensmuir National Nature Reserve, Scotland includes grazing with Limousin cattle to help reduce tree, scrub growth and open the thicker grassy sward (Houston and Durkin 2010).

The Dexter grazes dune vegetation on several sites in the United Kingdom, Murlough Dunes National Nature Reserve, Northern Ireland, Dawlish Warren, Devon, Saltfleetby and Gibraltar Point, Lincolnshire (Fig. 9.8). Grazing cattle have an added advantage that they can also create openings in the dune vegetation. The bare sand then provides habitat for specialist invertebrates and may help establish suitable conditions for the rare Natterjack toad.

Sandscale Haws, National Nature Reserve, Cumbria use cattle, mainly modern breeds of Friesian and Hereford in combination with sheep. About 300 sheep graze an area of 282 ha in winter, with about 150 in summer. They maintain their condition well and require little or no supplementary feeding. They have catholic tastes, although at this level of stocking numbers are too low to suppress scrub (Burton 2001). Herdwicks graze the extensive dune system of the Ainsdale Sand Dunes and Cabin Hill National Nature Reserves, Merseyside in winter. Here they help provide control of several woody species, including *Salix arenaria*, *Hippophaë rhamnoides*, *Betula pubescens* and *Rubus* spp.

9.5.6 Caveats to the Use of Grazing Animals on Sand Dunes

Despite grazing being the preferred management strategy on many sand dune sites, there are a number of important caveats. These especially apply when grazing relies on modern animal breeds:

1. The requirement for supplementary feeding can result in the stock remaining close to the feeding locations, causing localised damage from trampling and eutrophication, as well as under-grazing coarser vegetation elsewhere on the site (Ross Links, Fig. 7.4);
2. Reduced value of fleece because of entrapped sand;
3. Fluctuating rabbit populations make determining stocking levels difficult (Burton 2001).

There are no hard and fast rules for determining the most appropriate grazing regime. Individual breeds may be more suited to the conditions on one site than on another. In some locations, availability of animals is an overriding consideration and compromises may be required in relation to the breed or even period of grazing. It is generally better to graze than not to graze at all, at least in the sites in northwest Europe. However, some consider the restoration of sand dynamics is one of the most important management tools in Europe (Hoffmann 2008; Chap. 12).

9.5.7 Mowing

Mowing can provide an alternative to grazing by livestock. It can simulate grazing, aiding the removal of course grasses and scrub from overgrown dune grassland. It also obviates the need for expensive fencing, often required with the introduction of grazing animals for stock management. When applied to relatively small areas it can be effective, as for example in the Voorne dunes in Holland where regular mowing of a dune slack helped retain a species-rich community (Boorman 1977; Fig. 9.9). It is, however, very labour intensive and is not as effective in maintaining short species-rich swards typical of calcareous dunes for four main reasons:

1. It is non-selective and all species are ‘grazed’ to the same height at the same time;
2. Invertebrates may be adversely affected as the ‘patchiness’ of grazed swards disappears;
3. It can result in a topographically uniform surface;

Fig. 9.9 Mowing dune slack vegetation Voorne dunes, the Netherlands, August 1987

4. Unless removed the cut material remains in situ, providing a potential source of enrichment.

9.5.8 Managing the Rabbit Oryctolagus cuniculus

Rabbit grazing may be an essential component of sand dune management, if for no other reason than they have played such an important part if the development of sand dune vegetation in Europe. Certainly, in their absence (Sect. 7.3.2) many of the more valuable aspects of the flora and fauna of temperate sand dunes have been lost in the last 50 years or so. Two situations are considered:

1. Areas where rabbit grazing has become so intense that destabilisation threatens the survival of vegetated dune surface;
2. Areas where following their disappearance due to myxomatosis in the 1950s, the sand dunes became clothed in coarse grassland and scrub.

The first requires control, whilst the latter may involve re-introduction. Methods of control are not the subject of this book but there are numerous references in the literature; see for example the summary by Wray (2006). There appears to be only one reported case where there have been deliberate attempts to reintroduce rabbits to a sand dune system for nature conservation management (Whatmough 1995).

Despite the success of this introduction programme, scrub encroachment, notably by *Hippophaë rhamnoides* continues to be a significant management problem (Sect. 8.3). Thus, although moderate rabbit population levels help sustain the diversity of sand dune vegetation communities (Zeevalking and Fresco 1977) mechanical scrub control may still be necessary.

9.6 Managing People and Caring for Dunes

Recreational activities often appear to have a detrimental impact on sand dunes. The perception of the dune manager may sometimes overemphasise their impact especially in causing erosion in densely populated areas. Locally heavy visitor use can cause significant destabilisation and erosion requiring remedial action. However, recreational use is an important cultural service. Managing this use is often a significant part of any nature conservation plan.

9.6.1 Signage and Walkways

Access control on sand dunes may be by signage (Fig. 9.2) or by the provision of walkways. The latter are particularly useful for facilitating access from tourist accommodation such as caravans and chalets or from car parks. On Ynyslas Dunes in west Wales, for example, a dune separated a caravan site from the beach, and dune erosion occurred along the line of a path, which linked the two (Fig. 7.7). Tackling the problem by the construction of a raised walkway to facilitate access to the beach and the use of brushwood and fencing to re-establish the dunes was largely successful.

Similar approaches are common in many other sand dunes sites. These may include pathways facilitating access, including wheelchairs, whilst at the same time 'protecting' the dune. This type of structure may be anathema to those seeking a natural landscape. The absence of grazing (incompatible with such recreational use) may exacerbate the problems of bracken and scrub encroachment (Fig. 9.10) requiring more costly control. However, this may be an acceptable trade-off on larger sites, given the recreational benefits particularly for disabled people.

The need for such costly constructions away from the main access points may not be required. In fact, moderate pressure by pedestrians elsewhere may cause little damage. Trampling may even help reduce the growth of coarse grasses and herbs and can be a management tool in some circumstances (Sect. 7.6.1).

Fig. 9.10 Walkway at Murlough National Nature Reserve, Northern Ireland October 2010. Note *Pteridium aquilinum* invasion in areas grazed only by rabbits

9.6.2 Zonation

A second approach is to adopt a policy of zonation. Tourists often require a wide range of facilities, which cater for the traditional beach holiday, as well as more active pursuits such as walking, running and natural history. Visitor management is a key to helping avoid the more damaging effects of trampling and other activities on dunes.

A key aspect of the control of large-scale tourist access is by the location of car parks. At Meijendel Dunes in the Netherlands, the location and size of car parks helped control visitor numbers and their overall impact on the dune habitat. By closing car parks and reducing available spaces, it was possible to increase the number of people arriving by bicycle from 10 to 50% in the years between 1973 and 1993. Correspondingly, the number arriving by car dropped from 63% and 73% in 1963 and 1973 respectively, to less than 50% in the 1990s (Bakker 1997). Signage, walkways and a visitor centre in the central part of the site also helped manage visitor use.

The natural configuration of a site may provide adequate control, especially on sand spits. Dawlish Warren on the Exe Estuary on the south Devon coast, England provides one example. Here access to the beach is from a large car park situated at the anchor point of the spit to the mainland. This is also where the visitor attractions are located. Access to the site is by foot only. Consequently, other than the pressure on the foredune (Sect. 6.3.4; Fig. 6.5) recreational use on the rest of the sand dune nature reserve is relatively low.

9.7 Deforestation

Deforestation within the context of this book, involves the removal of plantations, usually of nonnative pines, and restoration of dune habitat. Forests of this type cover many coastal sand dunes (Sect. 2.3.2) and have had a major adverse impact on the flora and fauna of many sites. In theory simply removing trees, either as a commercial operation or for nature conservation purposes and not replanting them, may appear to be all that is required.

However, creating a more open and dynamic approach to management and restoration through the removal of trees is not an easy option. Factors that militate against the removal of trees include their perceived value for sea defence (unfounded) and economic values of the forest itself (for timber) are principal amongst these. Other interests that have built up as the forest matures, in Great Britain at least, include the presence of bats, Red Squirrel or Pine Martin *Martes martes*. Recreational use is also important. However, today there is increasing recognition that the extensive planting programme of past decades may have harmed the conservation value of many dune systems including the impact on the hydrological system (Sect. 7.5).

Restoration may be easier on an already degraded system. In Spain, in the Llobregat delta, near Barcelona, a sand dune planted with *Pinus pinea* lies adjacent to a large expanding conurbation. Dam construction reduced the sediment delivered to the coast resulting in erosion of the delta. Water abstraction for the burgeoning population caused salt-water intrusion to the dune aquifer. Second homes, campsites and other tourist developments encroached onto the dune. Attempts to restore this degraded dune landscape included removal of planted trees, reinstatement of dune dynamics and the re-creation of a characteristic succession. Based on an assessment of neighbouring 'reference' areas this included sclerophyllous shrubs, *Juniper* and *Pinus* spp. woodland (Aronson and Vallejo 2006).

9.7.1 Value for Sea Defence

There is a perception that pine plantations provide a more robust sea defence capability, even if it does not. The forest may help control blowing sand inland, but once the high water reaches the base of the sand dune, it will erode (Fig. 9.11).

On the coast of the Aquitaine region of France, threats from blowing sand to a commercial dune forest and urban areas resulted in construction of an artificial dune rampart along the coast. The Office National des Forêt, whose remit in the past has been to prevent sand blow in forested areas allows some sand dune movement, except when habitation is threatened. Thus, despite the importance of the dunes to coastal

Fig. 9.11 Eroding sand dunes at the edge of the Culbin Forest, northeast Scotland, July 1982

defence, its maintenance involves a more dynamic approach. The Conservatoire du Littoral, which owns many sites, also accepts sand mobility where it results from natural processes. (Paskoff 2001).

9.7.2 Overcoming Objections

The sand dunes of the Sefton Coast, northwest England are approximately 2,000 ha of vegetated and blown sand lying between the Mersey and Ribble estuaries. They include 130 ha dominated by pine (mostly *Pinus nigra* ssp. *laricio*) planted between 1897 and 1960 in an attempt to emulate Les Landes in south-west France, and turn the 'wasteland' into a more productive estate. By the early 1970s, the woodlands covered about one third of a National Nature Reserve, contributing to loss of nature conservation values on the site as the water table dropped and the dunes became more stable. In 1983, there were proposals for their removal along the dune front. During lengthy consultation, the local authority expressed concern about the plan to remove the plantations partly because of the implications for sea defence. The presence of one of the few remaining colonies of Red Squirrel in England also caused a great deal of local opposition to scrub control and removal of some of the woodland areas (Edmonson and Velmans 2001).

Despite these reservations a phased programme of felling of the 'frontal woodlands' commenced on the Sefton Coast in 1992 (Fig. 9.12). The approach included a sequence of events designed, in part, to overcome local objections (Simpson 2001). The work continued into 1999 under the Sefton Coast Life Project (Rooney 2001). In 2010, the emphasis was on consultation and community involvement through a Sefton Coast Partnership (Nolan 2010 http://www.seftoncoast.org.uk/partnership.html).

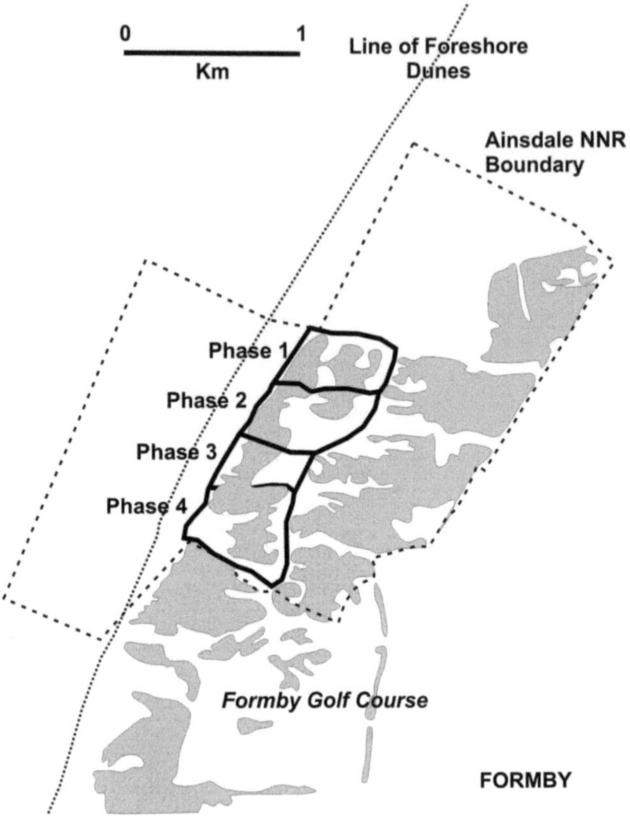

Fig. 9.12 Proposals for the phased removal of the 'frontal woodlands' within the Ainsdale National Nature Reserve from 1992. Plantations are shaded in *grey*

One of only two significant Red Squirrel colonies in Wales occupies the conifer plantations on Newborough Warren. Introduced in 1996 it has thrived to reach a sustainable population level (Shuttleworth 2003). Proposals by the Government Agency, Countryside Council for Wales to fell a large area of the alien pines to restore open sand dune, met with considerable local opposition, largely because of their presence.

Whatever the merits of the nature conservation case it is clear that there must be extensive consultation with local people when such emotive issues are involved with management operations.

9.7.3 Restoring Acid Dune Grassland and Dune Heath – Projects in the European Union

The European Commission LIFE Nature projects include three aimed at restoring threatened, "fixed grey dunes" and "decalcified fixed dunes with *Empetrum nigrum*" both listed in the Habitats Directive. These involved removal of pine, mainly *Pinus mugo* planted in the mid 1880s to help stabilise the dune systems in order to restore natural dynamics. The projects all in Denmark were:

1. "Restoration of dune habitats along the Danish West Coast", LIFE02 NAT/DK/8584. From 2001 to 2005 the project included clearing 264 ha non-native conifers and restoration of dune heath;
2. "Re-establilshing lichen an coastal heaths in the Anholt Desert", LIFE 94NAT/DK/492;
3. "Protection of grey dunes and other habitats on Hulsig Hede/Hulsig Heath", LIFE96 NAT/DK/ 3000;

The first follows on from a programme of deforestation, which began in 1992. It involved the large-scale removal of exotic conifers and the re-establishment of natural mosaic vegetation through burning, grazing and cutting. By 1999, restoration of 850 ha of open dune had taken place and during the following 5–8 years about 90 ha a year was proposed to be clear-felled (Ovesen 2001).

The second project took place on the sand dunes on Anholt Island, originally covered with native *Pinus sylvestris*. In the sixteenth century, the forest provided fuel for a lighthouse. By 1631, the forest had all but disappeared. Subsequent use for agriculture, including turf cutting, overgrazing and haymaking all contributed to the 'desertification' of the dunes. Afforestation began in 1885 over a wide area with *Pinus mugo* and *P. sylvestris* to prevent sand movement. These forests expanded as self-sown plants colonised areas outside the plantations (Christensen and Johnsen 2001a). The pines, felled manually and burnt on location or chipped for use in private small-scale, chip-fired heating plants on the island, were a key to the restoration of rare dune lichen-heath. The results of the LIFE-Nature project were highly successful. However, *P. mugo* in particular remains a threat to the rich lichen flora (Christensen and Johnsen 2001b) and its expansion must be contained in the longer-term.

The third project on Hulsig Hede (Hulsig Heath) included the restoration of mobile grey dunes and other habitats in north Jutland. It focused on the large-scale removal of conifers from 1,680 ha of the site. The varied techniques, included removal (Fig. 9.13), burning and crushing. The project was a success with pine cleared from 574 ha (99%) of the targeted area. The future control of self-seeding pines remains a challenge and continuous removal (by hand) of seedlings will probably be essential for the operation to be completely successful.

These projects served as pilot studies for a larger-scale pine clearance programme. The LIFE-Nature project on the restoration of dune habitats (LIFE02NAT/DK/8584), launched in 2002, covered 11 sites along the western coast of Jutland. The project planned restoration actions on a total net area of 5,675 ha – covering 65% of the total sand dune. The project provides an indication of the scale and wide-ranging nature of the management and restoration requirements for sand dunes in Europe. The principal objectives included:

- Removing 264 ha of non-native pine plantation;
- Clearing 542 ha of scrub;
- Removal of tree encroachment on 3,452 ha;
- Management of 2,800 ha to combat nutrient enrichment and over stabilisation;

Fig. 9.13 Removal of self-sown pine trees from sand dunes in North Jutland, June 1992. The mature plantation is visible in the background

- Restoration of natural hydrology at three sites;
- Improving habitat for reptiles and amphibians;
- Improving breeding localities for bird species such as Golden Plover *Pluvialis apricaria* Wood Sandpiper *Tringa glareola* and Crane *Grus grus*.

Summary guidelines suggest that management takes place as soon as pine seedlings appear. In more heavily infested areas, clearing plantations and scrub should include the removal of all above ground biomass, including needles and cones. Follow up treatment with mosaic burning (a new approach), removal by hand of self-sown pine seedlings on at least a 3 year cycle, and in some areas introduction of extensive grazing particularly by sheep, but also cattle (Danish Forest and Nature Agency http://ec.europa.eu/ourcoast/download.cfm?fileID=1120).

On the Curonian Spit, Lithuania there has been large-scale erosion, including loss of villages in the past. This resulted in reforestation projects, which commenced in 1825. Today remedial work involving pine removal, has been successful in re-establishing mobile dunes. As a result, moves to destabilise some other areas of the Curonian Spit are underway (Povilanskas et al. 2009).

European Union programmes in the Netherlands included equally wide ranging methods. The restoration of 4,700 ha of sand dunes between 2005 and 2010 included the erection of almost 60 km of fencing, the removal of 1,200 t of trees and shrubs, infilling of almost 8 km of ditches and the removal of 25,000 truckloads of nutrient rich soil from several dune valleys. Grazing regimes involved the introduction of 15 Exmoor ponies, 100 Dutch land goats, 47 Galloway cattle, 60 Shetland ponies and

15 Hereford cattle to various grazing areas. In addition, turf cutting, mowing and burning to help rejuvenate vegetation also took place. Perhaps more significantly, there was no attempt to control aeolian sand movement stimulated by these activities. The project included three stretches of dunes along the Dutch coast including:

1. Wadden Islands: Texel, Vlieland, Terschelling;
2. Mainland Coast: Hollands Duin (Noordwijk, Coepelduynen, Wassenaar);
3. Delta Area: Kop van Schouwen, Manteling van Walcheren (Zevenberg 2010).

9.8 Creating New Sand Dunes

There are a few examples of areas where the creation of new sand dunes has taken place, where none existed before. They are dependent on the presence of abundant sediment, and opportunities for accumulation to take place above high water creating a beach/foredune. The subsequent development of inland dune depends on there being a continuing supply of sediment to create new foredunes, as described in previous chapters. Three examples are from Ireland and France (created unintentionally), and Denmark (designed).

9.8.1 North Bull Island – The Development of a New Sand Dune

The need to prevent the accumulation of silt in Dublin Bay and hence allow continuing access to Dublin Port, resulted in the construction of a breakwater on the south side of the harbour in 1730. A further wall on the north side in 1825 helped form a self-dredging channel. Around this time, sediment deposited to the north and outside the sea walls resulted in the development of a sand dune and saltmarsh system (Fig. 9.14). The growth of this area (North Bull Island) includes a complex sequence of habitats with mud and sand flats, sand dune and saltmarsh. These form a 300 ha nature reserve of considerable importance and extending for approximately 5 km outwards from the northern sea wall (Jeffrey 1977).

9.8.2 Baie d'Audierne, Southwest Brittany

The destruction of a protective shingle barrier followed a long period of shingle extraction. In 1966, a wide breach opened up because of wave action, facilitating the influx of a large quantity of sand. This provided the material for new dunes to form landward of the ridge, soon after the breach occurred. These developed over a period of only 10 years or so, and between 1968 and 1978 the sand was colonised by vegetation. The dunes showed periods of erosion and accretion and in 1991 were still growing (Guilcher et al. 1992).

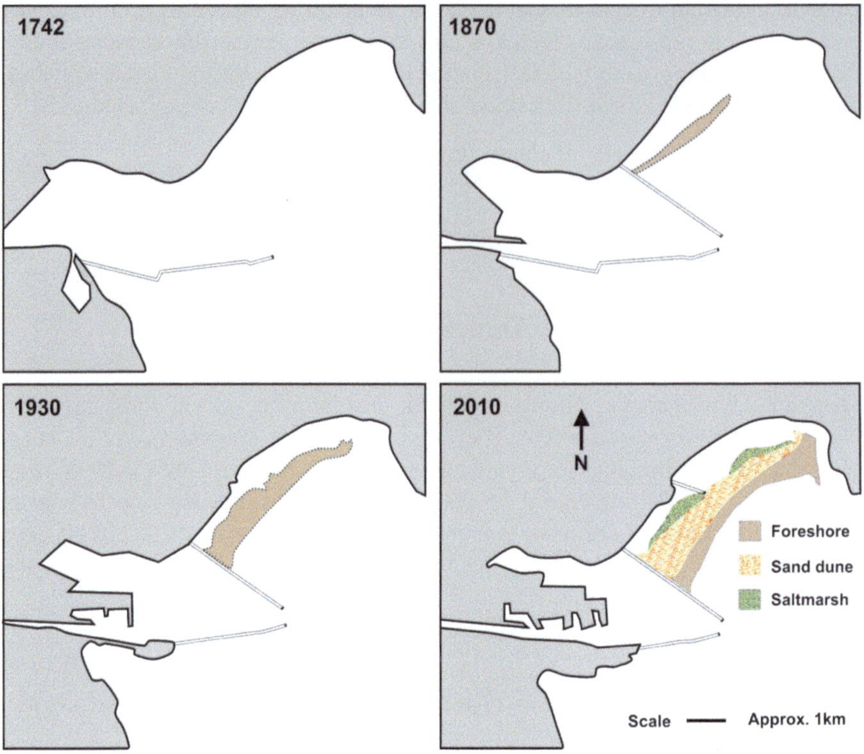

Fig. 9.14 *Sketch map* of the growth and development of Bull Island, Dublin Bay (After Jeffrey 1977). *Grey areas* are built areas of Dublin and Dublin Harbour

9.8.3 Køge Bay Beach Park, Denmark

Here an artificial barrier on a low natural sand bar provided the basis for dune development. Stabilisation by pioneer dune forming species followed the importation of new sand. The new dune grew through natural succession to develop a near natural appearance. This extended into the upper beach allowing dune grassland to colonise the artificial dune (Vestergaard and Hansen 1992).

The development worked well providing flood protection and recreational space, including a 5,000 ha dune area and 8 km of sandy bathing beach. Furthermore, the dune areas have a natural appearance with forms modified by wind and colonising vegetation rather than human interference. This site illustrates how it is possible to combine coastal defence measures and tourism development (Sistermans and Nieuwenhuis undated).

In summary, these examples illustrate a number of key points of importance when considering the artificial creation of sand dunes:

1. To be sustainable in the medium to long term, there must be an adequate and continuing supply of sediment or room for the sand dune ridges to migrate;

2. Dunes are 'opportunists' and foredunes can develop spontaneously;
3. Changes can be quite rapid;
4. Dunes are not fragile and vulnerable, but are adaptable and will react to sea level rise and storms to create new habitat.

9.9 Monitoring

Undertaking any form of conservation management should have a defined purpose. It is important to establish if the action taken has had the desired effect. For this, monitoring is an essential tool. Satellite and other remote sensed images can provide an overview suitable for the design and assessment of macro-level planning. Detailed work is required to assess some of the micro-scale issues. Remote Sensing (RS) and Geographical Information Systems (GIS) help create baseline inventories for mapping and monitoring coastal resources, including identifying their location, size and any changes that have taken place.

Establishing the position of the beach/foredune using permanent and temporary markers or using fixed-point photography can be simple and effective. Coupled with knowledge of the local sea level trend, this will provide the basis for establishing the movement of the dune landward or seaward. Historical artefacts (Sect. 6.1) can also give an indication of the relative movement of the foreshore. It can be time consuming to establish seasonal change here as this requires more frequent monitoring. Wind, rainfall and temperature data all help set the results in context and establish the trend and reasons for it.

Assessing the quality of the vegetation, however, requires a more intimate study of the species and vegetation types. Identifying the presence of rare species (plants and animals) requires knowledge of the species and field study. Details of the vegetation succession along a transect line provides an easy method of monitoring changes in the sequence of vegetation. It was more reliable than measurements using quadrates. Annual surveys from the beach to the foredune and into the first inland dune ridge provided an overall assessment of the effectiveness of sand dune restoration. Coupled with data on elevation, this provided a way of assessing both structural and vegetation development (Rozé and Lemauviel 2004).

Monitoring vegetation change on inland dunes, using a land-cover hierarchy identifiable on false colour aerial photographs, proved to be cost effective and accurate. This helped in the assessment of the success of cattle grazing in reducing grass encroachment (by *Calamagrostis epigejos*) in Mijendel dunes, the Netherlands (van der Hagen et al. 2008). Aerial surveillance (Aerial Digital Photographic System, ADPS) was effective in assessing the impact of visitor pressure. It used a self contained and transportable camera, easily mounted on a light aircraft (Curr et al. 2000). A sequence of aerial photographs, corrected for any distortions in older photographs, can also provide a means of assessing change when scrub encroachment takes place.

At Penhale dunes in North Cornwall, attempts to control recreational use and introduce grazing stock to manage and enhance the conservation significance of this large (1,070 ha) dune system have been accompanied by a comprehensive monitoring programme (Crummay et al. 2001). Techniques included:

- National Vegetation Classification survey (Rodwell 2000) – to be repeated every 5 years;
- Fixed-point photography – to record changes in physical features;
- Selected species monitoring – according to taxonomic groups with well-established methodologies. To include indicator species that are particularly sensitive to changes in environmental conditions, nationally scarce and local species and easily identified species.

The Handbook of Ecological Monitoring (Clarke 1986) brings together information on three important techniques: ground surveys, low-level aerial reconnaissance and remote sensing from high-flying aircraft and satellites. It is a valuable introduction to the subject, providing practical guidance for the design of ecological monitoring programmes, with particular reference to the sources of statistical bias, which can arise. In Great Britain, a Common Standards Monitoring Guidance is available for Sand Dune Habitats (Joint Nature Conservation Committee 2004).

9.10 Conclusion

This chapter has described the management methods employed to move the sand dune from one state to another. This has looked at specific parts of the succession, and included methods of preventing erosion and promoting accretion at the beach/foredune interface. There are examples of successful dune stabilisation, and even the creation of new sand dunes. In the medium to long term attempting to prevent landward movement in locations with a sediment deficit is probably neither sustainable nor desirable.

Management of inland dunes revolves around the internal relationships associated with vegetation succession. The individual approaches described here allow fine-tuning of management to suit specific requirements, especially for dune grassland and heath. Most of these activities, to a greater or lesser degree reflect human use, such as changes in grazing intensity by domesticated stock or afforestation with alien trees. However, it is clear whatever the remedial treatment in the short term, the natural forces of succession and stabilisation will require continued management in the long term.

Destabilising inland dunes may help sustain some features of nature conservation significance. However, even the most severe disturbance to the dune surface can become stabilised in only a few decades. During this period, it may only support significant sand dune plant and animal communities for a relatively short period. The implications for the long term viability of coastal sand dunes, is considered in Chap. 12.

The next two chapters illustrate integrated approaches to management, which accommodate recreational and cultural activities as well as helping to conserve wildlife.

References

Anderson P, Romeril MG (1992) Mowing experiments to restore a species-rich sward on sand dunes in Jersey, Channel Islands, GB. In: Carter RWG, Curtis TGF, Sheehy-Skeffington MJ (eds) Coastal dunes: geomorphology, ecology and management for conservation. A.A. Balkema, Rotterdam, pp 219–234

Aronson J, Vallejo R (2006) Challenges for the practice of ecological restoration. In: van Andel J, Aronson J (eds) Restoration ecology: the new frontier. Blackwell Science Ltd, Malden, pp 234–247

Avis AM (1995) An evaluation of the vegetation developed after artificially stabilizing South African coastal dunes with indigenous species. J Coast Conserv 1(1):41–50

Bacon J (ed) (2003) The scrub management handbook: guidance on the management of scrub on nature conservation sites. The Forum for the Application of Conservation Techniques, (FACT), with the assistance of English Nature. Source: http://publications.naturalengland.org.uk/publication/72031. Accessed 20 Aug 2012

Bakker ThWM (1997) Fifty years of recreational planning in the Mijendel dunes near the Hague. In: Drees JM (ed) Coastal dunes recreation and planning. European Union for Coastal Conservation, Leiden, pp 11–17

Bakker C, de Graaf H, Ernst WHO, van Bodegom PM (2005) Does the seed bank contribute to the restoration of species-rich vegetation in wet dune slacks? Appl Veg Sci 8:39–48

Bergin DO, Kimberley MO (1999) Rehabilitation of coastal foredunes in New Zealand using indigenous sand-binding species. Sci Conserv, 122, Department of Conservation, Wellington, New Zealand. Source: http://www.doc.govt.nz/upload/documents/science-and-technical/Sfc122.pdf. Accessed 13 Jan 2012

Boorman LA (1977) Sand-dunes. In: Barnes RSK (ed) The coastline. Wiley, London, pp 161–197

Boorman LA, Fuller RM (1982) Effects of added nutrients on dune swards grazed by rabbits. J Ecol 70(1):345–355

Brooks A, Agate E (2005) Sand dunes: a practical handbook. British Trust for Conservation Volunteers Doncaster, 109 pp.

Burton P (2001) Grazing as a management tool and the constraints of the agricultural system: a case study of grazing on Sandscale Haws Nature Reserve, Cumbria, Northwest England. In: Houston JA, Edmondson SE, Rooney PJ (eds) Coastal dune management, shared experience of European conservation practice. Liverpool University Press, Liverpool, pp 80–85

Christensen SN, Johnsen I (2001a) The lichen-rich coastal heath vegetation on the isle of Anholt, Denmark – description, history and development. J Coast Conserv 7(1):1–12

Christensen SN, Johnsen I (2001b) The lichen-rich coastal heath vegetation on the isle of Anholt, Denmark – conservation and management. J Coast Conserv 7(1):13–22

CIRIA (1996) Beach management manual. Construction Industry Research and Information Association, (CIRIA), Report 153

CIRIA (2010) Beach management manual, 2nd edn. Construction Industry Research and Information Association (CIRIA), Report C685B

Clarke R (1986) The handbook of ecological monitoring. Oxford University Press, Oxford, 350 pp

Corbett KF, Beebe TJC (1975) The disappearing Natterjack. Oryx 13:47–49

Cosyns E, Degezelle T, Demeulenaere E, Hoffmann M (2001) Feeding ecology of Konik horses and donkeys in Belgian coastal dunes and its implications for nature management. Belg J Zool 131(2):111–118

Crummay S, Neil C, Spalding A (2001) Monitoring the changing ecology of sand dunes near Gear Sands SSSI in conjunction with management and access, recreation and wildlife. In: Houston JA, Edmondson SE, Rooney PJ (eds) Coastal dune management: shared experience of European conservation practice. Proceedings of the European symposium, coastal dunes of the Atlantic biogeographical region, Southport, northwest England, September 1998. Liverpool University Press, Liverpool, pp 429–439

Curr RHF, Koh A, Edwards E, Williams AT, Davies P (2000) Assessing anthropogenic impact on Mediterranean sand dunes from aerial digital photography. J Coast Conserv 6(1):15–22

Doody JP (2001) Coastal conservation and management: an ecological perspective, vol 13, Conservation biology series. Kluwer Academic Publishers, Boston, 306 pp

Dugan JE, Hubbard DM (2010) Loss of coastal strand habitat in southern California: the role of beach grooming. Estuaries Coasts 33(1):67–77

Edmonson SE, Velmans C (2001) Public perception of nature management on a sand dune system. In: Houston JA, Edmondson SE, Rooney PJ (eds) Coastal dune management, shared experience of European conservation practice. Liverpool University Press, Liverpool, pp 206–218

European Commission (2007) Interpretation. Natura 2000. European Commission, DG Environment, Nature and Biodiversity, Brussels. Source: http://ec.europa.eu/environment/nature/legislation/habitatsdirective/docs/2007_07_im.pdf. Accessed 10 Jan 2012

FitzGibbon C, Albon S, Robinson P (2005) The effects of a mixed grazing regime on sand dune vegetation communities at Braunton Burrows, Devon. English Nature Research Reports, 637, p 2

Greipsson S, Davy AJ (1997) Responses of *Leymus arenarius* to nutrients: improvement of seed production and seedling establishment for land reclamation. J Appl Ecol 34(5):1165–1176

Greipsson S, El-Mayas H (1996) The stabilisation of coastal sands in Iceland. In: Jones PS, Healy MG, Williams AT (eds) Studies in European coastal management, EUCC. Samara Publishing, Cardigan, pp 93–100

Grootjans AP, Geelen HWT, Jansen AJM, Lammerts EJ (2002) Restoration of coastal dune slacks in the Netherlands. Hydrobiologia 478(1–3):181–203

Grootjans AP, Adema EB, Bekker RM, Lammerts EJ (2004) Why coastal dune slacks sustain a high biodiversity? In: Martinez ML, Psuty N (eds) Coastal dunes – ecology and conservation, vol 171, Ecological studies. Springer, Berlin, pp 85–101

Guilcher A, Hallégouët B, Meur C, Talec P, Yoni C (1992) Exceptional formation of present-day dunes in the Baie d'Audierne, southwestern Brittany, France. In: Carter RWG, Curtis TGF, Sheehy-Skeffington MJ (eds) Coastal dunes: geomorphology, ecology and management for conservation, proceedings of the third European Dune Congress, Galway, Ireland, 17–21 June 1992. A.A. Balkema, Rotterdam, pp 15–23

Herrier J-L, Van Nieuwenhuyse H, Deboeuf C, Deruyter S, Leten M (2005) Sledgehammers, cranes and bulldozers: restoring dunes and marshes by removing buildings and soil. In: Herrier J-L et al (eds) Proceedings 'Dunes and Estuaries 2005': international conference on nature restoration practices in European coastal habitats, Koksijde, Belgium 19–23 September 2005, vol 19, VLIZ special publication. Vlaams Instituut voor de Zee, Oostende, pp 79–94

Hewett DG (1985) Grazing and mowing as management tools on dunes. Plant Ecol 62(1–3):441–447

Hoffmann M (2008) Pros and cons of grazing management for coastal dune biodiversity: a review of what we know, don't know and should know. In: Hoffmann, M. Zendingsverslag van deelname aan de Internationale Conferentie over "Monitoring Biodiversity in Europe. Volunteers, Efficiency and Costs", Duitsland, Leipzig". UFZ, 28–30. Source: http://www.inbo.be/files/bibliotheek/65/179165.pdf, Accessed 05 Jan 2012.

Hoffmann M, Cosyns E, Deconinck M, Lamoot I, Zwaenepoel A (2001) Donkey diet in a Flemish coastal dune area in the first year after introduction. In: Houston JA, Edmondson SE, Rooney PJ (eds) Coastal dune management, shared experience of European conservation practice. Liverpool University Press, Liverpool, pp 95–107

Holmstrom RM, Etterson JR, Schimpf DJ (2009) Dune restoration introduces genetically distinct American beachgrass, *Ammophila breviligulata*, into a threatened local population. Restor Ecol 182:426–437

Houston J, Durkin C (2010) Sand dune and shingle network, Ninth Newsletter. Liverpool Hope University. Source: http://www.hope.ac.uk/coast/newsletters/. Accessed 05 Jan 2012

Jamieson SL (2010) Sand dune restoration in New Zealand: methods, motives, and monitoring. A thesis submitted for the partial fulfilment for the degree of Master of Science in Ecological

Restoration, Victoria University of Wellington, School of Biological Sciences. Source: http://researcharchive.vuw.ac.nz/bitstream/handle/10063/1229/thesis.pdf. Accessed 12 Jan 2012

Jeffrey DW (ed) (1977) North Bull Island, Dublin Bay – a modern coastal natural history. The Royal Dublin Society, Dublin, 158 pp

Joint Nature Conservation Committee (2004) Common standards monitoring guidance for sand dune habitats, Version August 2004. JNCC, Peterborough. Source: http://www.jncc.gov.uk/pdf/CSM_coastal_sand_dune.pdf. Accessed 12 Jan 2012

Jones L (ed) (2011) Coastal margins. In: The UK National ecosystem assessment technical report. UK National Ecosystem Assessment, UNEP-WCMC, Cambridge. Source: http://uknea.unep-wcmc.org/. Accessed 16 Feb 2012

Jones MLM, Norman K, Rhind PM (2010) Topsoil inversion as a restoration measure in sand dunes, early results from a UK field-trial. J Coast Conserv 14(2):139–151

Ketner-Oostra R, Sýkora KV (2007) Long term vegetation change (1966–2006) on calcium-poor coastal sand dunes in the Netherlands related to nitrogen deposition, the evaluation of restoration management. International conference on management and restoration of coastal dunes, Santander, Spain, 3–5 October 2007. Source: http://edepot.wur.nl/43650. Accessed 05 Jan 2012

Lemoine G (2005) La gestion des dunes flamandes: de l'éco-jardinage aux perturbations mécaniques à coups de bulldozers! In: Lemoine G (ed) Dunes flamandes: Joyau naturel du Nord. Service Espaces Naturels Sensibles-Département du Nord, pp 18–25

Lowday JE, Marrs RH (1992) Control of bracken and the restoration of heathland. I. Control of bracken. J Appl Ecol 29(1):195–203

Marrs RH (1987a) Studies on the conservation of lowland *Calluna* heaths. I. Control of birch and bracken and its effect on heath vegetation. J Appl Ecol 24(1):163–175

Marrs RH (1987b) Studies on the conservation of lowland *Calluna* heaths. II. Regeneration of *Calluna*, and its relation to bracken infestation. J Appl Ecol 24(1):177–189

Marrs RH, Lowday JE (1992) Control of bracken and the restoration of heathland. II. Regeneration of the heathland community. J Appl Ecol 29(1):204–211

McCue J, Pye K, Wareing A (2010) Beach nourishment operations in Wales and likely future requirements for beach nourishment in an era of sea level rise and climate change – a pilot study. CCW Science report, no. 928, Countryside Council for Wales, Wales

Moore P, Davis A (2004) Marram grass *Ammophila arenaria* removal and dune restoration to enhance nesting habitat of Chatham Island oystercatcher *Haematopus chathamensis*, Chatham Islands, *New Zealand.* Conserv Evid 1:8–9

New South Wales Government (1990) NSW coastline management manual. Source: http://www.environment.gov.au/archive/coasts/publications/nswmanual/index.html. Accessed 04 Jan 2012

Nordstrom KF (2004) Beaches and dunes of developed coasts. Cambridge University Press, Cambridge, 352 pp

Nordstrom KF, Lampe R, Vandemark LM (2000) Reestablishing naturally functioning dunes on developed coasts. Environ Manage 25(1):37–51

Ovesen CH (2001) Management of Danish dunes today: theory and practice. In: Houston JA, Edmondson SE, Rooney PJ (eds) Coastal dune management: shared experience of European conservation practice. Proceedings of the European symposium, coastal dunes of the Atlantic biogeographical region, Southport, northwest England, September 1998. Liverpool University Press, Liverpool, pp 302–305

Packham JR, Willis AJ (2001) Braunton Burrows in context, a comparative management study. In: Houston JA, Edmondson SE, Rooney PJ (eds) Coastal dune management: shared experience of European conservation practice. Liverpool University Press, Liverpool, pp 65–79

Paskoff P (2001) Dune management on the Atlantic coast of France: a case study. In: Houston JA, Edmondson SE, Rooney PJ (eds) Coastal dune management, shared experience of European conservation practice. Liverpool University Press, Liverpool, pp 34–40

Peterson CH, Bishop MJ (2005) Assessing the environmental impacts of beach nourishment. Bioscience 55(10):887–896

Phillips CJ, Willetts BB (1978) A review of selected literature on sand stabilisation. Coast Eng 2:133–147
Pickart AJ (2008) Restoring the grasslands of northern California's coastal dunes. Calif Nativ Grassl Assoc, 18/1:4–9. Source: http://www.fws.gov/humboldtbay/pdfs/grasslandsarticle.pdf. Accessed 10 Jan 2012
Pickart AJ, Barbour MG (2007) Beach and dune. In: Barbour M, Keeler-Wolf T, Schoenherr AA (eds) Terrestrial vegetation of California, 3rd edn. University of California Press, Berkely, pp 155–179
Pilkey OH, Young RS, Bush DM, Thieler ER (1994) Predicting the behaviour of beaches: alternatives to models. In: Littoral´94, proceedings, Eurocoast, Lisbon, Portugal. pp 26–29. Source: www.vliz.be/imisdocs/publications/54902.pdf. Accessed 10 June 2011
Pizzey JM (1975) Assessment of dune stabilisation at Camber, Sussex, using air photographs. Biol Conserv 7:275–288
Plassmann K, Jones M, Laurence M, Edwards-Jones G (2009) Effects of long-term grazing management on sand dune vegetation of high conservation interest. Appl Veg Sci 13(1):100–112
Povilanskas R, Baghdasarian H, Arakelyan S, Satk nas J, Taminskas J (2009) Secular morphodynamic trends of the Holocene dune ridge on the Curonian Spit (Lithuania/Russia). J Coast Res 25(1):209–215
Ranwell DS, Boar R (1986) Coast dune management guide. Institute of Terrestrial Ecology, HMSO, London, 105 pp
Rodwell JS (ed) (2000) British plant communities. Volume 5, maritime communities and vegetation of open habitats. Cambridge University Press, Cambridge, 512 pp
Rooney PJ (2001) The Sefton coast life project: a conservation strategy for the sand dunes of the Sefton coast, northwest England. In: Houston JA, Edmondson SE, Rooney PJ (eds) Coastal dune management: shared experience of European conservation practice. Proceedings of the European symposium, coastal dunes of the Atlantic biogeographical region, Southport, northwest England, September 1998. Liverpool University Press, Liverpool, pp 243–261
Rozé F, Lemauviel S (2004) Sand dune restoration in North Brittany, France: a 10-year monitoring study. Restor Ecol 12(1):29–35
Scottish Natural Heritage (2000) A guide to managing coastal erosion in beach/dune systems. Source: http://www.snh.org.uk/publications/on-line/heritagemanagement/erosion/sitemap.shtml. Accessed 31 May 2011
Seliskar DM (1995) Coastal dune restoration: a strategy for alleviating dieout of *Ammophila breviligulata*. Restor Ecol 3(1):54–60
Shuttleworth CM (2003) A tough nut to crack: red squirrel conservation in Wales. Biologist 50:231–235
Simpson DE (2001) Towards best practice in the sustainable management of sand dune habitats: 1. The restoration of open dune communities at Ainsdale sand dune National Nature Reserve. In: Houston JA, Edmondson SE, Rooney PJ (eds) Coastal dune management: shared experience of European conservation practice. Proceedings of the European symposium, coastal dunes of the Atlantic biogeographical region, Southport, northwest England, September 1998. Liverpool University Press, Liverpool, pp 255–261
Sistermans P, Nieuwenhuis O No date EURosion case Study, Køge Bay Beach Park. Source: http://copranet.projects.eucc-d.de/files/000110_EUROSION_Koge_bay.pdf. Accessed 06 Jan 2012
Smith PH (2006) Changes in the floristic composition of sand-dune slacks over a twenty-year period. Watsonia 26:41–49
Terlouw L, Slings R (2005) Dynamic dune management in practice – remobilisation of coastal dunes in the National Park Zuid-Kennemerland in the Netherlands. In: Herrier J-L, Mees J, Salman A, van Nieuwenhuyse H, Dobbelaere I (eds) Proceedings 'Dunes and Estuaries 2005', vol 19, VLIZ special publication., pp 211–217
van der Hagen HGJM, Geelen LHWT, de Vries CN (2008) Dune slack restoration in Dutch mainland coastal dunes. J Nat Conserv 16(1):1–11
van der Laan D, van Tongeren OFR, van der Putten WH, Veenbaas G (1997) Vegetation development in coastal foredunes in relation to methods of establishing marram grass (*Ammophila arenaria*). J Coast Conserv 3:179–190

van der Putten WH, Peters BAM (1995) Possibilities for management of coastal foredunes with deteriorated stands of *Ammophila arenaria* (marram grass). J Coast Conserv 1:29–39

van Dijk HWJ (1992) Grazing domestic livestock in Dutch coastal dunes: experiments, experiences and perspectives. In: Carter RWG, Curtis TGF, Sheehy-Skeffington MJ (eds) Coastal dunes: geomorphology, ecology and management for conservation, proceedings of the third European Dune Congress, Galway, Ireland, 17–21 June 1992. A.A. Balkema, Rotterdam, pp 235–250

Vestergaard P, Hansen K (1992) Changes in morphology and vegetation of a man-made beach-dune system by natural processes. In: Carter RWG, Curtis TGF, Sheehy-Skeffington MJ (eds) Coastal dunes: geomorphology, ecology and management for conservation, proceedings of the third European Dune Congress, Galway, Ireland, 17–21 June 1992. A.A. Balkema, Rotterdam, pp 165–176

Whatmough JA (1995) Case study, grazing on sand dunes: the re-introduction of the rabbit *Oryctolagus cuniculus* L. to Murlough NNR, Co. Down. In: Bullock DJ, Harvey HJ (eds) The National Trust and nature conservation 100 years on. Academic, London, pp 39–43

Williams MJ (2007) Native plants for coastal restoration: what, when, and how for Florida. USDA, NRCS, Brooksville Plant Materials Center, Brooksville, FL. p 51. Source: http://www.fl.nrcs.usda.gov/programs/pmc/flplantmaterials.html. Accessed 10 Jan 2012

Wray S (2006) A guide to rabbit management. Construction Industry Research and Information Association (CIRIA), Report C645

Zarnetske PL, Seabloom EW, Hacker SD (2010) Non-target effects of invasive species management: beachgrass, birds, and bulldozers in coastal dunes. Ecosphere 1(5):1–20

Zeevalking HJ, Fresco LFM (1977) Rabbit grazing and species diversity in a dune area. Plant Ecol 35(3):193–196

Zevenberg J (2010) LIFE dunes. Report on five years of dune restoration in the Netherlands. Source: Staatsbosbehee. http://www.staatsbosbeheer.nl/Nieuws%20en%20achtergronden/Dossiers/LIFE%20Nature/~/media/00%20PDF/Actueel/Dossiers/LIFE/Publieksrapport_EN.ashx. Accessed 05 Jan 2012

Chapter 10
Integrated Action – Golf Course Management

Abstract This chapter sets out to consider the relationship between recreational golf and the conservation values of sand dunes. The importance of this lies in the origins of golf, on sand dunes (links) in Scotland. Golf is and continues to be a major use of sand dunes, especially in the United Kingdom and Ireland. It has expanded into other European countries, North America and around the world. Reconciling golfing with nature conservation objectives is an important issue. It considers the merits and problems associated with this recreational activity by reference to specific case studies in the United Kingdom. These include two new links courses in Scotland (Machrihanish Dunes and Foveran Links) which illustrate very different approaches to dune conservation in modern times.

10.1 Introduction

The name golf may have come from the old Scots verb "to gowff" meaning to "strike hard". King James II in 1457, concerned that playing 'golfe' resulted in neglect of the vital sport of archery, demanded that *"fute-ball and golfe be utterly cryed down and not to be used."* The sand dunes (links courses) at St Andrews in Scotland have been in use since around the 1400s. Links courses are preferred localities for those seeking the development of a traditional course even today.

Links golf courses number about 200 out of the approximately 6,000 courses in Europe (WorldGolf.com 1997–2009 http://www.europegolf.com/). Of the 100 top golf courses in the world about 25 % are on coastal sand dunes and classified as 'links' courses (Top 100 Golf Courses http://www.top100golfcourses.co.uk/index.asp). This is therefore an extensive and valuable recreational resource. The next sections deal with the trade-offs between the negative effects caused by building the course in the first place and the opportunities for nature conservation management.

J.P. Doody, *Sand Dune Conservation, Management and Restoration*, Coastal Research Library 4, DOI 10.1007/978-94-007-4731-9_10,

10.2 Habitat Loss

In Great Britain, approximately one third of the larger sand dunes provide the location for golfing activities. Although the proportion worldwide is probably less than this, there are still significant areas of sand dune with this form of development, and new ones continue to be proposed. There is no doubt that building a new golf course on a sand dune destroys many surface features and interferes with the hydrology and coastal dynamics. The extent to which this adversely affects the nature conservation value of the sand dune depends on the area converted to intensively used greens, tees, fairways, and management of the surrounding areas of 'rough'.[1] Whatever the losses, some golf courses provide oases of biodiversity and can hold a relatively high level of species richness, in comparison to surrounding land. This difference is most marked when urban or agricultural land surrounds the golf course, much less so in areas where anthropogenic use is limited (Colding and Folke 2009).

10.2.1 St Andrews

One of the oldest golf courses in the world is the famous St Andrews, in Scotland '*the home of golf*'. The location for the Royal and Ancient, a private golf club, it has helped nurture and develop the game for 600 years. Initially, there was little interference with the sand dunes. In the early 1900s, grazed by sheep and rabbits they were biologically diverse. Lichens were abundant and the dune grassland included Moonwort *Botrychium lunaria*, English Catchfly *Silene anglica* and Mountain Everlasting *Antennaria dioica*. Thyme *Thymus serpyllum* was said to be extremely plentiful (Wilson 1910). A picture of the links from this period, suggests the dunes had much less Gorse *Ulex europaeus* scrub than today (Wilson 1910; Plate IV, Fig. 12). However, even then the conversion of the rough moor into more modern golf links by trampling, rolling, mowing and grazing was taking place (Wilson 1910).

Today St Andrews sand dunes support seven golf courses, which have largely destroyed the open sand dune grassland and heath. Intensively managed greens, tees and fairways now cover much of the surface area (Fig. 10.1.). They are mostly devoid of any significant nature conservation value because of the use of herbicides and fertilisers, regular watering and mowing. There are progressively potentially richer areas outside these intensively managed zones in the 'semi-rough' and 'rough'. However, over the years dense stands of gorse vegetation have come to dominate these areas. Small pockets of dune heath do occur but these are very fragmented. Good populations of the nationally scarce Curved

[1] A good general definition of 'Roughs' is "undesired playing areas where grass is maintained above 1.5 in. in height". They lie adjacent to the main areas of play, notably alongside fairways.

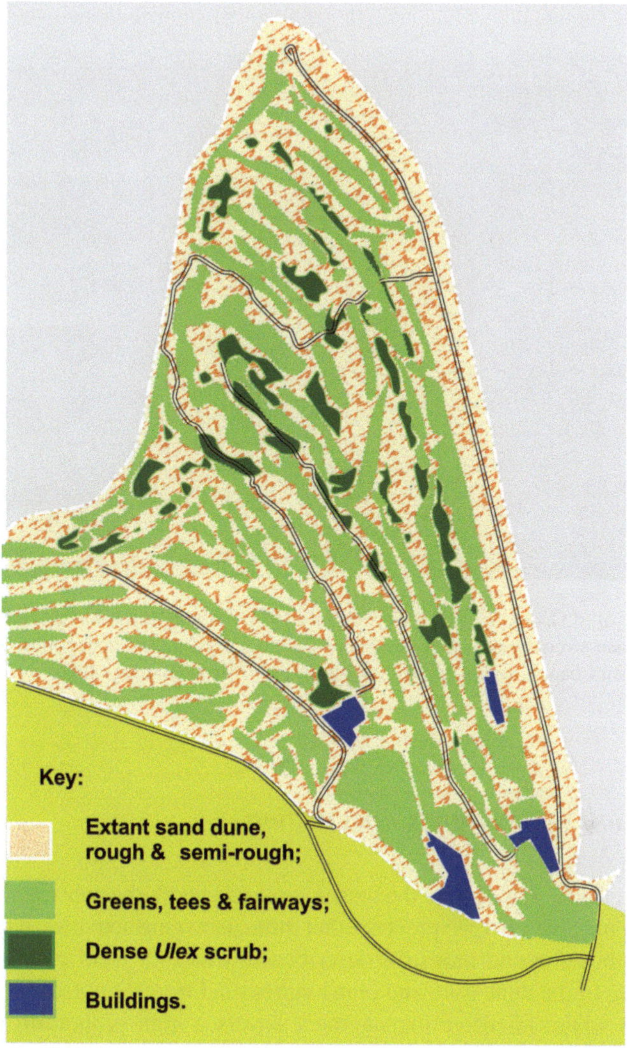

Fig. 10.1 Habitat modification on St Andrews golf course showing the extent of development of greens, tees and fairways in relation to the area of blown sand (Interpreted from Google Earth imagery December 2006)

Sedge *Carex maritima*, surviving in wet fairways in 1981 have been lost since then (Pearman and Lockton 2007).

This is typical of the many golf courses built on sand dunes to a greater or lesser degree. For example, a study on the island of Jersey showed that semi-natural dune vegetation contained 30–40 species per 25 m^2 compared with 5–10 species per 25 m^2 on fairways on St. Ouen's Bay, which have received fertiliser and herbicide applications as well as irrigation (Ranwell 1975). However, as we will see there are many sites where features of nature conservation value remain.

Fig. 10.2 Royal St George's golf course, 9th July 1985, 6th hole in the foreground. The sites for *Himantoglossum hircinum* and other rare plants were out of bounds to both golfers and spectators during the Open Championship, which took place from 18th to 21st July 1985

10.2.2 Golf and Nature Conservation

It is perhaps ironic that despite the losses associated with the construction of a golf course, they may have helped prevent other more damaging activities, notably development for housing and industry (described in Chap. 2) from taking place. The extent of loss of the dune flora and fauna varies on different courses. St Andrews is an example where intensive management covers a high proportion of the sand dune. All golf courses contain areas of 'rough'. In some cases, intensive management restricts the diversity of vegetation. On other sites relatively large and intact areas of sand dune vegetation remain, which may support important plant and animal communities.

In the United Kingdom, some courses retain sufficient interest to fall within a Site of Special Scientific Interest (SSSI). Some links courses are in this category and support rare species including the Sand Crocus *Romulea columnae* on Dawlish Warren, Devon and Natterjack Toad and Sand Lizard at Royal Birkdale in Lancashire. One of the most celebrated examples is Royal St George's in Kent, a course that hosted the Open Championship of golf in 1985, 1993, 2003 and 2011 (Fig. 10.2). In the undisturbed areas of the course, there are 11 species of orchid with the Lizard

Orchid *Himantoglossum hircinum* especially common (Gange et al. 2003). As with many other links courses, these sand dunes are ungrazed. Management centres on mowing with the removal of the cut material. This helps to ensure the survival of the valuable vegetation including the orchids.

A review of the status of the network of SSSIs by English Nature in 2004 revealed that 57 % of the surface area was in 'favourable condition'[2] from a wildlife perspective. This figure rose to 66 % for all those SSSIs associated with golf courses. However, on sand dunes (and inland grasslands) the percentage surface area in 'favourable condition' was much lower with many in an 'unfavourable'[3] condition due to scrub development (Tew 2004).

As with sand dunes more generally, one of the key issues is grazing. Control of rabbit populations and reduction or cessation of grazing by domesticated stock is common on most links courses. Erosion control, eutrophication and water abstraction for irrigation are also important. Sympathetic management can do much to protect wildlife. Without it, considerable damage can result.

10.3 Grazing – Trends and Trade-offs

It is probable that grazing took place on many sand dunes as golf developed in the late nineteenth century. However, problems associated with grazing animals, include dunging and the hindrance they may cause during play. As a result, there appears to have been a tendency to remove animals from many courses by 'buying out' the Common Grazing Rights.[4] In addition to this, as with grazing by domestic stock elsewhere, there has been a reduction in stock levels and grazing intensity on marginal land, such as sand dunes. The conservation agencies in the United Kingdom see this as a significant issue for sand dune conservation. Common Standards Monitoring guidance for sand dune habitats, specifically says "the absence of stock and strict control of rabbit populations on golf courses has led to rapid successional change and widespread loss of dune grassland to mesotrophic swards" (Joint Nature Conservation Committee 2004).

The following sections describe a number of case studies from England and Wales, which highlight some of the issues in securing the appropriate levels of management. The sites discussed are sand dunes that have golf courses lying within existing Sites of Special Scientific Interest.

---

[2] Favourable condition means that special habitats and features are in a healthy state and conserved for the future by appropriate management.

[3] Unfavourable condition is the term used to describe sites where the Special Features of a site are not adequately conserved, or are being lost.

[4] Right to pasture cattle, horses, sheep or other animals on common land.

Fig. 10.3 Dense 'wind-pruned' Blackthorn *Prunus spinosa* mixed with Bramble *Rubus fruticosus* scrub, Aberdovey dunes, Wales July 2007

10.3.1 Aberdovey Dunes – A Site in Transition

The Aberdovey Golf Club own part of Aberdovey Dune system, which includes extant exercised Common Grazing Rights. The area forms part of the Dovey Site of Special Scientific Interest and falls within the Pen Llyn A'R Sarnau Special Area of Conservation. This grazing has a long history, with Common Rights extending over a significant part of the golf course and Pen-llyn marshes behind. A 1995 management agreement specifies continuation of grazing by domestic stock (29 cattle 31st May–31st October and a number of sheep during the winter months).

The economic return on the relatively small number of stock, and the difficulties of control, make the continuation of this form of management less attractive to the small number of local graziers. As a result, there has been a reduction in stock grazing such that there is a noticeable deterioration in the dune grassland vegetation and its nature conservation value. Vegetation maps are available for 1991 (Ashall et al. 1994) and from a survey in 2004, of the land occupied by the golf course.

Comparison between these maps and a visit by the author in 2007 provides a picture of change. The distribution of the main scrub areas in 2004 was similar to 1991. However, by 2007 there had been a considerable extension in their area and density. The most extensive areas lay just outside the mown semi-roughs. *Prunus spinosa* (Fig. 10.3.) occurred in dense patches and *Crataegus monogyna* was present throughout the site. There was also evidence of further scrub development by a variety of additional species, ranging from *Acer pseudoplatanus*, to Cedar *Cedrus* spp. and Willow *Salix* spp. This, together with the extensive and expanding *Rosa pimpinellifolia*, is a clear indication that grazing levels were much lower than formerly.

In addition, in recent years the Aberdovey Golf Club has identified issues relating to the conflict between stock grazing, golfers and other people using the site (there is a footpath across the golf course providing access to the beach). These include:

- Animals may impede play. As a result fencing of greens and tees is required;
- Dung has to be cleared from the course every morning before play starts, for health and safety and aesthetic reasons;
- Issues of safety can arise, especially from cows during the calving season;
- In the UK where there is a legal requirement for 'Common' grazing to take place, if local graziers stop their traditional practices the club may have to take on the role of stock management.

Whilst the mown 'roughs' retain open sand dune grassland on other parts of the site, where grazing is the only form of management, the communities are deteriorating. For example, the more typical sand dune slack communities are becoming taller mire vegetation. The lack of grazing pressure and increased water logging of the soils appear to be responsible for this.

This process is likely to continue at current stocking levels. Any further reduction or cessation of grazing, of either cattle or sheep, would result in the coarsening of the vegetation and scrub development accelerating in the unmanaged parts of the course. This would have a further adverse impact on the nature conservation values of the site.

10.3.2 Romney Warren, Kent, Southeast England

Romney Warren is an extensive area of sand dune in Kent. It is a Site of Special Scientific Interest and includes two 18-hole golf courses forming the Romney Warren Golf Club. The more recent course, on the west of the site, has wide fairways and relatively limited areas of 'rough'. A review of the management of the site in relation to its declared nature conservation interest took place in 2005. This provides an indication of the effects of cessation of grazing by domestic stock on the quality of the vegetation.

The greater part of Romney Warren includes the Littlestone Warren Golf Club, founded in 1888. As with many other sand dunes in Europe, as the name implies, the cultivation of rabbits took place here. Domesticated stock also grazed the site, but this ceased in the 1920s. Grazing by rabbits continued until 1953 when myxomatosis decimated the population. The combined effect was to favour the growth of coarse grasses and scrub at the expense of dune grassland and heath.

The communities present in 2005 were largely the same as those recorded during the Sand Dune Vegetation Survey of Great Britain (Hedley 1990). Their location

Fig. 10.4 Ungrazed sand dune grassland, overgrown with coarse grasses, Romney Warren, Kent, October 2005

and distribution was also comparable. However, there was a major difference in that they appeared to be consistently ranker, with coarse grasses such as Yorkshire Fog *Holcus lanatus* frequent and sometimes dominant in the sward (Fig. 10.4). Today the management of the grassland depends on mowing, although this is restricted to the fairways and semi-rough. As a result, much of the Site of Special Scientific Interest has become stabilised and overgrown, with loss of open low-growing vegetation typically associated with sand dunes.

Reintroduction of grazing by sheep has taken place in two grazing enclosures in the north of the site to test this approach to management. However, in the grazed paddocks the levels of grazing are insufficient to open up the denser swards. Increasing both the extent of grazing and density of stock remain the preferred management strategy. Although rabbits were present, there was little evidence they made any significant inroads on the dense vegetation on or off the course.

10.3.3 Rye Bay Dunes

The dunes to the west of Rye Bay include the Rye Golf Club, established in 1894. There is no record of grazing having taken place on the dunes. None of the pictures

Fig. 10.5 General view of Rye Golf course, Kent November 2005. The intensively managed greens, tees and fairways have 'rough' undulating dunes regularly cleared of invading scrub

or references to the site in the early stages of development indicates the presence of any domesticated stock, although it seems likely that rabbits grazed the dunes in the past. The golf course lies within the Camber Sands and Rye Saltings Site of Special Scientific Interest. In "Views About Management" for the Dungeness, Romney Marsh and Rye Bay Site of Special Scientific Interest (Natural England http://www.sssi.naturalengland.org.uk/special/sssi/sssi_details.cfm?sssi_id=2000533) it states: "Selective scrub management and grazing or mowing may be necessary, especially where dunes have become over stabilised."

Intensive management of the greens, tees and fairways leaves little, if anything of nature conservation value. The fairways have a narrow band of semi-rough and wider areas of rough. The open vegetation adjacent to some of the fairways includes good lichen-rich grassland cover and Sea Campion *Silene uniflora* is prominent in the sward.

Scrub clearance is an ongoing management activity, which controls the invasion of *Ulex europaeus* and *Hippophae rhamnoides*. As a result, scrub cover is less than 5 % and the site retains a good representation of dune grassland. The intensive management of the greens, tees and especially the fairways is offset by the generally 'favourable condition' of the intervening roughs, including the undulating sand dunes. Regular mowing and raking helps to retain many of the dune grassland features, in the absence of rabbits or other grazing animals (Fig. 10.5).

10.4 Other Issues

Erosion control, water abstraction and eutrophication are issues that have relevance to golf courses. These centre on the need for stability to prevent the loss of parts of the golf course. The availability of water and use of fertilisers and herbicides is also necessary to maintain the greens, tees and fairways.

10.4.1 Erosion

The layout of greens, tees and fairways creates a fixed, specifically designed landscape. Green keepers will not tolerate erosion and sand movement on the inland parts of the golf course. This restricts the options for management of degraded sites, from a nature conservation perspective. Beach or foredune erosion can threaten important features such as a tee, a green or even an entire hole.

Given that eroding sandy beaches far outnumber those that are accreting, it is perhaps not surprising that problems occur. The northwest flank of the Jubilee Course towards the mouth of the Eden Estuary at St. Andrews suffered from erosion. Buried gabion baskets provided the 'last line of defence' in front of a tee. 'Recharging' the dune face with sand and replanting *Ammophila arenaria* helped maintain the natural characteristics of the sand dune. Five years on, the area appeared to be healthy and even showed signs of accretion. More generally and in relation to climate change, protecting isolated areas with hard structures may not be effective and can lead to further loss in adjacent areas (Windows 2004).

A visit to Royal Troon golf course on the west coast of Scotland in January 1991, revealed a series of structures placed to 'protect' a tee from erosion. These included plastic barrels and concrete pipes filled with sand, chestnut paling fences and gabions filled with stone. These had failed and in some cases appeared to have exacerbated the erosion (Fig. 10.6).

Northam Burrows lies on top of a pebble spit stretching along the North Devon coast of England from Westward Ho! to the mouth of the Taw-Torridge Estuary. Erosion of the pebble foreshore in recent years has threatened at least two holes of England's oldest links golf course, the Royal North Devon Golf Club established in 1864. This is nothing new and Spearing (1884), reported that in the previous 9 years the beach eroded at a rate of about 30 ft (approximately 9 m) per year. Erosion continued intermittently during the twentieth century.

The sand dunes and the pebble ridge form part of a Site of Special Scientific Interest. In order to allow 'natural' development to take place Natural England, the Government Agency responsible for protecting the site, has indicated that erosion should continue. This will inevitably result in the loss of the eighth hole, which in 2008 was only 18 m from the sea. Erosion threatens at least one other hole with others to follow. In the short-term Natural England agreed to consider the resumption of minor repair work to the sea defences, providing relocation of the threatened holes

Fig. 10.6 Foreshore along the southwest facing sand dune of the Royal Troon, January 1991. The picture shows erosion outflanking the concrete pipes filled with sand, installed to prevent loss of part of the golf course

takes place in the medium to long-term. The pebble ridge, which protects the sand dune, no longer receives material from cliff erosion to the southwest, and the whole system is effectively fossilised. In addition to compromising the integrity of the geomorphology of the site, protecting the eroding ridge is difficult and likely to be costly. Note that the sand dunes are heavily grazed (Fig. 5.2) but this has no impact on the erosion of the pebble beach or the inland dune.

10.4.2 Water Relationships

Sand dunes often have a domed water table fed by rainwater, surface water run-off from adjacent land or groundwater. The aquifer provides a readily accessible supply of water for the irrigation of greens, tees and fairways. Although the amount of water taken from the reservoir is limited (<1 % of the aquifer recharge in the case of the golf courses on the Sefton coast), it can have a localised impact on the dune vegetation. This is also true more generally on European Natura 2,000 sites, although it is less significant than drainage schemes (Houston 2008). Dune slacks are disproportionately affected and are highly sensitive to both water abstraction and drainage (Davy et al. 2006).

Nutrient inputs to greens and tees can cause damage to adjacent 'roughs' as fertilisers are leached into the surrounding areas. A loss of quality results as communities change from the more species-rich sand dune grassland types to ranker mesotrophic grassland. Changes in hydrology resulting from irrigation can also cause a similar effect.

10.5 Management Options

Introduction or reintroduction of grazing is the preferred management strategy on sand dunes in temperate regions generally (Sect. 9.5). However, the particular restrictions imposed by the needs of golfing militate against this on most sand dunes used for this purpose. Alternative forms of management to control the growth of rank vegetation and scrub are therefore required. This section considers grazing and some of the alternative management strategies employed.

10.5.1 Grazing

Introduction or reintroduction of grazing is not common practice on links courses. Grazing stock and the presence of high rabbit numbers can hinder play, and cause aesthetic and health issues. In a few sites in the United Kingdom existing grazing 'rights' prevent removal of domesticated stock as at Northam Burrows (Sect. 5.2.1) and Aberdovey (Sect. 10.3.1). Fencing the greens and tees can help. However, this may be unsightly and inconvenient for the golfer. Introducing or reintroducing grazing animals, more generally to links golf courses is unusual but see Sect. 10.6.2.

10.5.2 Mowing

In the absence of grazing animals, mowing provides an alternative. It already takes place in the semi-rough and in the rough on many sites as part of normal course management. Extension of mowing regimes for conservation purposes enhances the potential value of golf courses to wildlife. In Scotland, for example, the Gullane Golf Club has been working closely with Scottish Natural Heritage (SNH) and the Scottish Golf Environment Group (SGEG). The aim is to transform areas of tussocky grasses such as Cock's-foot *Dactylis glomerata* and *Holcus lanatus* into vegetation with finer grasses more typical of sand dunes, by mowing. This benefits both the vegetation from a golfing point of view as well as improving the biodiversity of the sward. Mowing once a year and importantly removing the cuttings, helps reduce fertility favouring the less robust species (Windows 2004). Forage harvesters are probably the best mowing equipment from a nature conservation point of view, as they can both cut and collect the vegetation.

10.5.3 Management Plans

The golf courses on the Sefton coast may have prevented development, notably for housing (Sect. 2.8.2), on some dune landscapes. Today they occupy over a quarter of the dune area, some 550 ha out of a total of approximately 2,000 ha. The modified greens, tees and fairways represent 25 % of the golf course land (Simpson et al. 2001). In the 1970s and 1980s following the decimation of the rabbit population in 1953, coarse grassland and scrub rapidly invaded the dunes, as it had elsewhere, compromising many nature conservation values.

In 1996, a golf and nature initiative began on the Sefton Coast. A partnership between the conservation organisations and the golf clubs drew up integrated management plans. Included in this was a combination of activities designed to help maintain the open dune habitat. These included scrub management, mowing, turf stripping on dune ridges, dune stabilisation, heather restoration and water management including the creation of scrapes. Modifying existing management regimes also helped. Changes included mowing the rough every 2 years rather than one and mowing later in the season to allow seed set. 'Turf-stripping' to patch up worn areas of greens and tees also provided an opportunity for rare invertebrates by creating bare sand. Taken from areas with limited nature conservation value, these can have other positive benefits by creating pockets of instability (Simpson et al. 2001).

10.6 Developing New Links Golf Courses

From the above it is clear that the nature conservation value of existing golf courses largely depends on their current management. The manicured lawns of courses in the United States of America Golf became fashionable in the mid to late 1900s. The American champion Sam Snead reportedly said in 1946 of the Old Course at St Andrews "down home, we plant cow beets on land like that" (Sailer 2005).

Despite the comments attributed to Sam Snead, golfers look upon links courses more favourably today. As already indicated above, they include a high proportion of the world's favourite courses, with 10 of the top 20 golf courses being on sand dunes. The following advertisement provides an illustration: "Located on the rolling sand dunes of the Mornington Peninsula, in the area known as the "Cups Country", the Dunes Golf Club forms an exquisite 'links-style' golf course as in the original Irish or Scottish links." This may be why Donald Trump, in seeking to establish a new European course of "World class", settled on sand dunes on the Aberdeenshire coast of Scotland. This damaging development lies in sharp contrast to another new golf course Machrihanish Dunes, also in Scotland.

10.6.1 Foveran Links, Balmedie, Aberdeenshire

The Trump International Golf Links Scotland web site boasts, "I have never seen such an unspoilt and dramatic seaside landscape and the location makes it perfect for our development". The location includes a large part of a Site of Special Scientific Interest. It has a championship golf course suitable for hosting major international events with a second 18-hole golf course proposed. Adjacent to this will be an eight-storey "iconic hotel set amidst an array of luxury holiday homes; and a residential village" with almost 1,000 holiday homes and 500 private houses.

The proposal came before an inquiry held from June 10th to July 4th 2008 in Scotland. There were objections from a wide variety of statutory and voluntary nature conservation bodies. The 'Reporters' to the inquiry submitted their report to Scottish Ministers in October the same year. It concluded, "The development would have a significant adverse impact on the southern third of the Foveran Links SSSI. Here the dynamism that underpins the designation of the SSSI would, for the majority of the holes in the back nine of the championship course, be halted." Despite this they recommended approval of the outline proposal because, "The development has the potential to deliver major benefits against economic and social objectives at national, regional and local level."

In November 2008, Scottish Ministers said they would "grant outline planning permission for a golf course and resort development on land at Menie House, Balmedie, Aberdeenshire". Work began in 2011, when finished it will have destroyed a large proportion of the sand dune with considerable loss of wildlife values.

10.6.2 Machrihanish Dunes, Kintyre, Scotland

Machrihanish dunes Site of Special Scientific Interest (SSSI) represent a considerable contrast to the development on the east coast. Close collaboration between wildlife experts and the golf course designers, made it possible to overcome many of the problems associated with building a new course. Survey provided the developers with information that ensured the greens, tees and fairways avoided environmentally sensitive areas. As a result, only 3 ha, less than 3 % of the natural dune landscape has been lost. The restricted use of fertilisers and pesticides to greens and tees, avoids damage to sensitive vegetation elsewhere. Mowing the fairways and grazing by a flock of Hebridean sheep will keep the vegetation of the roughs under control, allowing the rare plant life to flourish. Buildings, roads and a car park are all outside the SSSI boundary.

Advertised as the first true links course on the west coast of Scotland for more than 100 years, Machrihanish Dunes Golf Club opened to the public in 2009. It is debatable whether this course, or indeed the development of the "World Class course" proposed by Donald Trump, is really such a positive step forward. There are already many golf courses on sand dunes in Scotland. Most of the unaffected areas are over stable from a nature conservation perspective, and the requirements of golf will inevitably lead to more habitat loss and stabilisation.

10.6.3 The Future

Today there is increasing recognition of the value of combining golf course management with nature conservation (e.g. Dodson 2000; Wood 2004). In Florida, a manual for planning strategies for wildlife habitat management includes a Chapter on "Planning Wildlife Friendly Golf Courses" (Pennington 2010).

The success of working with conservationists at Machrihanish has lead to other proposals for new links courses. The team behind the Machrihanish development (Kintyre Property Trust) together with Credential Holdings Company propose a new development near the world-famous Royal Troon Golf Club, on the Ayrshire coast in Scotland (e-golf Business Ltd. 2000–2012 http://www.golfbusinessnews.com/news/courses/the-ayrshire-a-multi-million-pound-investment/). Commenting on the Machrihanish development the team said "A links course is much cheaper to build and maintain, therefore making the game more accessible to people" (Paul Kelbie 2008 http://www.telegraph.co.uk/earth/earthnews/3353152/Eco-friendly-golf-course-to-open-in-Scotland.html).

This could result in proposals for new links courses on many of the remaining untouched sand dunes, not only in the United Kingdom but also worldwide. It is arguable that the creation of an 'environmentally friendly' golf course can sustain wildlife values. However, there is an equally valid argument for keeping sand dunes that are in a more or less 'natural' condition, free from such development. This is despite policies, which help sustain features of ecological value, described by Gange et al. (2003). Whatever the response to proposals for new links courses, developing dialogue with golf course managers on existing courses, is a key requirement for the future.

10.7 Conclusion

It is interesting to speculate what would have happened to coastal sand dunes in the absence of the game of golf. How many existing sand dunes would be lost to housing or industrial development, sand extraction or used for agriculture? It is impossible to tell. Constraints on management for wildlife because of the golfing need mean that there will always be a trade off between the two. However, it is possible to reconcile some of these differences by adopting appropriate management strategies. There is an increasing willingness on the part of the golf course manager to do so.

These may never fully meet the nature conservation need. Constraints on grazing by domestic stock or introducing instability to the inland dune will remain. However, there are approaches to management that can have nature conservation benefits. Scrub control and mowing are two. The construction of any new links golf course will destroy some sand dune habitat, however carefully planned. It will also restrict those forms of management that are equally difficult to reconcile within existing courses.

References

Ashall J, Duckworth J, Holder C (1994) Sand dune survey of Great Britain, Site report no. 113, Dunes between Tywyn and Aberdovey, Meirionydd, Wales. 1991. Joint Nature Conservation Committee Report (JNCC), no. 81, Peterborough

Colding J, Folke C (2009) The role of golf courses in biodiversity conservation and ecosystem management. Ecosystems 12(2):191–206

Davy AJ, Grootjans AP, Hiscock K, Peterson J (2006) Development of eco-hydrological guidelines for dune habitats – phase 1. English Nature Research reports no 696, Peterborough. Source: http://publications.naturalengland.org.uk/publication/61018. Accessed 20 Aug 2012

Dodson RG (2000) Managing wildlife habitat on golf courses. Wiley, New York, 177 pp

Gange AC, Lindsay DE, Schofield JM (2003) The ecology of golf courses. Biologist 50(2):63–68

Hedley SM (1990) Romney Warren, Kent. Site report no 77, Chief Scientist Directorate, Nature Conservancy Council, Peterborough

Houston JA (2008) Management of Natura 2000 habitats. 2190 humid dune slacks. European Commission. Source: http://ec.europa.eu/environment/nature/natura2000/management/habitats/pdf/2190_Humid_dune_slacks.pdf. Accessed 04 June 2011

Joint Nature Conservation Committee (2004) Common standards monitoring guidance for sand dune habitats. ISSSN 1743–8160 (online). Source: http://jncc.defra.gov.uk/pdf/CSM_coastal_sand_dune.pdf. Accessed 26 Feb 2012

Pearman D, Lockton A (2007) Progress with the *Carex maritima* survey. BSBI Rec 11:11–12

Pennington D (ed) (2010) Wildlife habitat planning strategies, design features and management practices for Florida communities and landowners. http://www.floridahabitat.org/wildlife-manual/florida%20wildlife%20manual%20All%20Pgs%20final2%20Lo%20Res.pdf. Accessed 07 Jan 2012

Ranwell DS (1975) The dunes of St. Ouen's Bay, Jersey: an ecological survey. Annu Bull Soc Jersiaise 21:381–391

Sailer S (2005) From Bauhaus to golf course: the rise, fall, and revival of the art of golf course architecture. Source: http://www.isteve.com/golf_art.htm. Accessed 04 Mar 2012

Simpson DE, Rooney PJ, Houston JA (2001) Towards best practice in the sustainable management of sand dune habitats: 3. Management for golf and nature on the Sefton Coast. In: Houston JA, Edmondson SE, Rooney PJ (eds) Coastal dune management: shared experience of European conservation practice. Proceedings of the European symposium, coastal dunes of the Atlantic biogeographical region, Southport, northwest England, September 1998. Liverpool University Press, Liverpool, pp 271–280

Spearing HG (1884) On the recent encroachment of the sea at Westward Ho, north Devon. Q J Geol Soc Lond 40:474

Tew T (2004) Golf courses and wildlife sites. The Golf Club Secretary, Magazine, January edition. Broadside Publishing Ltd., Essex

Wilson JH (1910) Nature study rambles around St Andrews. WC Henderson and Sons, University Press, St Andrews, 258 pp

Windows R (2004) Climate change and Scottish golf courses. Scottish Golf Environment Group Limited. Source: http://www.sgeg.org.uk/documents/Publications%20and%20Case%20Studies/Climate%20Change%20and%20Scottish%20Golf%20Courses%20%28SGEG%202004%29.pdf. Accessed 04 June 2011

Wood M (2004) Nature conservation and golf course development: best practice advice, Scottish Golf Environment Group (SGEG). http://www.sgeg.org.uk/documents/Publications%20and%20Case%20Studies/Nature%20Conservation%20and%20Golf%20Course%20Development%20%28SGEG%202004%29.pdf

Chapter 11
Integrated Action – Machair, Human History and Nature Intertwined

Abstract Machair is a special type of sand dune occurring as a sandy plain. It is restricted to northwest Scotland and western Ireland. This chapter describes a landscape shaped by natural forces associated with the geology of the islands, glacial history, geographical position, climate and geomorphological development. In considering this formation, it is important to understand that whilst the geomorphological processes govern where the machair exists, human activity has had a major influence on the system and its nature conservation values. These considerations form the basis for discussion in this chapter.

11.1 Origins and Definition

Machair comes from a Gaelic word meaning 'plain' and refers to an extensive low-lying flat, fertile area behind foredunes (where these exist). The origin of the machair sand derives from glacial material deposited by the retreating ice at the end of the last glaciation. Mixed with this are large quantities of calcareous shells, remnants of animals living in the sea. This material, subsequently moved onshore by wave action and then blown inland, provides the basis for the development of machair. Machair is a complex system where there is a close interplay between geomorphological processes and agricultural activity (Angus 1997, 2001a; Love 2009). Cycles of severe erosion associated with overgrazing and intensive cultivation may have helped form the machair in the Outer Hebrides (Ritchie 1979). It occurs predominantly on the west coasts of Scotland and Ireland (Fig. 11.1), where exposure to North Atlantic depressions with high rainfall and a high frequency of strong westerly winds is crucial to its development.

Note, although it is generally considered to be restricted to Scotland and Ireland, a similar ecosystem is also reported in New Zealand (Wilson et al. 1993).

It forms part of a wider 'machair system', which encompasses a sequence of habitats from strandline and foredune, the 'machair plain' and transitions to saline

J.P. Doody, *Sand Dune Conservation, Management and Restoration*, Coastal Research Library 4, DOI 10.1007/978-94-007-4731-9_11,

Fig. 11.1 Machair distribution in Scotland and Ireland

lagoons and saltmarsh, or to calcareous lochs, acidic grasslands, fens, heath or bog. A typical functional sequence includes the 'machair grassland' within the 'machair system' (Fig. 11.2).

A typical machair has the following characteristics:

1. A base of blown sand, which has a significant percentage of shell-derived materials;
2. Lime-rich soils with pH values normally greater than 7.0;

Fig. 11.2 A typical cultivated machair site in the Western Isles, on the island of Barra. Beach, dune (mostly absent here), machair grassland and hill machair, the simplest sequence. A picture taken from the same location on the Pastoral web site http://www.efncp.org/hnv-showcases/scottish-hebrides/machair/system/ shows a remarkably similar situation to this picture taken in June 1983. The 'system' often includes additional elements such as marsh, loch and blackland; saltmarsh, brackish loch and blackland (Blackland refers to the inland transitional area between the machair and moorland); saltmarsh, tidal sand flats and blackland (Angus 2001a)

3. A level or low-angle smooth surface at a mature stage of geomorphological evolution;
4. Sandy grassland vegetation, which lacks taller dune binding grasses such as *Ammophila arenaria*. Core plants are *Festuca rubra*, Birdsfoot Trefoil *Lotus corniculatus*, White Clover *Trifolium repens*, Yarrow *Achillea millefolium*, *Galium verum*, Ribwort Plantain *Plantago lanceolata*, Eyebright *Euphrasia officinalis* agg., Daisy *Bellis perennis* and the moss *Rhytidiadelphus squarrosus*;
5. Biotic interference, such as is caused by heavy grazing, sporadic cultivation, trampling and sometimes artificial drainage should be a detectable influence within the recent historical period;
6. An oceanic location with a moist, cool climatic regime;
7. Machair plains flood or are at least marshy in winter (Ritchie 1976, 1979).

However, the machair 'system' is only part of a wider landscape and cultural unit. Providing a model of the machair at a landscape scale supplements the definition provided above (Angus 2006).

Although ancient agricultural use of the machair began with Neolithic peoples, there is evidence to suggest that humans played a significant role in its development much earlier. Occupation in Scotland may date back more than 5,000 years ago to the Mesolithic Period (Edwards et al. 2005; Gregory et al. 2005) and in Ireland, possibly up to 8,000 years ago (Basset and Curtis 1985). During this period, in Scotland, burning vegetation and scrub clearance may have contributed to aeolian activity in causing deflation of the sand dune surface (Edwards et al. 2005).

11.2 Agricultural Management

Land management falls into two broad categories, which split conveniently into cultivated and uncultivated ground. Historically the machair in both Scotland and Ireland was cultivated. Today in Scotland, grazing takes place in both cultivated areas between periods of crop rotation when the land lies 'fallow' and in uncultivated areas amongst the sand dunes. Note that grazing is by both cattle and sheep. In Ireland, there is little or no cultivation and grazing takes place year round across the whole machair plain.

11.2.1 Traditional Machair Cultivation in the Outer Hebrides

Traditionally cultivation in Scotland involved the use of the machair to graze domesticated stock in the winter months, before being removed in May and taken to the hills for summer grazing, or in more recent years 'fenced out'. Cultivation, especially in the Outer Hebrides, involves a 'crofting'[1] agricultural system with traditional crops such as potatoes and cereal. Local varieties of cereal, Oats *Avena strigosa*, Rye *Secale cereale* and an indigenous form of Barley *Hordeum* spp. provided fodder for the wintering animals. These species are more tolerant of the highly alkaline, nutrient deficient soils than other faster growing species grown on the mainland such as *A. sativa*. In the islands of Uist (Outer Hebrides) these species sown in mixtures, produce a sparse crop within which there is bare ground providing soils suitable for a variety of agricultural weeds.

This is probably representative of an ancient cropping rotation and involves:

- Small scale scattered cultivation mainly in dunes immediately inland from the foredune, and usually consisting of small scale potato patches;
- Traditional small-scale strip cultivation, characterised by a variety of weed species in the cultivated strips and a variety of dominants in the areas of fallow over a 3 year or longer cycle (Fig. 11.3). Often there are no added artificial fertilisers or herbicides.

The diversity in crop rotation, and presence of these old agricultural 'weeds' in the Hebridean machair is a feature representative of an agricultural system virtually absent from anywhere else in Europe.

[1] Crofting is a land tenure system in Scotland, which arose because of the land clearances. The first crofting act of 1886 provided security of tenure to crofters, protecting them from further forced removal from the land. Crofts are small agricultural units owned, or leased from a landlord. Within these units stock (sheep and cattle) are the main product, although winter keep (cereals for cattle) and potatoes are also grown (Scottish Crofting Federation http://www.crofting.org/index.php/home).

Fig. 11.3 Traditionally cultivated machair showing fallow land in the foreground, ploughed land planted with potatoes middle distance (*left*) and cereal cultivation middle distance (*right*) in the Outer Hebrides, Scotland. Note the shallow plough, used to reduce the effects of erosion, July 1983

11.2.2 Machair in Ireland

Irish machair is morphologically similar to the Scottish machair. In historical times, (following its introduction to Europe in the 1600s) the main machair crop was potatoes. Potato cultivation took place on a much smaller scale in the Western Isles. Depopulation due to the potato famine, which occurred from 1845 to 1851, probably altered the management considerably (Bassett and Curtis 1985). Today grazing is predominantly by sheep and tends to occur throughout the year; the machair normally remains uncultivated.

The structure of the machair 'system' is also different in that the transition is to species-poor heath, bog or acid grassland, rather than to wetlands and marshes present in Scotland (Bassett and Curtis 1985).

11.3 Nature Conservation Value

The presence of the calcareous material helps to make the machair relatively rich in species, when compared to the more acid upland blackland areas. The long history of management, involving low input agriculture, has helped to sustain a rich and varied flora and associated fauna in both Scotland (Angus 2001a) and Ireland (Bassett and Curtis 1985). The differences in management between the two have helped to create different nature conservation interests, described next.

11.3.1 Vegetation of Uncultivated Sand Dunes

The early stages in sand dune formation (strandline and foredune) are relatively uncommon in the Outer Hebridean machair. This reflects the overall erosional nature of the western dune/machair systems on their seaward margins. There is usually, an area of dry base-rich dune grassland with uncommon species such as Hoary Whitlowgrass *Draba incana*, *Ophioglossum vulgatum* and *Coeloglossum viride*. This vegetation type normally extends into the machair plain proper. In more modern parlance, this is the *Festuca rubra-Galium verum* fixed dune grassland (SD8) according to the British National Vegetation Classification (Rodwell 2000). This dune grassland type divides into two sub-communities, 'dry' and 'damp' machair.

The Outer Hebridean 'dry' machair grassland is the most extensive. This, the *Bellis perennis-Ranunculus acris* sub-community (SD8d), is widespread and in summer colourful with clovers in addition to Daisy and Meadow Buttercup that characterise the vegetation. Interspersed within this, in lower-lying ground is the *Prunella vulgaris* sub-community (SD8e). In this vegetation, a number of the less common orchids including *Coeloglossum viride* and Early Marsh Orchid *Dactylorhiza incarnata* ssp. *coccinea* occur, together with other Hebridean races and unusual hybrids. This community together with a slightly more acid, damp grassland help form the wet machair. In a few areas, the influence of salt water can result in species of upper saltmarsh such as Sea milkwort *Glaux maritima* and the Baltic Rush *Juncus balticus* uncommon in the British Isles. For a more detailed review of the machair grassland, see Angus (2001a), pp. 205–214.

In Ireland, three principal machair communities are recognised:

1. Dry machair community one, which lies close to the seaward side of the habitat. As with the vegetation in Scotland, this has a virtual absence of the dune building *Ammophila arenaria*. The presence of the salt-tolerant plants *Plantago maritima*, *Elytrigia juncea* and *Armeria maritima* is indicative of the close proximity to the sea;
2. Dry machair community two is much more frequent and typical of the machair plain. The salt-tolerant species present in community one are absent. *Ammophila arenaria* can be locally frequent and there is also a greater abundance of the mosses *Rhytidiadelphus squarrosus* and *Brachythecium rutabulum*;
3. Wet machair has a variety of moisture loving herbs such as *Potentilla anserina*, Marsh Pennywort *Hydrocotyle vulgaris*, Common Sedge *Carex nigra*, Jointed Rush *Juncus articulatus*, Common Spike Rush *Eleocharis palustris*, Bog Pimpernel *Anagallis tenella*, Water Mint *Mentha aquatica*, Marsh Bedstraw *Galium palustre* and the moss *Calliergonella cuspidate*.

These communities are broadly similar to those found in Scotland in terms of both communities and species (Gaynor 2006). The mainland machair-like grasslands,

Fig. 11.4 Species-rich sandy grasslands at Strathy Bay, Sutherland, June 1987. The wind blows sand over exposed cliffs covering the steep slopes. Cowslip *Primula veris* visible in the foreground is amongst a variety of species which includes the rare (in United Kingdom) *Oxytropis halleri*

such as those associated with the rocky coasts of Sutherland; include arctic-alpine species *Dryas octopetalla* in Sandwood Bay and Purple Oxytropis *Oxytropis halleri* at Bettyhill. Here, windblown calcareous sand covers the slopes to create steep sandy grassland (Fig. 11.4).

11.3.2 Vegetation of Cultivated Machair

Rotational ploughing and cropping of machair has probably taken place for centuries, especially in the Outer Hebrides. Use of natural fertiliser such as seaweed, and a traditional cropping pattern allowed expression of an older and richer flora. Extant cereal crops can include 80 species or more and have a seed bank of arable 'weeds', many long-since eliminated from farmland elsewhere and now rare in the British Isles.

Species include those surviving ploughing and regenerating from vegetative propagules (the majority). Other species colonise because of the exposure of buried seed, and a few come from outside because of seed-rain (Owen et al. 2000). Ruderals Scarlet Pimpernel *Anagallis arvensis*, Red Poppy *Papaver rhoeas*, Charlock Mustard *Sinapis arvensis*, Field Forget-me-not *Myosotis arvensis* and Field Pansy *Viola arvensis* were most frequently associated with cereal crops as in dry machair at

Fig. 11.5 Cereal cultivation in Western Isles with *Chrysanthemum segetum* prominent, September 1981

Drimsdale, South Uist (Crawford 1990). Additional species included Corn marigold *Chrysanthemum segetum*, which can be a significant visual component in the crop (Fig. 11.5).

Left to lie fallow, a further sequence of visually dominant species develops. In the first year of fallow the ruderals are replaced by Heartsease *Viola tricolor*, Creeping Buttercup *Ranunculus repens*, *Myosotis arvensis*, Sea Mouse-ear *Cerastium diffusum* and Common Stork's-bill *Erodium cicutarium*. By the second year all the ruderal species of the newly-ploughed and 1 year fallow were absent and replaced by a further suite of species including *Festuca rubra*, Ragwort *Senecio jacobaea*, *Bellis perennis*, Common Selfheal *Prunella vulgaris*, *Trifolium repens* and Yarrow *Achillea millefolium* (Owen et al. 2000).

By contrast, potatoes crops are less rich with only 37 species recorded. This appears to be partly due to the smaller size of the patches and absence of the 'arable' species component of the cereal crop (Owen et al. 2000).

11.3.3 Birds

The variety of plant species supports a wide range of invertebrates, and provides food and shelter for breeding birds. Partly because of this and the absence (until recently) of predators, the machair is especially important for ground-nesting birds. Amongst these the most numerous are:

- Lapwing *Vanellus vanellus*;
- Dunlin *Calidris alpina*;
- Redshank *Tringa tetanus*;
- Snipe *Gallinago gallinago*;
- Oystercatcher *Haematopus ostralagus*;
- Ringed Plover *Charadrius hiaticula.*

The regularly occurring migratory breeding wader species *Calidris alpina* and *Charadrius hiaticula* can reach densities of 200 pairs per ha in the most favourable habitats, though 100 per ha is more usual. These densities are far greater for this group of birds than for habitats where they nest in other parts of Europe (Fuller et al. 1986). In 1983 and 1984, surveys of breeding waders in the Outer Hebrides showed that there was a high nesting density throughout most of the machair and blackland of the larger islands.

For two Annex 1 species (EU Species Directive), namely: Corncrake (*Crex crex*) and Chough (*Pyrrhocorax pyrrhocorax*) the machair system supports significant populations.

In Ireland the same suite of wader species breed on the machair as in Scotland, in order of abundance they are:

- Lapwing *Vanellus vanellus*;
- Dunlin *Calidris alpina*;
- Ringed Plover *Charadrius hiaticula*;
- Snipe *Gallinago gallinago*;
- Oystercatcher *Haematopus ostralegus*;
- Redshank *Tringa tetanus*;

In addition

- Common Sandpiper *Actitis hypoleucos*;
- Golden Plover *Pluvialis apricaria*;
- Red-necked Phalarope *Phalaropus lobatus* are also recorded (Nairn and Sheppard 1985).

11.3.4 Invertebrates

The invertebrate fauna of machair is not particularly rich (Welch 1979). The range of habitats helps support a wider range of species than is normally associated with dry grasslands. The fact that the machair is relatively undisturbed also means some characteristic species take a relatively long time to complete their life cycle and can grow into large specimens. There are also characteristic species that because of the varied conditions occur in large numbers. Together these are an important food resource for insectivorous birds described above (McCracken 2009). The Great Yellow Bumblebee *Bombus distinguendus*, which has shown a dramatic decline in Britain, has disappeared from much of mainland Great Britain recently. The species still occurs in some numbers in the Outer Hebrides and is notable as it highlights the decline of bumblebees more generally.

11.4 Changes in Management

The presence of the rich and varied plant and animal communities depends on the continuation of traditional forms of management. Modern, more intensive management includes reseeding with agricultural grasses. When coupled with artificial fertilisers, amalgamation of strips and fencing, these result in a considerable reduction in nature conservation values.

11.4.1 Effects on Vegetation – Cultivation

Over the last few decades, management of the machair has changed from 'traditional' cultivation (small-scale planting of potatoes and cereal on a cycle of several years using seaweed as a fertiliser and soil stabiliser, followed by a period of fallow) to a more intensive rotation. This usually involves planting with cereal only, and using artificial fertilisers and herbicides as a means of boosting productivity. This 'modernisation' occurs in the following sequence:

1. **Traditional** small-scale scattered cultivation immediately inland from the foredunes, which usually consists of small potato patches. Small-scale strip cultivation characterised by a variety of ancient 'weed' species in the cultivated strips with a variety of dominants in areas of fallow;
2. **Simplification** involving a 3-year rotation, but within amalgamated units. The use of artificial fertiliser and/or herbicides resulting in some areas retaining characteristics of the 'weedy' traditional cultivations whilst others are virtually weed free;
3. **Intensive** cereal cultivation involving a 2-year rotation with artificial fertiliser and herbicides, fewer crofters and larger parcels of land. Within these areas, there are virtually no 'weed' species either in the cultivated or fallow periods. Reseeding with agricultural grasses may also occur in 2 and 3 above.

The management regimes progressively reduce the occurrence of the rare agricultural 'weeds' and also the visual impact of the fallow swards. There is a tendency for the management to harvest the crops earlier in the season, reducing the ability of the arable weeds to set seed. Herbicide and fertiliser treatment, and deeper ploughing using modern machinery, further reduces the variety of herbs in the sward. Under-sowing areas of arable crop with grass seed, especially Perennial Ryegrass *Lolium perenne* have a similar effect.

Overall, there is a reduction in vegetation diversity and elimination of the period of fallow when colourful machair grasslands develop. In some cases, the adjacent species-rich wet grassland, which is also a feature of the machair plains of North and South Uist, is lost. Scottish Natural Heritage's Site Condition Monitoring shows that losses of biodiversity in the crop and fallow plants, although variable, affected nearly 50% of 34 sites surveyed (Angus 2009).

11.4.2 Effects on Vegetation – Grazing

Cattle and sheep graze the machair grassland of the Outer Hebrides. The former due to their unselective grazing, can improve the quality of the grassland for other grazers as well as for wildlife. However, cattle are marginally economic and declining, resulting in the greater use of sheep. Fenced all year, they can cause overgrazing, with a loss of ecological value and greater susceptibility to soil erosion.

The opposite problem occurs on some uninhabited offshore islands. In the past loading and offloading animals to graze in the summer months were regular occurrences. However, welfare issues and the effort of moving the stock resulted in cessation of grazing on some islands. Here the vegetation freed from the control of grazing animals became rank, with a loss of plant species diversity.

11.4.3 Effects on Birds

Corn bunting *Emberiza calandra* historically bred within traditionally cultivated areas in the Western Isles. However, a change in cereal harvesting which involves cutting earlier in the season before the seed is ripe reduces the availability of seed suitable for corn buntings. It appears this has led to a decline in the breeding pairs of 62% between 1995 and 2005 (Wilson et al. 2007). The population of wintering Barnacle Goose *Branta leucopsis* appears to have declined on some of the outer islands because of the decline in use for grazing stock. The rank grassland is less palatable for the geese and soon becomes unsuitable for breeding waders. Note however, that movement towards more intensive farming, and goose management schemes, have attracted geese to alternative sites in recent years (Mitchell et al. 2008).

In Ireland there was a significant loss of breeding birds between 1985 (Nairn and Sheppard 1985) and 1996 (Madden et al. 1998) with declines of approximately 30% for comparable sites. A resurvey of 55 sites, surveyed in either 1985 or 1996, within Counties Donegal, Sligo, Mayo and Galway, showed this loss had continued and by 2009 there had been a further significant reduction in breeding pairs from mainland sites (Suddaby et al. 2010). Enclosure of formerly open common land, sub-divided by fencing to increase grazing levels, is amongst the reason for this decline. Drainage (Fig. 11.6) affects 44% of sites, and together with increased grazing pressure this has led to shorter turf. For species such as *Calidris alpina* this is unsuitable for nesting. In addition, it exposes nests, eggs and young of ground-nesting birds, posing a greater risk of predation from foxes and Hooded Crow *Corvus cornix* (Suddaby et al. 2010).

Fig. 11.6 Heavily grazed machair western Ireland July 1991, with a recently excavated drainage channel

11.5 Human Activity and Deflation

Historically, cycles of sand dune erosion and accretion occur in many areas, and in Outer Hebrides, northwest Scotland there is a clear relationship with human activity. Mean annual wind speeds on very exposed sites in the area reach 6.2–8.0 m/s. Sand grains move with wind speeds around 4–5 m/s, so the exposed sites in Outer Hebrides are susceptible to sand movement for half the year (Angus and Elliot 1992). During severe storms, therefore, the land is vulnerable to blowing sand. Changes in human use have exacerbated this.

Seaweed, formerly gathered as fertiliser for the machair in the Outer Hebrides, had the additional effect of acting as a stabilising agent. However, in 'kelp manufacture' seaweed was burnt to provide products included soda (alkali) for the soap, glass and textile industries, and later iodine for use in medicine and photography for a period of nearly 100 years up to 1822 (Forsythe 2006). During this period, neglect of the land combined with accelerated loss to the sea increased erosion and sand drift (Angus 1997).

By the mid nineteenth century and into the twentieth century the machairs became more stable. The use of seaweed, dung combined with *Ammophila arenaria* planting helped bind the sand. However, erosion was still a problem with a large proportion

Fig. 11.7 Eroding and under-cutting of the machair surface on the west coast of Ireland. In this location relative sea level rise is implicated. Heavy grazing, recreational activity, sand extraction and exposure combine to cause loss of surface vegetation and erosion of the sand. The sand cliff in the foreground is approximately 3 m high, July 1991

(c25%) of blown sand in the Highlands and Islands occurring in the Outer Hebrides (Richie and Mather 1977) and this continued into the 1990s (Angus 1997).

Sheep grazing is one of the main factors on the west coast of Ireland contributing to machair surface deflation. When combined with the extensive undercutting of the surface occurring along the seaward machair front (Fig. 11.7) erosion can be severe. A survey of 24 machair sites in 1996 along the west coast of Ireland from Galway to Donegal recorded that "all the systems encountered in the survey were retreating and eroding…" (Crawford et al. 1998).

11.6 Alien Mammals

Islands are areas where some predatory animals are absent. As a result, they can become a haven for ground-dwelling species. The high densities of ground-nesting waders in the Hebridean islands are a manifestation of this. The European Hedgehog *Erinaceus europaeus*, feral ferrets and American Mink *Mustela vison* are three introduced animals, which prey on ground-nesting birds. They have had a significant adverse impact on the breeding success of waders since the 1970s.

11.6.1 Breeding Waders and Hedgehogs

In 1974, hedgehogs, introduced to South Uist in an attempt to control garden pests spread rapidly. At the same time, there was a reduction in the nesting population of the waders, due to hedgehog predation, with animals eating the eggs of ground-nesting birds (Jackson and Green 2000). Experimental removal of hedgehogs from fenced plots showed that waders had 2–4 times more breeding success in their absence (Jackson 2001). Further studies showed that as the hedgehog population expanded northwards egg predation increased. Between 1983 and 2000 on South Uist and Benbecula, nesting Dunlin, Ringed Plover and Snipe fell by 57%, Redshank by 41% and Lapwing by 31%. By contrast, in North Uist where the hedgehog population was low, the decline in number was less severe and several breeding waders increased. Thus, overall whilst wader numbers fell by 22%, in areas with hedgehogs, the decline was greater at 39% (Jackson 2003; Jackson et al. 2004). The key to the success of hedgehogs in the Western Isles appears to lie in the absence of predators, the availability of food and in recent years increasing temperatures, favouring their survival through the winter (Jackson 2006).

The Uist Wader Project, which began in 2000 set out to restore the breeding success of the nesting birds through culling the hedgehog population. Experimental removal showed that this could be successful (Jackson 2001). Initially, humane killing of trapped live animals took place. However, this procedure met with considerable opposition from animal welfare groups. As a result, from 2007 the Uist Hedgehog Rescue translocated trapped animals to the Scottish mainland. The results have been encouraging, with a significant reduction in the hedgehog population and an increase in breeding waders (Scottish Natural Heritage http://www.snh.gov.uk/land-and-sea/managing-wildlife/uist-wader-project/).

11.6.2 Machair and Mink

A second predator introduced to the Outer Hebridean islands is the non-native American mink. The first escapees were from fur farms in Lewis during the 1950s, and they have since spread southwards reaching South Uist by 2001. These aggressive predators caused significant losses of adult nesting birds, chicks and eggs on Harris and Lewis, which coincided with widespread breeding failures of arctic terns, gulls and lapwings (European Union Birds Directive Annex 1 birds). They had the potential to have a significant impact on the three most important species breeding on the machairs of South Uist (Corncrake, Dunlin and Ringed Plover).

In 2001, because of the threat to the breeding species in the south a 5 year European Union LIFE project (LIFE00 NAT/UK/007073) aimed at eliminating the animal from the islands of North Uist, Benbecula and South Uist. It also set out to reduce significantly their numbers in Harris. The last mink trapped was in 2005 and the animal it appears, is now eradicated from the southern islands (Anon 2006).

11.7 The Need for Conservation Action

Both cultivated and uncultivated (grazed) machair are under threat. These threats derive not only from the specific problems associated with 'modernisation' the increased use of artificial fertilisers, herbicides and non-native seeds as described above, but also from other issues, such as climate change.

11.7.1 Machair and Climate Change

There has been a general assumption that in Scotland, isostatic emergence will compensate for rising sea levels around the coast. Maps thus show Scotland continuing to emerge slowly from the sea, as postglacial rebound outpaces sea level rise. However, recent investigations show that the rate of uplift in Scotland is modest and less than rising sea levels (Rennie and Hansom 2011).

This seems to confirm anecdotal evidence that some of the Scottish islands, such as the Outer Hebrides may have been experiencing relative sea level rise for some time. The glacial ice sheets were thicker over the central parts of Scotland than around the margins. Consequently, the thinner ice mass caused less crustal depression, resulting in a lower rate of rebound when the ice retreated (Angus 2001a).

On the sheltered side of some small machair islands there are signs of lost cultivated land, which is not due to direct wave attack or sediment depletion. This also suggests that relative sea level rise must play a part. The combination of rising sea levels, higher winter precipitation and increased storminess pose a particular threat to the machair. This may be especially damaging to the cultivated machair of the islands of south and north Uist. Overtopping of foredunes, flooding and restricted drainage inland could also have serious consequences for both their economic and nature conservation values (Angus and Hansom 2004).

11.7.2 Restoring Stability

In some areas of both Scotland and Ireland, extensive erosion both at the seaward margin and on the machair plain, is significant. Whilst many of the mechanisms described in Chap. 9 to combat erosion are applicable, their likely cost and effectiveness militate against their use. Such work as has been tried, as for example in Eoligarry on Barra, involved *Ammophila arenaria* planting and the use of geotextiles (Fig. 11.8). These appeared to be effective at least in the short term (Angus and Elliot 1992), however in 2012 the problems continued to require active intervention.

In Scotland, many of the machair erosion sites result from disturbance due to access (both cars and people), overgrazing or intensification of crop management.

Fig. 11.8 Remedial action to control erosion at Eoligarry, Barra, June 1983. In 2012 extensive areas of bare sand remained (Google Earth). Nearly 30 years later, local action continues in an attempt to control the extent of blowing sand through a "Coastal Care" project (Coast Hebrides 2012 http://www.coasthebrides.co.uk/index.php?option=com_content&view=article&id=66&Itemid=88)

In almost all instances, removing the destabilising factor is the best way of combating erosion. The same is probably true in Ireland, where increasing grazing and recreational activity already implicated in erosion, are also ecologically damaging. The resulting bare sand increases the risk from storms, requiring controls on livestock density, the length of the grazing season and recreation activities to facilitate vegetation recovery (Cooper et al. 2005).

Recommendations for Irish machair management and restoration include:

- Reduce or remove sheep grazing to allow regeneration of the vegetation;
- Graze seasonally with cattle at low to medium stocking rates;
- Reduce fencing on township land;
- Cease sand extraction;
- Control location, expansion and use of caravan sites (Crawford et al. 1998).

The list also includes the comment "Plan golf course management with nature conservation in mind". As with sand dunes elsewhere, this is an activity affecting several machair systems. The previous chapter deals with these issues.

The recommendations are equally applicable to the machair in Scotland. A project begun in 2010 encompasses the need for adaptation and local involvement. It seeks to demonstrate that traditional crofting practices have a sustainable future, to the benefit of the conservation of this exceptional ecosystem, (Machair Life http://www.machairlife.org.uk/ and CoastAdapt 2010–2011 http://www.coastadapt.org/Outer-Isles/index.html). Most conservationists recognise that traditional agricultural management must be able to evolve, and compromises may be required. The future of these grasslands in Scotland depends ultimately on the active and skilled management by the crofters (Angus 2001b).

11.8 Conclusion

The machair is a special form of sand dune where cultural, economic and nature conservation values are closely interlinked. The move away from traditional agricultural practices has changed the value of the system, especially from a nature conservation perspective. In addition, the introduction of mammals absent from these offshore islands has posed a serious threat to the breeding wader populations, which are amongst the most significant in Europe. Management strategies, aimed at reversing these trends must engage local people in the decision-making process. The various initiatives, including those concerned with erosion control, appear to be moving in the right direction.

References

Angus S (1997) The shaping of the islands. The Outer Hebrides. White Horse Press, Cambridge, 264 pp

Angus S (2001a) Moor and machair. The Outer Hebrides. White Horse Press, Cambridge, 339 pp

Angus S (2001b) The conservation of machair in Scotland: working with people. In: Houston JA, Edmondson SE, Rooney PJ (eds) Coastal dune management, shared experience of European conservation practice. Liverpool University Press, Liverpool, pp 177–191

Angus S (2006) De tha machair? Towards a machair definition. Sand dune machair 4:7–22. Source: http://www.snh.gov.uk/docs/B100728.pdf. Accessed 07 June 2011

Angus S (2009) Dé tha cearr air a'mhachaire?Biodiversity issues for Scottish machair: an initial appraisal. Glasg Nat 25(Suppl):53–62

Angus S, Elliott MM (1992) Erosion in Scottish machair with particular reference to the Outer Hebrides. In: Carter RWG, Curtis TGF, Sheehy-Skeffington MJ (eds) Coastal dunes: geomorphology, ecology and management for conservation. A.A. Balkema, Rotterdam, pp 93–112

Angus S, Hansom JD (2004) Tir a'mhachair, tir nan loch? Climate change scenarios for Scottish machair systems: a wet future? In: Proceedings of the vol.2, Littoral 2004. Delivering sustainable coasts: connecting science and policy. pp 565–569

Anon (2006) Mink control to protect important birds in the SPAs in the Western Isles. Technical final report, LIFE00 NAT/United Kingdom/007073, Scottish Natural Heritage. Source: http://www.snh.org.uk/pdfs/scottish/wisles/Hebridean%20Mink%20Project%20-%20Final%20Technical%20Report%20-%20Final%20version%20October%202006.pdf. Accessed 15 Mar 2012

Bassett JA, Curtis TGF (1985) The nature and occurrence of sand dune machair in Ireland. Proc R Ir Acad 85(1):1–20

Cooper A, McCann T, Ballard E (2005) The effects of livestock grazing and recreation on Irish machair grassland vegetation. J Plant Ecol 181(2):255–267
Crawford IC (1990) Agriculture, weeds and the Western Isles, machair. Trans Bot Soc Edinb 45:483–492
Crawford I, Bleasdale A, Conaghan J (1998) Biomar survey of Irish machair sites, 1996. Vol. 1, site information. Ir Wildl Man 3, 127 pp
Edwards KJ, Whittington G, Ritchie W (2005) The possible role of humans in the early stages of machair evolution: palaeoenvironmental investigations in the Outer Hebrides, Scotland. J Archaeol Sci 32(3):435–444
Forsythe W (2006) The archaeology of the kelp industry in the northern islands of Ireland. Int J Naut Archaeol 35(2):218–229
Fuller RJ, Reed TM, Buxton NE, Webb A, Williams TD, Pienkowski MW (1986) Populations of breeding waders *Charadriidae* and their habitats on the crofting lands of the Outer Hebrides, Scotland. Biol Conserv 37:333–361
Gaynor K (2006) The vegetation of Irish machair. Biol Environ: Proc R Ir Acad 106B(3):311–321
Gregory RA, Murphy EM, Church MJ, Edwards KJ, Guttmann EB, Simpson DDA (2005) Archaeological evidence for the first mesolithic occupation of the Western Isles of Scotland. Holocene 15:944–950
Jackson DB (2001) Experimental removal of introduced hedgehogs improves wader nest success in the Western Isles, Scotland. J Appl Ecol 38(4):802–812
Jackson DB (2003) Waders, hedgehogs and machair: research and conservation lessons from the Outer Hebrides. Wader Study Group Bull 100:14–19
Jackson DB (2006) Factors affecting the abundance of introduced hedgehogs (*Erinaceus europaeus*) to the Hebridean island of South Uist in the absence of natural predators and implications for nesting birds. J Zool 271(2):210–217
Jackson DB, Green RE (2000) The importance of the introduced hedgehog (*Erinaceus europaeus*) as a predator of the eggs of waders (Charadrii) on machair in South Uist, Scotland. Biol Conserv 93(3):333–348
Jackson DB, Fuller RJ, Campbell ST (2004) Long-term population changes among breeding shorebirds in the Outer Hebrides, Scotland, in relation to introduced hedgehogs (*Erinaceus europaeus*). Biol Conserv 117(2):151–166
Love JA (2009) Oh, dear! What can the machair be? Glasg Nat 25(Supplement. Machair Conservation: Successes and Challenges):3–10
Madden B, Cooney T, O'Donoghue A, Norriss DW, Merne OJ (1998) Breeding waders of machair systems in Ireland in 1996. Ir Birds 6:177–190
McCracken DI (2009) Machair invertebrates: the importance of 'mosaiciness' Glasg Nat 25(Supplement. Machair Conservation: Successes and Challenges):29–30
Mitchell C, Walsh A, Hall C, Crowe O (2008) Greenland Barnacle Geese *Branta leucopsis* in Britain and Ireland: results of the international census, spring 2008. Wildfowl and Wetlands Trust, Slimbridge. Source: http://monitoring.wwt.org.uk/pdf/Greenland%20Barnacle%20Goose%20census%20report%202008.pdf. Accessed 17 Apr 2012
Nairn RGW, Sheppard JR (1985) Breeding waders of sand dune machair in north-west Ireland. Ir Birds 1:160–170
Owen NW, Kent M, Dale MP (2000) Ecological effects of cultivation on the machair sand dune systems of the Outer Hebrides, Scotland. J Coast Conserv 6:155–170
Rennie AF, Hansom JD (2011) Sea level trend reversals: land uplift outpaced by sea level rise on Scotland's coast. Geomorphology 125:193–202
Ritchie W (1976) The meaning and definition of machair. Bot J Scotl 42(4):431–440
Ritchie W (1979) Machair development and chronology in the Uists and adjacent islands. Proc R Soc Edinb 77B:107–122
Ritchie W, Mather AS (1977) The beaches of the highlands and islands of Scotland. Commissioned by the Countryside Commission for Scotland 1977. Reprinted 2005 by Scottish Natural

Heritage as Commissioned report no 048. Source: http://www.snh.org.uk/pdfs/publications/commissioned_reports/ReportNo048.pdf. Accessed 15 Mar 2012

Rodwell JS (ed) (2000) British plant communities. Volume 5, maritime communities and vegetation of open habitats. Cambridge University Press, Cambridge, 512 pp

Suddaby D, Nelson T, Veldman J (2010) Resurvey of breeding wader populations of machair and associated wet grasslands in North-west Ireland. Irish Wildlife Manuals, No. 44. National Parks and Wildlife Service, Department of the Environment, Heritage and Local Government, Dublin, Ireland. Source: http://www.npws.ie/publications/irishwildlifemanuals/IWM44.pdf

Welch RC (1979) Survey of the invertebrate fauna of sand dune and machair sites in the Outer Hebrides during 1976. Proc R Soc Edinb 77B:395–404

Wilson JB, Watkins AJ, Rapson GL, Bannister P (1993) New Zealand machair vegetation. J Veg Sci 4(5):655–660

Wilson JD, Boyle J, Jackson DB, Lowe B, Wilkinson NI (2007) Effect of cereal harvesting method on a recent population decline of Corn Buntings *Emberiza calandra* on the Western Isles of Scotland: capsule population decline since 1995 is associated with the harvesting of cereals as arable silage. Bird Study 54:362–370

Chapter 12
Present Threats and Future Prospects

Abstract This chapter looks at what we have learnt about sand dune management, conservation and restoration. It seeks to summarise the trends in human activity, setting their conservation requirements in a wider historical context. In particular, it looks at the way human use has constrained and changed the processes necessary for sand dune development, along the active coastal margin, and inland. It considers whether this and the extent of habitat loss have contributed to a protectionist philosophy, and a preoccupation with sand stabilisation. Have the timescales for assessing change and management need been too short? Learning from the past and looking to the future, it proposes a more dynamic approach for sustainable management and restoration for sand dune conservation.

12.1 Trends in Human Activity

Sand dune species, habitats and ecosystems are all adapted to coastal change both spatially and temporally. Human activity has sought to restrict the movement associated with this natural change, to protect assets established along the coastal margin from erosion and/or blowing sand. These assets include agricultural land (often derived from coastal habitats including sand dune), the built environment (ports and harbours, housing and industry) and coastal defence structures. As the coastal habitats are 'squeezed' from the land (Sect. 2.8), and sea by rising sea levels they have become less resilient to the forces acting upon them. Thus storms, where change can occur over minutes to hours, and even daily cycles of tidal inundation, cause erosion and drive the dunes landward. Hemmed in by housing, roads or other infrastructure and truncated by agriculture or forestry some sand dunes may eventually disappear altogether.

Fortunately, there are still substantial areas of coastal sand dune surviving around the world. What happens to them in the future largely depends on the availability of new sediment, whether or not they have 'room to move' and their management. Future trends in human use and the implications of climate change, particularly sea

J.P. Doody, *Sand Dune Conservation, Management and Restoration*, Coastal Research Library 4, DOI 10.1007/978-94-007-4731-9_12,

level rise, also have a significant bearing on this. It is clear that understanding the geomorphological context in which the development of a dune system takes place is essential when making management decisions. This applies to management and restoration in the face of changes taking place at a local level, as well as those related to regional weather patterns and worldwide climatic trends.

Several thousand years of historical human intervention have modified the sand dunes that survive in temperate regions today. In the past, we have perhaps not fully appreciated our role in influencing their physical structure and vegetation development. We now have a much better understanding of the processes that helped to create and nurture them including the influence, for better or worse, of human activity. At the beach/foredune interface changes in sediment availability are a key, helping to determine the response to exposure (waves, wind and rain) and sea level change (Chaps. 4 and 6). The analysis of the vegetated states described above (Chaps. 5 and 7) gives us an important insight into the role of grazing management in some parts of the world. The extent of mobile sand is another indicator of the impact of external factors such as sediment availability, exposure and human activity.

12.1.1 Agriculture and Forestry

The history of sand dune development is one of exploitation and habitat loss. Forestry plantations and agricultural development have been major factors in reducing the extent of open dune landscapes. New large-scale coniferous forestry planting on dune areas seems unlikely. However, the historical legacy of existing forests includes lowering of the water table (Sect. 7.5.2) and regeneration of an unplanned forest outside the planted area. These will continue to pose a threat so long as the forest remains in situ. However, in Denmark removal of large areas of forest, originally planted to combat erosion, is taking place. Despite opposition from some amenity interests, this is an ongoing activity both here and elsewhere in Europe with positive benefits for nature conservation (Sect. 9.7).

It is likely that most areas suitable for conversion to agricultural use are already in cultivation. Opportunities for further development are limited. Of more significance from a nature conservation perspective in northwest Europe, is the continuing move away from traditional farming enterprises and with it the reduction in grazing on many sites. This continues to pose a long-term threat. Atmospheric nitrogen deposition exacerbates the growth of coarse grasses and scrub encroachment as grazing intensity drops. At a smaller number of sites, the opposite is true and high stock levels cause loss of structural and species diversity, which in some instances can result in severe erosion. Thus, a key issue for the future lies in determining the appropriate stocking regime to meet nature conservation objectives, highlighted in Chaps. 7 and 9. This is particularly important where rabbits are no longer present or unwilling to graze unpalatable vegetation.

12.1.2 Recreation and Tourism

In the last 50 years, tourist development and recreational activity have had a continuing and increasing influence on the dune landscapes of Europe, notably in the Mediterranean, and in North America. Loss of dune areas has been on a large scale and cumulative. Urbanisation in Western Europe probably reached its zenith in Spain in the 1960s and 1970s. The pattern of development continued eastwards, along the Mediterranean coast and into the Black Sea, causing further loss of sand dune habitat in the European Union accession countries Romania and Bulgaria. Climate change could help mitigate some of these effects. Rising sea levels threaten low-lying coastal areas, whilst the increasing temperatures associated with climate change will require adaptive measures to be implemented (Nicholls and Hoozemans 1996). An analysis suggests by 2100 there could be a reduction in tourism in the Mediterranean, as summer visits become increasingly uncomfortable. However, despite this, tourism will continue to develop and tourist numbers grow, even if this growth may be substantially slower because of climate change (Bigano et al. 2008).

12.1.3 European Wars

Severe disruption took place on sand dunes such as those near Antwerp as they became part of the battlefield in World War I. During the Second World War, a further wave of damage occurred. The beaches of Dunkirk saw some of the greatest disturbance, with the evacuation of troops from the British Expeditionary Force in May/June 1940. In the months leading up to the allied invasion of Europe, sand dunes became places for military training. During this period, extensive physical damage took place, with destabilisation of many foredunes resulting from practice landing on the beach (Sect. 6.3.6).

Infantry training and the use of ordnance and other hardware on inland dune caused destabilisation on sites such as Braunton Burrows (Sect. 12.3.2). In the vast Łeba dunes, (Slowinski National Park) Poland, Field Marshal Erwin Rommel, Germany's most famous military leader, apparently practiced desert warfare. Many other dune areas in mainland Europe suffered from military activities; not least the sand dunes on the Normandy coast where the D Day invasion by Allied Forces took place on the 6th June 1944.

These disturbances set the succession back many years. There was no intent to re-create mobile dunes as part of a process of restoration. However, it provided the foundation for the development of the systems so highly valued today for nature conservation, especially in northwest Europe. What would sand dunes look like if the disturbance had not taken place? It is likely there would be much less dune grassland or heath, and many more inland sand dunes would be clothed in scrub or woodland.

12.1.4 The Military and Management

Since the Second World War many sites used for training or directly affected by invasion or other military action, continue to support military use. Habitat loss occurs with military installations such as buildings or runways. At others the exclusion of grazing animals, because of the threat from ordnance, has resulted in a return to stability. When combined with the presence of alien species introduced to control erosion (e.g. Eskmeal Dunes *Hippophaë rhamnoides*, Sect. 8.3.1) large areas of open sand dune habitat can be lost to scrub encroachment.

However, military activity requires large areas of land as safety zones for live firing and/or the movement of soldiers on foot or in tanks and other vehicles. These activities do not obliterate the sand dunes but can cause significant disturbance. The precise effect varies from site to site. As an indication of the contribution to nature conservation management made by military use in Europe, the following are examples, although not all are on sand dunes:

- The Netherlands, 50% of the total military estate of 30,000 ha and all firing ranges lie within Natura 2,000 sites;
- Belgium, of the 26,000 ha of the total military estate, 70% is included in Natura 2,000 sites; 9,400 ha (12 sites) in Flanders and 8,000 ha (3 sites) in Wallonia;
- Denmark, 45% of military areas (which total 32,000 ha) are included in Natura 2,000 sites (Gazenbeek 2005).

Some specific projects of positive benefit to nature conservation on coastal sand dunes occur in a range of European Countries. In Denmark, for example, a LIFE-Nature project involves a sand dune area used by the military, at Oxbøl. Here the Danish Armed Forces clear encroaching trees (including non-native trees, primarily *Pinus mugo*), and carry out "mosaic burning" in a few places (LIFE02NAT/DK/008584). In addition to this, they have also been engaged in the restoration of amphibian habitats and excavation of new ones, for the benefit of Natterjack Toad and Moor Frog *Rana arvalis* (Annex IV species in Habitats and Species Directive). Natterjack Toad is restricted to one of only three main locations in Denmark, the islands of Fanø and Rømø, the dynamic parabolic dune of the Råbjerg Mile and in the military training area of Oxbøl. Natural dune dynamics provide the necessary habitat for the species on the islands and the Råbjerg Mile. However, tank movement helps keep breeding and foraging habitats open in the military training areas.

At Vattaja in Finland, there is a comprehensive management programme, initiated to help protect the Natura 2,000 site, one of the largest Boreal (northern) sand dune systems in Europe. It also forms part of a European Union LIFE Programme study where military live firing and training go hand in hand with nature conservation aims (Koskela and Sievänen 2009).

In Spain a few of the more important and large populations of Maritime Juniper *Juniperus oxycedrus* spp. *macrocarpa* still survive in natural or semi-natural situations within protected or military areas (Muñoz-Reinoso 2003).

In Germany four inland military training areas not near the coast, are amongst the last large remnants of sparse, dry, sandy grasslands in Europe. For some specialist plants and animals, these are important refuges. Two species, Blue-winged Grasshopper *Oedipoda caerulescens* and Northern Dune Tiger Beetle *Cicindela hybrida* protected species in Germany depend on semi-stable sandy soils. Adults of the former require between 60% and 100% surface disturbance, corresponding to 50–70% plant cover, depending on the location. Adults of the latter preferentially occupy areas with >40% disturbance, or an average of 61% plant cover (Warren and Büttner 2008). Although these are not coastal sandy areas, they illustrate the significance of disturbance due to military activity for the conservation of such specialist animals.

In the United Kingdom, the Ministry of Defence has statutory and non-statutory obligations for sustainable development. To this end, it has 120 Conservation Groups covering its establishments at home and abroad. A yearly magazine "Sanctuary" provides a "Spotlight" on the activities of these groups. Several of these include information on the management of coastal sand dunes where they help in a wide range of conservation management activities, such as:

- Removal of scrub and less mature trees to help restore areas of dune heath at RAF Woodvale. Followed by chemical treatment, this included new stock fences and gates to link the area with the current grazing managed by Lancashire Wildlife Trust;
- Introduction of grazing on large fenced headlands on Ministry of Defence (MOD) land and controlling willow re-growth in the dune slacks at Penhale Dunes with the Cornwall Wildlife;
- At Eskmeals, Cumbria work includes *Hippophaë rhamnoides* control and reintroduction of grazing;
- The Ministry of Defence erected 12 km of livestock fencing to facilitate the reintroduction of traditional grazing on the sand dunes of Magilligan and Ballykinler in Northern Ireland.

On the 208 ha Altcar Rifle Range estate, within the Sefton Coast Merseyside, northwest England, work included mowing to maintain orchid-rich grasslands and dune slack vegetation. Management of a series of shallow pools for Natterjack Toad and the creation of new wet slacks, were also part of the programme (Gazenbeek 2005).

12.2 Climate Change

Climate change will have important consequences for the physical and biological attributes of sand dunes and nature conservation values. To some extent, the precise effect will depend on sediment availability, the location and movement of the beach/foredune and the size of the inland sand dune. Larger systems with adequate sediment to create new dunes will be more robust than smaller ones (Doody 2004). However, changes in relative sea level are likely to have the greatest impact, especially for the beach/foredune. The potential for change on inland dunes is more difficult to assess.

12.2.1 Sea Level Rise and the 'Sand Dune Squeeze'

Average global sea level is rising at approximately 3.2 mm/year nearly twice that experienced for most of the twentieth century. Some predictions suggest a rise of between 18 and 79 cm by the end of the twenty first century (Church and White 2006, 2011). When combined with an increase in the intensity and frequency of storms this is likely to have a significant impact on the shoreline, contributing to an escalation of global erosion trends (Brown and McLachlan 2002). However, there is considerable regional variation in sea level mainly due to differences in ocean thermal expansion (Cazenave and Llovel 2010).

Predictions of global eustatic rise do not tell the whole story. The interplay between isostic and eustatic movement is important in assessing the relative movement of sea level in relation to the land at a given location. Even in areas generally thought to be safe as glacial rebound outpaces risings seas, the coast is not necessarily secure. Isostatic uplift appears to contribute little towards mitigating the effect of relative sea level rise on the Scottish coast (Rennie and Hansom 2011). This may be even more significant on some of the outlying islands where the glacial cover was relatively thin when compared to the mainland and there is a relative rise in sea level (Sect. 11.7.1). The relative movement at the coast then depends on local conditions including sediment availability and coastal resilience. Areas with a reduced sediment supply (Sect. 4.2.1) may be particularly vulnerable.

In some areas, the forces are such that there are no foredunes and the eroding beach cuts directly into the inland dune (Physical State 1). In others, the beach/foredune may retain its characteristics (Physical State 2) as it moves landward (Carter 1991; Psuty and Silveira 2010). However, in both situations there is net landward migration, which inevitably results in the loss of dune habitat inland. The extent of this loss and the resulting 'sand dune squeeze' depends on the size of the dune and the presence of barriers such as infrastructure or other human assets (Sect. 2.5) to sand movement. Building on sand dunes close to the shore will result in the ultimate 'sand dune squeeze' (Fig. 12.1).

In this situation, narrowing the zone occupied by colonising plants causes a breakdown of successional processes (Feagin et al. 2005). This will have knock-on effects for other interests. For example, a 0.5 m rise in sea levels could result in the loss of up to 32% of beaches in the Caribbean due to the vulnerability of coastal zones. This loss could have serious consequences for nesting sea turtles (Fish et al. 2005) and other interests present in the beach/foredune (Sect. 4.3).

Fig. 12.1 A classic example of sand dune squeeze, the New Jersey shoreline, May 1996. The foredune has virtually disappeared as sea level rise, storms and sediment depletion combine to promote landward movement of the dune. The sand is kept in place, at least in the short term by 'protective' sand-trapping fences

12.2.2 Climate Change – Biological Effects

In temperate regions, vegetation plays a major role in halting or slowing down the movement of sand, when driven onshore by drying winds over an exposed beach. It is, therefore, important to consider the way in which a change in climate might affect the biological processes within the primary dune vegetation. *Ammophila* spp. and other sand-binding species respond to burial by sand. The extent to which an increase in temperature and wetness might affect the ability of these species to build dunes is difficult to assess. However, since both are especially adapted to overcome sand inundation it seems likely this factor will dominate any change in status. Any alteration of the balance of the species due to climatic factors per se is likely to be minor and difficult to detect.

It is also difficult to predict what factors might be most important for more stable inland dune vegetation. Drier summers and a rise in temperature resulting in drought conditions will cause changes in the structure and species composition, at least in the short term. However alternatively, greater winter precipitation could increase the rate of scrub development and help stabilise the system. Experience from the drought of 1976 in Great Britain suggests two principal affects on dune vegetation. Firstly, in dune slacks, there was a marked adverse impact on plants of special conservation interest. Secondly, there was also a significant lowering of the water table more generally (Rodda and Marsh 2011). The drought appears to have accelerated the gradual process of drying, which had been taking place for several years previously. This resulted in a loss of typical dune slack plants. It had an especially adverse impact on the rare Natterjack Toad as the dune slacks, important for breeding in the spring, dried out. Fire also causes damage, and dry summers may increase the incidence of burning of *Ammophila arenaria* and fixed dune grassland. In addition to causing changes in species composition, it can lead to the exposure of bare sand and severe erosion, with loss of sand from the system.

Species at the limits of their geographical range and with special habitat requirements may be the most responsive. Invertebrates, which are dependent on open dry, sandy substrates, could show the most rapid change. Hotter, drier summers, which lead to greater dune mobility, will tend to favour these species. Warmer wetter winters may on the other hand bring about more closed vegetation, causing a decline in the same species. Distinguishing this effect from the impact of changes in grazing regimes (Sects. 7.3 and 7.4) or nitrogen enrichment (Sect. 7.7.2) will be difficult. In order to separate climatic effects from anthropogenic impacts, historical information and contemporary management will need to be scrutinised.

12.3 Stabilisation – Too Much of a Good Thing?

Stabilising sand dunes as part of a restoration process is perhaps understandable when there is a threat to human assets from sand inundation. It is less obvious in areas where such interests are not threatened. However, in the 1950s extensive areas of many sand dunes had lost their surface vegetation, restoration became a priority and several sites in England are illustrative of the process. These sites show how ultimately this led to loss of nature conservation values and helped bring about a change in attitude to the role of sand dune mobility as part of the restoration process, especially on stable inland sand dunes.

12.3.1 Camber Sands – Recreation Pressure

Camber Sands is a narrow dune lying in front of Camber village in Kent. Erosion has been a feature of this site since the 1930s. Human pressure from vehicle access and public use caused destabilisation of the sand dunes prior to 1940. Following on from this, military use in the 1940s caused the destruction of large parts of the foredune. Due to the threat to the village from blowing sand and flooding, between 1947 and 1955 the erection of sand fences and planting *Ammophila*

Fig. 12.2 Fencing at Camber Sands, November 2005. The fence in the foreground delimits an access route to the beach

arenaria took place helping to stabilise the dune. However, continuing heavy pedestrian and vehicle use reversed the process (Pizzey 1975). By the 1960s extensive erosion was again present along the dune front (Countryside Commission 1969). In the early 1970s, further remedial measures involving the application of hydraulically sown seeds, coupled with fencing to regulate pedestrians and trap windblown sand, proved more successful (Ranwell and Boar 1986). Although sand continued to accumulate through to 1994, by 2005 it became necessary to erect replacement fencing to protect the village from blowing sand once again (Fig. 12.2).

During this period, other changes took place. In 1994 there was a considerable amount of *Hippophaë rhamnoides* invading the site. As with many other areas in the United Kingdom, this has continued to expand in the absence of physical control or grazing pressure into the dune inland from the foredune. This site illuminates a number of issues related to visitor pressure:

1. Concentrated human use causes erosion which requires control when adjacent infrastructure is threatened;
2. The presence of car parks close by exacerbates these impacts, by facilitating easy access to the dunes and beach;
3. Remedial action can be costly;
4. Focusing on visitor control can result in other damaging changes such as alien scrub invasion.

Fig. 12.3 An extensive area of destabilised sand dune planted with *Ammophila arenaria* at Braunton Burrows in 1956/1957

12.3.2 Braunton Burrows – From Instability to Stability and Back Again

Braunton Burrows provides an insight into the importance of taking a wider view and longer-term perspective of management requirements when seeking to protect valuable wildlife assets. Extensively studied by ecologists in the last 50 years, it shows the way attitudes to stability have changed over the years. The American army trained for the D-Day landings here in 1944. This had a major impact on the stability of the foredune, where practice amphibious landings took place. Troop movements including the use of flamethrowers, mortars and vehicles such as tanks (Bass 1992) caused major destabilisation of inland dunes.

By the early 1950s the impact of uncontrolled rabbit grazing and burrowing, combined with the effects of military use, resulted in extensive erosion. This included the foredune, and created extensive areas of bare sand inland, amounting to hundreds of hectares. Between 1953 and 1960, a large-scale programme of *Ammophila arenaria* planting took place to rehabilitate the inland, vegetated sand dune (Fig. 12.3).

This proved to be very successful and by the 1970s a considerable area of the site was clothed in vegetation. This included dense scrub with *Hippophaë rhamnoides* reintroduced to speed up the stabilisation programme in the 1950s, following an earlier unsuccessful attempt in 1937. The loss of the rabbit population through myxomatosis in 1953, combined with a lowering water table due to drainage, were important factors in the stabilisation process (Packham and Willis 2001).

Fig. 12.4 Braunton Burrows National Nature Reserve, stages in the development of stable sand dune grassland and scrub

From 1964 Natural England (formerly the Nature Conservancy Council and then English Nature) managed the site as a National Nature Reserve under a lease from the owners Christie Devon Estates. By the early 1970s it became clear that stabilisation was adversely affecting the biodiversity of the site with *Hippophaë rhamnoides* having a major influence. Despite an extensive *Hippophaë rhamnoides* control programme between 1985 and 1996, the natural stabilisation of the dune system and the growth of coarse grasses and scrub continued (Fig. 12.4).

In the late 1980s, introduction of sheep grazing to several small enclosures attempted to demonstrate the value of this form of management in arresting scrub encroachment and restoring dune grassland. This had limited success and in the early 1990s, discussion about the introduction of cattle took place with the owners. These failed, and by September 1996 scrub encroachment was even further advanced and with it considerable loss of nature conservation interest.

Negotiations took place, but the nature conservation value of a large area of the site continued to decline. Between 1997 and 2003, mixed grazing by cattle and sheep took place in a small (27 ha) plot, confirming the value of this form of grazing in reversing the encroachment of coarse grasses and scrub and helping to re-establish species rich grassland (FitzGibbon et al. 2005). By 2009, the owners agreed to the reintroduction of cattle and began fencing three large enclosures to accommodate them. Military training continues.

This site shows how quickly vegetation can develop on inland sand dunes. It took only 20 years for most of the extensive areas of bare sand to become recolonised. Within a further 10 years, the adverse impact on the original nature conservation

Fig. 12.5 Chestnut paling fencing on Blakeney Point, a mobile shingle and sand dune complex on the North Norfolk coast, England August 1986

values, including species rich dune slacks, became apparent. It also confirms the importance of grazing as a means of controlling scrub and restoring dune grassland.

12.3.3 Blakeney Point – A Dynamic Spit

Blakeney Point is a shingle spit on the east coast of England with a series of recurves, partly covered by sand dunes at its distal end. Its inherent dynamism described in the early 1930s showed how it responded to changes in tides, storms and sediment movement (Oliver and Salisbury 1913). Bought by Charles Rothschild in 1912 at the suggestion of F.W. Oliver and given to the National Trust in the same year, it was the first nature reserve in the county of Norfolk.

Blowouts in the sand dune occurred in the 1930s and whilst some remained more or less stable, others showed signs of rapid change in the following years. The use of fencing to control sand movement occurred from time to time, including in the 1980s (Fig. 12.5). Given the inherent dynamics of the site, it is questionable whether such interference was appropriate. In 2011, the same area had a good cover of vegetation, the result of processes that would probably have occurred without fencing.

12.4 Management Options

It is important when considering the management need to make a distinction between the beach/foredune sand sharing system and the inland dune. In the former mobile sand, ephemeral vegetation along the strandline and early dune-forming vegetation are inherently unstable. Many of the nature conservation attributes depend on this instability. In temperate regions, the inland dune develops into more stable forms, and vegetation plays a key role in the process (Chap. 1). Here bare sand occurs because of deposition of sand blown inland from the beach or foredune, or through damage or loss of surface vegetation often caused by human activity (Chap. 2). From a nature conservation perspective, the two systems appear to require different approaches. In the first, the best representation of nature conservation values occurs when the system remains active. The second depends on stability within the system for the full development of dune grassland or heath. However, as we have seen, too much stability can result in loss of nature conservation values.

12.4.1 The Beach/Foredune Interface

Eroding shorelines, i.e. those with limited supply of sediment or in locations with a relative rise in sea level, may elicit a response involving erosion control. On beaches with little or no vegetation and an eroding dune face (Figs. 4.2 and 11.7) or on those where the vegetated foredune is shifting landwards (Fig. 6.1) the underlying cause may lie outside the control of the local conservation manager. These include activities such as offshore sediment extraction (Sect. 6.3.1), sand mining from the foreshore (Sect. 6.3.2) or river damming (Sect. 6.3.3). Erosion may be partially reversible when governments are prepared to curtail these activities and/or initiate remedial action to restore the sediment supply, for example by beach nourishment (Sect. 9.2.7).

The nature conservation manager should consider whether erecting barriers to sand movement is appropriate. In most cases, it will treat the symptoms not the cause, and will probably be ineffective in the medium to long term. In many situations, the beach/foredune will continue to move landward whatever the local action. Maintaining the active dune front may be the preferred option, even if this is at the expense of the more stable inland dune. The specialist plants and animals will continue to exist within the dynamic matrix of bare, mobile sand and open sparsely vegetated dune. This will erode and/or overwhelm the inland dune (Fig. 12.6) with important consequences for the values of the dune grassland or heath associated with it.

In the Netherlands where dune preservation is important for sea defence, since 1990, 'dynamic preservation' of the coast has become an integral part of policy. In this context beach nourishment, allowing natural dynamics to create beach/foredune barriers (Sect. 9.2.7) is an important management tool. Trade-offs associated with the creation of a more dynamic foredune system need not compromise

Fig. 12.6 The relationship between, physical State 1 eroding dune, physical State 2 mobile foredune and vegetated State 2/3 sand dune, Skagen, northern Denmark in September 1997. The eroding dune front and the shifting foredune diminishes the area of stable vegetated inland sand dune

their sea defence function (van der Meulen and van der Maarel 1989; van Bohemen and Meesters 1992).

12.4.2 Inland Dunes

In all but a few locations where local conditions result in accreting physical State 3 foredunes, those responsible for nature conservation often preside over a diminishing asset, as described above. Even on larger systems where losses are small relative to the size of the dune, there are significant management issues. These include control and removal of alien species (Chap. 8), controlling scrub encroachment (Sect. 9.4), establishing the most appropriate grazing regime (Sect. 9.5) and managing recreational use (Sect. 9.6) including golf course development and management (Chap. 10).

Of these, in temperate regions at least, scrub invasion at the expense of dune grassland and heath is probably the most significant. Four of the states of vegetated inland sand dune relate directly to the level of grazing pressure. The most frequently encountered on sand dunes of high nature conservation values, are those where grazing is absent or where it is at too low a level to control scrub development (vegetated State 4).

In the past both planners and managers viewed mobile sand as a threat. In the Netherlands, for example, large parts of the country are 'protected' from flooding

Fig. 12.7 *Ammophila arenaria* planting, used to stabilise bare sand, the Netherlands 1983

from the sea by massive sand dune barriers (Fig. 6.7). The absence of grazing animals, and *Ammophila arenaria* planting to stabilise even the smallest patches of bare sand (Fig. 12.7) in the 1970s and 1980s, left most of the dune landscape with little or no natural dynamics.

The reduction in nature conservation value, resulting from stability of these inland dunes, called into question the reliance on sand stabilisation as the primary policy for sand dune management. From the 1980s, coastal scientists began to suggest adopting a more flexible dynamic approach to sand movement, not only immediately above the beach in the foredune, but also inland (Doody 1989, 2001).

Since then there has been a change in attitude to valuing mobile features within the inland dune. This has meant that rather than expending energy on barriers to sand movement, the modern manager is more likely to embrace it as a feature in its own right. However, when is erosion a threat to the biological diversity of a sand dune? Of course, it all depends! A small site with little or no additional sediment might appear to be at risk of total destruction as the beach/foredune migrates inland, overwhelming the stable grassland or heath. This will be particularly acute for dunes backed by rising ground or infrastructure. On larger sites without such restrictions, the losses of vegetated dune may appear to be less significant.

In areas where there is an adequate supply of sediment for the formation of foredunes, the beach will continue to move seawards and with it the strandline and foredune. Locations with falling sea levels relative to the land are more likely to experience this form of progradation. As the inland dunes are 'left behind', in the absence of grazing they quickly become stabilised and scrub covered. There is a value judgement to be made as to the extent the vegetation succession should be allowed to revert to scrub and eventually become woodland (Sect. 9.3.2). However, even where grazing is present, there may still be a loss of important areas of grassland, heath or dune slacks with increased stability. In these circumstances, active promotion of instability may ultimately provide a more sustainable and cost effective way of promoting nature conservation. In terms of practical management, this could mean cutting down trees rather than planting them, and creating blowouts rather than preventing them.

These solutions, however, will require a more complete understanding of the dune system and the historical, physical and climatic context in which development has taken place. Clearly the larger the dune system the more opportunities there are for encouraging the full range of forms from bare sand, dune slacks, grassland and/or heath, scrub and woodland.

12.4.3 A Recreational Experience

Where once mobile sand appeared as a threat to human activity, it has become part of a landscape and cultural experience. Large migrating dunes can be a recreational magnet, such as la Dune du Pyla (Dune de Pilat) Arcachon Bay in France, possibly the highest dune in Europe. Also in Europe is the mobile dune Råbjerg Mile (Fig. 14.102 in Doody 2001) situated near Skagen in northern Denmark. This remained unplanted even during the major periods of afforestation. Today it serves as an example of a natural phenomenon and as an aid to understanding sand drift. Moving at a speed of about 15 m per year depending on climate conditions, its height and mobility have also become a recreational attraction. The public has free access to the area by foot, which helps maintain the moving dune front. Other examples include the Rubjerg Knude lighthouse (Fig. 12.8) in Denmark and the Łeba sand bar in Poland (Fig. 7.58 in Doody 2001).

12.5 Conclusions

Past human activities have resulted in widespread habitat modification and irreversible loss of sand dunes both big and small. At many locations, the surviving areas represent a depleted resource. There is legislation designed to protect both beach/foredune and inland sand dune from further damage and destruction. In the case of the former, this often recognises their contribution to sea defence. For the latter, nature conservation and/or recreational values provide the focus. Losses continue, and preventing further harmful development requires a more robust attitude by those responsible for implementing legislation affecting the coastal zone than hitherto.

At the beach/foredune interface, the lack of sediment is already significant. Due to the other factors involved, notably rising sea levels, a reversal of the predominance

Fig. 12.8 Rubjerg Knude lighthouse, a popular visitor attraction on the Danish coast, June 1992 when the buildings were still accessible. Sand has since overwhelmed the buildings and only the lighthouse tower was visible in 2009

of erosion over accretion is unlikely without major intervention to restore sediment delivery to the coast. On a few coastlines, economic considerations make beach nourishment worthwhile. In many others, even some with developed shorelines, such intervention does not take place. The result is a narrowing foreshore, breakdown of successional processes and loss of vegetation and associated animals (Feagin et al. 2005). Where there are undeveloped inland dunes, the beach and/or foredune may continue to migrate landwards.

For dunes inland from an eroding foredune, the situation is different. Here the migrating beach/foredune occurs at the expense of the vegetated inland dune. This may not matter so long as the dune is large in comparison to the migrating dune, or has itself room to migrate. Difficulties arise on smaller sand dunes and those where infrastructure development, afforestation or agricultural use inhibits landward migration. The resulting sand dune squeeze (Sects. 2.8 and 12.2.1; Fig. 12.9) makes management of the surviving areas of habitat even more important.

Whatever the status of the beach/foredune interface, the vegetated inland dune may continue to require prescriptive management. This will include scrub control, removing alien invaders, reintroducing grazing by domesticated stock and catering for visitors. Without such management, in Europe at least, most of the surviving areas of dune grassland and heath will continue to deteriorate, as stabilisation degrades their biological diversity. The most appropriate forms of management for nature reserves and other 'protected' areas, described in Chap. 9 continue to evolve.

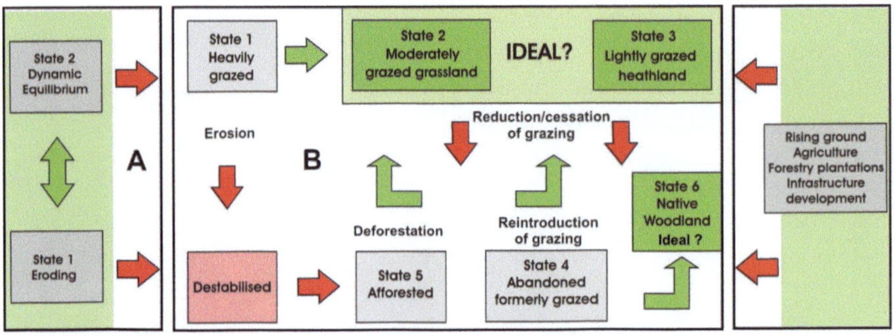

Fig. 12.9 Relationship between (**a**) the beach/foredune Physical State Evaluation Model (Sect. 6.2.4), (**b**) the inland dune Vegetated State Evaluation Model (Sect. 7.2.6) and some of the factors preventing migration of the sand dune landward. The inland dune is 'squeezed' into a narrowing zone. *The Ideal? States represent* situations that are especially important for nature conservation values

Based at Liverpool Hope University, the European Sand and Shingle Network facilitates exchange of experience amongst dune managers. Management can be labour intensive, costly and often restricted to individual problems. However, if we look to the past and consider the way sand dunes respond to environmental and human perturbations, alternative management options become apparent.

"The dune complex is a restless maze" so said Cowles (1899). It seems that in the intervening 100 years or so we have forgotten this fact and sought to stop blowing sand and control dune mobility. Ancient Acts of Parliament and Laws forbidding removal of *Ammophila* spp. and otherwise carrying out activities that cause destabilisation were relatively common. Forestry plantations, usually of non-native conifers stabilised areas of inland sand dune (Sect. 2.3).

The local demand for a 'seaside holiday' and the burgeoning foreign tourist industry in the latter part of the twentieth century meant cleaning beaches and creating car parks, access points, paths and tourist facilities (Sect. 6.3.4). Local planners and coastal managers failed to recognise the impact on beaches and inland dunes or, if they did, viewed the economic advantages as outweighing any environmental damage. They were more likely to provide finance to prevent sand dune erosion than control the activities that helped to create it. Chestnut paling fencing was, and continues to be, a means of controlling erosion along the coastal foredune (Sect. 9.2.4), around car parks and access points to the beach (Fig. 12.2).

Nature conservation organisations were similarly concerned with preventing sand movement. Sand fences became the first response to blowing sand, even in nature reserves where access was strictly controlled. Some nature conservation bodies continued to use stabilisation techniques such as fencing and *Ammophila* spp. planting (Fig. 12.5) until quite recently. However, as this book has demonstrated, in temperate regions vegetation is an extremely effective stabilising agent. There are far more examples of scrub infested inland sand dunes than mobile ones.

It is also true that destabilisation provides opportunities for habitat regeneration. Prof. Bill Carter said in 1990 at a conference in Sefton, England dedicated to dune conservation "these habitats are not sensitive but robust and designed to accommodate changes in tides, tidal energy and sediment availability. It is our desire to prevent them from moving that has caused so much damage", personal recollection (Carter 1990). We sometimes forget that nature has a way of restoring itself if left alone. It may not be desirable to embark on the level of destabilisation in Europe, brought about during two world wars (especially the Second World War, Sect. 12.1.3). However, if it had not been for this disturbance, it is likely many more sand dunes would be clothed in dense species-poor scrub or woodland.

Thus, the tendency has been for the nature conservation movement to adopt the same approach as those concerned with protecting land and property. It was, and to some extent still is common practice to plant *Ammophila arenaria*, to thatch eroding dunes with brushwood or otherwise to stabilise bare sand using the techniques described in Sect. 9.2. During the 1980s, however, there was an increasing recognition that instability of inland dunes, as part of a natural process, plays a vital part in conserving sand dune biodiversity (e.g. Doody 1989). Conferences of the then newly formed European Union for Dune Conservation included further consideration of the theme (van der Meulen et al. 1989; Carter et al. 1992).

We must continue to manage some special areas to prevent loss due to stabilisation. However, we should also direct our efforts towards giving beaches and inland sand dunes, room to move. We must actively resist building new structures on inland sand dunes. Taking tentative steps to clear vegetation, reintroduce grazing and other relatively small-scale management at first, there is an increasing recognition of the value of techniques involving destabilisation. Although this will not be on the scale of earlier unplanned activities, nevertheless it does offer a longer term solution to the conservation of many of the important natural attributes of this highly prized habitat.

Increased storminess and other effects brought on by global warming will increase the erosive forces acting upon sand dune systems. We must resist the temptation to direct our efforts towards protecting them with traditional stabilisation techniques in either the foredune or inland, vegetated dune. A change in attitude, which recognises blowing sand as integral to the healthy functioning of a sand dune system, whether in the foredune or inland dune (Fig. 12.6) must continue.

Allowing greater mobility will help reverse some of the problems associated with stabilisation. In one sense, grazing is a secondary ecosystem/succession process, which has modified the sand dune species composition. Sediment movement, climate and soil forming processes are the basis for sand dune development. The restoration of sand dune dynamics could therefore, be an important management tool for restoring this habitat. Human use has been a significant agent of change in the past and could be in the future. Despite opposition to removing forests planted with alien species or remobilising stable dunes, these management options may provide a more enduring prescription for conserving this valuable habitat.

References

Bass RT (1992) Spirits of the sand. The history of the U.S. Army Assault Training Centre. Lee Publishing, Woolacombe, 170 pp

Bigano A, Hamilton JM, Tol RSJ (2008) Climate change and tourism in the Mediterranean, FNU-157, Hamburg University and Centre for Marine and Atmospheric Science, Hamburg. Source: http://www.fnu.zmaw.de/fileadmin/fnu-files/publication/working-papers/mediterraneanwp.pdf. Accessed 28 Mar 2011

Brown AC, McLachlan A (2002) Sandy shore ecosystems and the threats facing them: some predictions for the year 2025. Environ Conserv 29:62–77

Carter RWG (1990) Coastal zone management: comparisons and conflicts. In: Houston J, Jones C (eds) Planning and management of the coastal heritage. Sefton Metropolitan Borough Council, Southport, pp 45–49

Carter RWG (1991) Near-future sea level impacts on coastal dune landscapes. Landsc Ecol 6(1–2):29–39

Carter RWG, Curtis TGF, Sheehy-Skeffington MJ (eds) (1992) Coastal dunes: geomorphology, ecology and management for conservation. Proceedings of the third European Dune Congress Galway, Ireland, 17th–21st June 1991. A.A. Balkema, Rotterdam, 533 pp

Cazenave A, Llovel W (2010) Contemporary sea level rise. Annu Rev Mar Sci 2:145–173

Church JA, White NJ (2006) A 20th century acceleration in global sea-level rise. Geophys Res Lett 33:L01602, doi:10.1029/2005GL024826. Source: http://naturescapebroward.com/NaturalResources/ClimateChange/Documents/GRL_Church_White_2006_024826.pdf. Accessed 19 Feb 2012

Church JA, White NJ (2011) Sea-level rise from the late 19th to the early 21st century. Surv Geophys 32:585–602

Countryside Commission (1969) Nature conservation on the coast. Special study report volume 2. HMSO, p 97 plus maps

Cowles HC (1899) The ecological relations of the vegetation on the sand dunes of Lake Michigan. Part 1 – Geographical relations of the dune floras. Bot Gaz 95:391

Doody JP (1989) Management for nature conservation. Proc R Soc Edinb, B 96:247–265

Doody JP (2001) Coastal conservation and management: an ecological perspective, vol 13, Conservation biology series. Kluwer Academic Publishers, Boston, 306 pp

Doody JP (2004) Coastal habitats to get bigger and better? In: Green DR et al (eds) Delivering sustainable coasts: connecting science and policy. Proceedings of Littoral 2004, Aberdeen, Scotland, UK, vol 1. Cambridge Publications, Cambridge, pp 84–89

Feagin RA, Sherman DJ, Grant WE (2005) Coastal erosion, global sea-level rise, and the loss of sand dune plant habitats. Front Ecol Environ 3(7):359–364

Fish MR, Côté IM, Gill JA, Jones AP, Renshoff S, Watkinson AR (2005) Predicting the impact of sea-level rise on Caribbean sea turtle nesting habitat. Conserv Biol 19(2):482–491

FitzGibbon C, Albon S, Robinson P (2005) The effects of a mixed grazing regime on sand dune vegetation communities at Braunton Burrows, Devon. English Nature Research Reports, 637, p 2

Gazenbeek A (2005) LIFE, Natura 2000 and the military. LIFE Focus, Journal of the LIFE III programme (2000–2006), p 86. Source: European Commission, Environment Directorate General. Source: http://ec.europa.eu/environment/life/publications/lifepublications/lifefocus/documents/military_en.pdf. Accessed 26 Mar 2011

Koskela K, Sievänen M (2009) Restoration, environmental management and monitoring in the Vattaja Dune LIFE Project, 2005–2009. Nature Protection Publications of Metsähallitus, Series A, 181, p 39. Source: http://julkaisut.metsa.fi/julkaisut/pdf/luo/a181.pdf. Accessed 26 Mar 2011

Muñoz-Reinoso JC (2003) *Juniperus oxycedrus* spp. *macrocarpa* in Spain: ecology and conservation problems. J Coast Conserv 9:113–122

Nicholls RJ, Hoozemans FMJ (1996) The Mediterranean: vulnerability to coastal implications of climate change. Ocean Coast Manag 31(2–3):105–132

Oliver FW, Salisbury EJ (1913) The topography and vegetation of Blakeney Point, Norfolk. Trans Norfolk Norwich Nat Soc 9:485–544

Packham JR, Willis AJ (2001) Braunton Burrows in context, a comparative management study. In: Houston JA, Edmondson SE, Rooney PJ (eds) Coastal dune management: shared experience of European conservation practice. Liverpool University Press, Liverpool, pp 69–75

Pizzey JM (1975) Assessment of dune stabilisation at Camber, Sussex, using air photographs. Biol Conserv 7:275–288

Psuty NP, Silveira TM (2010) Global climate change: an opportunity for coastal dunes?? J Coast Conserv 14(2):153–160

Ranwell DS, Boar R (1986) Coast dune management guide. Institute of Terrestrial Ecology, HMSO, London, 105 pp

Rennie AF, Hansom JD (2011) Sea level trend reversal: land uplift outpaced by sea level rise on Scotland's coast. Geomorphology 125(1):193–202

Rodda JC, Marsh TJ (2011) The 1975–76 drought – a contemporary and retrospective review. Centre for Ecology and Hydrology, Wallingford, 58 pp

Van Bohemen HD, Meesters HJN (1992) Ecological engineering and coastal defence. In: Carter RWG, Curtis TGF, Sheehy-Skeffington MJ (eds) Coastal dunes: geomorphology, ecology and management for conservation. A.A. Balkema, Rotterdam, pp 369–378

van der Meulen F, van der Maarel E (1989) Coastal defence alternatives and nature development perspectives. In: van der Meulen F, Jungerius PD, Visser J (eds) Perspectives in coastal dune management. SPB Academic Publishing, The Hague, pp 183–198

van der Meulen F, Jungerius PD, Visser J (eds) (1989) Perspectives in coastal dune management. SPB Academic Publishing, The Hague, 333 pp

Warren SD, Büttner R (2008) Active military training areas as refugia for disturbance-dependent endangered insects. J Insect Conserv 12(6):671–676

Index

J.P. Doody, *Sand Dune Conservation, Management and Restoration*,
Coastal Research Library 4, DOI 10.1007/978-94-007-4731-9,

H

I

K

L

M

N

MIX
Papier aus verantwortungsvollen Quellen
Paper from responsible sources
FSC® C105338
FSC
www.fsc.org

If you have any concerns about our products,
you can contact us on
ProductSafety@springernature.com

In case Publisher is established outside the EU,
the EU authorized representative is:
Springer Nature Customer Service Center GmbH
Europaplatz 3, 69115 Heidelberg, Germany

Printed by Libri Plureos GmbH
in Hamburg, Germany